Advances in Intelligent Systems and Computing

Volume 1067

The series "Advances in Intelligent Systems and Computing" contains publications on theory, applications, and design methods of Intelligent Systems and Intelligent Computing. Virtually all disciplines such as engineering, natural sciences, computer and information science, ICT, economics, business, e-commerce, environment, healthcare, life science are covered. The list of topics spans all the areas of modern intelligent systems and computing such as: computational intelligence, soft computing including neural networks, fuzzy systems, evolutionary computing and the fusion of these paradigms, social intelligence, ambient intelligence, computational neuroscience, artificial life, virtual worlds and society, cognitive science and systems, Perception and Vision, DNA and immune based systems, self-organizing and adaptive systems, e-Learning and teaching, human-centered and human-centric computing, recommender systems, intelligent control, robotics and mechatronics including human-machine teaming, knowledge-based paradigms, learning paradigms, machine ethics, intelligent data analysis, knowledge management, intelligent agents, intelligent decision making and support, intelligent network security, trust management, interactive entertainment, Web intelligence and multimedia.

The publications within "Advances in Intelligent Systems and Computing" are primarily proceedings of important conferences, symposia and congresses. They cover significant recent developments in the field, both of a foundational and applicable character. An important characteristic feature of the series is the short publication time and world-wide distribution. This permits a rapid and broad dissemination of research results.

** Indexing: The books of this series are submitted to ISI Proceedings, EI-Compendex, DBLP, SCOPUS, Google Scholar and Springerlink **

More information about this series at http://www.springer.com/series/11156

Miguel Botto-Tobar · Joffre León-Acurio ·
Angela Díaz Cadena · Práxedes Montiel Díaz
Editors

Advances in Emerging Trends and Technologies

Volume 2

 Springer

Editors
Miguel Botto-Tobar 🆔
Eindhoven University of Technology
Eindhoven, Noord-Brabant, The Netherlands

Joffre León-Acurio 🆔
Universidad Técnica de Babahoyo
Babahoyo, Ecuador

Angela Díaz Cadena
Universitat de Valencia
Valencia, Valencia, Spain

Práxedes Montiel Díaz
Centro de Investigación y Desarrollo
Profesional (CIDEPRO)
Babahoyo, Ecuador

ISSN 2194-5357   ISSN 2194-5365 (electronic)
Advances in Intelligent Systems and Computing
ISBN 978-3-030-32032-4   ISBN 978-3-030-32033-1 (eBook)
https://doi.org/10.1007/978-3-030-32033-1

This Springer imprint is published by the registered company Springer Nature Switzerland AG
The registered company address is: Gewerbestrasse 11, 6330 Cham, Switzerland

Preface

The 1st International Conference on Advances in Emerging Trends and Technologies (ICAETT) was held on the main campus of the Universidad Tecnológica Israel, in Quito–Ecuador from May 29 to 31, 2019, and it was organized jointly by Universidad Tecnológica Israel, Universidad Técnica del Norte, and Instituto Tecnológico Superior Rumiñahui, and supported by SNOTRA. The ICAETT series aims to bring together top researchers and practitioners working in different domains in the field of computer science to exchange their expertise and to discuss the perspectives of development and collaboration. The content of this volume is related to the following subjects:

- Technology Trends
- Electronics
- Intelligent Systems
- Machine Vision
- Communication
- Security
- e-Learning
- e-Business

ICAETT 2019 received 236 submissions written in English by 375 authors coming from 16 different countries. All these papers were peer-reviewed by the ICAETT 2019 Program Committee consisting of 222 high-quality researchers. To assure a high-quality and thoughtful review process, we assigned each paper at least three reviewers. Based on the peer reviews, 87 full papers were accepted, resulting in a 37% acceptance rate, which was within our goal of less than 40%.

We would like to express our sincere gratitude to the invited speakers for their inspirational talks, to the authors for submitting their work to this conference, and the reviewers for sharing their experience during the selection process.

May 2019   Miguel Botto-Tobar
                                                                              Joffre León-Acurio
                                                                          Angela Díaz Cadena
                                                                     Práxedes Montiel Díaz

Organization

General Chairs

Miguel Botto-Tobar — Eindhoven University of Technology, The Netherlands

Joffre León-Acurio — Universidad Técnica de Babahoyo, Ecuador

Organizing Committee

Miguel Botto-Tobar — Eindhoven University of Technology, The Netherlands

Joffre León-Acurio — Universidad Técnica de Babahoyo, Ecuador

Óscar Zambrano Vizuete — Universidad Técnica del Norte, Ecuador

Luis Chávez — Instituto Tecnológico Superior Rumiñahui, Ecuador

Paúl Baldeón Egas — Universidad Tecnológica Israel, Ecuador

Steering Committee

Miguel Botto-Tobar — Eindhoven University of Technology, The Netherlands

Joffre León-Acurio — Universidad Técnica de Babahoyo, Ecuador

Ángela Díaz Cadena — Universitat de Valencia, Spain

Práxedes Montiel Díaz — CIDEPRO, Ecuador

Publication Chair

Miguel Botto-Tobar   Eindhoven University of Technology,
                                      The Netherlands

Program Chairs

Technology Trends

Miguel Botto-Tobar   Eindhoven University of Technology,
                                      The Netherlands
Sergio Montes León   Universidad de las Fuerzas Armadas
                                      (ESPE), Ecuador
Hernán Montes León   Universidad Rey Juan Carlos, Spain
Jean Michel Clairand   Universidad de Las Américas, Ecuador
Ángel Jaramillo Alcázar   Universidad de Las Américas, Ecuador

Electronics

Ana Zambrano Vizuete   Escuela Politécnica Nacional, Ecuador
David Rivas   Universidad de las Fuerzas Armadas
                                      (ESPE), Ecuador
Edgar Maya-Olalla   Universidad Técnica del Norte, Ecuador
Hernán Domínguez-Limaico   Universidad Técnica del Norte, Ecuador

Intelligent Systems

Guillermo Pizarro Vásquez   Universidad Politécnica Salesiana,
                                      Ecuador
Janeth Chicaiza   Universidad Técnica Particular de Loja,
                                      Ecuador
Gustavo Andrade Miranda   Universidad de Guayaquil, Ecuador
William Eduardo Villegas   Universidad de Las Américas, Ecuador
Diego Patricio Buenaño Fernández   Universidad de Las Américas, Ecuador

Machine Vision

Julian Galindo   LIG-IIHM, France
Erick Cuenca   Université de Montpellier, France
Pablo Torres-Carrión   Universidad Técnica Particular de Loja,
                                      Ecuador
Jorge Luis Pérez Medina   Universidad de Las Américas, Ecuador

Communication

Óscar Zambrano Vizuete	Universidad Técnica del Norte, Ecuador
Pablo Palacios Jativa	Universidad de Chile, Chile
Iván Patricio Ortiz Garces	Universidad de Las Américas, Ecuador
Nathaly Verónica Orozco Garzón	Universidad de Las Américas, Ecuador
Henry Ramiro Carvajal Mora	Universidad de Las Américas, Ecuador

Security

Luis Urquiza-Aguiar	Escuela Politécnica Nacional, Ecuador
Joffre León-Acurio	Universidad Técnica de Babahoyo, Ecuador

e-Learning

Miguel Zúñiga-Prieto	Universidad de Cuenca, Ecuador
Verónica Fernanda Falconí Ausay	Universidad de Las Américas, Ecuador
Doris Macias	Universitat Politécnica de Valencia, Spain

e-Business

Angela Díaz Cadena	Universitat de Valencia, Spain
Praxedes Montiel Díaz	CIDEPRO, Ecuador

e-Government and e-Participation

Alex Santamaría Philco	Universidad Laica Eloy Alfaro de Manabí, Ecuador

Program Committee

Abdón Carrera Rivera	University of Melbourne, Australia
Abu Shakil Ahmed	Leading University, Bangladesh
Adrián Cevallos Navarrete	Griffith University, Australia
Ahmed Lateef Khalaf	Al-Mamoun University College, Iraq
Alba Morales Tirado	University of Greenwich, UK
Alejandro Ramos Nolazco	Instituto Tecnológico y de Estudios Superiores Monterrey, Mexico
Alex Cazañas Gordon	The University of Queensland, Australia
Alex Santamaría Philco	Universitat Politècnica de València, Spain/Universidad Laica Eloy Alfaro de Manabí, Ecuador
Alex Sevilla	Corporación Nacional de Telecomunicaciones CNT, Ecuador

Alexandra Elizabeth   Universidad de Cuenca, Ecuador
  Bermeo Arpi
Alexandra Shirley   Universidad de las Fuerzas Armadas
  Pérez Quisaguano   ESPE, Ecuador
Alexandra Velasco Arévalo   Universität Stuttgart, Germany
Alfredo Núñez   New York University, USA
Almílcar Puris Cáceres   Universidad Técnica Estatal de Quevedo,
                                  Ecuador
Ana Guerrero Alemán   University of Adelaide, Australia
Ana Núñez Ávila   Universitat Politècnica de València, Spain
Ana Santos Delgado   Universidade Federal de Santa Catarina
                                  (UFSC), Brazil
Ana Zambrano   Escuela Politécnica Nacional EPN,
                                  Ecuador
Andrea Mory Alvarado   Universidad Católica de Cuenca, Ecuador
Andrés Barnuevo Loaiza   Universidad de Santiago de Chile, Chile
Andrés Calle Bustos   Universitat Politècnica de València, Spain
Andrés Cueva Costales   University of Melbourne, Australia
Andrés Gonzalez   Universitat Politècnica de València,
                                  España
Andrés Jadan Montero   Universidad de Buenos Aires, Argentina
Andrés Molina Ortega   Universidad de Chile, Chile
Andrés Parra Sánchez   University of Melbourne, Australia
Andrés Robles Durazno   Edinburgh Napier University, UK
Andrés Vargas González   Syracuse University, USA
Angel Vazquez Pazmiño   Université Catholique de Louvain,
                                  Belgium
Ángela Díaz Cadena   Universitat de València, Spain
Angelo Vera Rivera   George Mason University, USA
Antonio Villavicencio Garzón   Universitat Politècnica de Catalunya,
                                  Spain
Arian Bahrami   University of Tehran, Iran
Audrey Romero Pelaez   Universidad Politécnica de Madrid, Spain
Bolívar Chiriboga Ramón   University of Melbourne, Australia
Byron Alejandro   Universidad Estatal de Campinas, Brazil
  Acuña Acurio
Carlos Barriga Abril   University of Nottingham, UK
Carlos Saavedra   Escuela Superior Politécnica del Litoral,
                                  Ecuador
Carlos Valarezo Loiza   Manchester University, UK
César Ayabaca Sarria   Escuela Politécnica Nacional (EPN),
                                  Ecuador
Cesar Mayorga Abril   Universidad Técnica de Ambato, Ecuador
Christian Báez Jácome   Wageningen University & Research,
                                  The Netherlands

Chrysovalantou Ziogou — Centre for Research & Technology Hellas, Greece

Cintya Aguirre Brito — University of Portsmouth, UK

Cristhian Flores Urgiles — Universidad Católica de Cuenca, Ecuador

Cristian Montero Mariño — University of Melbourne, Australia

Daniel Alejandro Guamán Coronel — Universidad Técnica Particular de Loja, Ecuador

Daniel Armijos Conde — Queensland University of Technology, Australia

Daniel Magües Martínez — Universidad Autónoma de Madrid, Spain

Daniel Silva Palacios — Universitat Politècnica de València, Spain

David Benavides Cuevas — Universidad de Sevilla, Spain

David Fernando Pozo Espín — Universidad de las Américas, Ecuador

David Rivera Espín — University of Melbourne, Australia

Diana Estefanía Cherrez Barragán — .

Diana Morillo Fueltala — Brunel University London, UK

Edgar Maya — Universidad Técnica del Norte, Ecuador

Edwin Guamán Quinche — Universidad del País Vasco, Spain

Efrén Reinoso Mendoza — Universitat Politècnica de València, Spain

Enrique Vinicio Carrera — Universidad de las Fuerzas Armadas ESPE, Ecuador

Eric Moyano Luna — University of Southampton, UK

Erick Cuenca Pauta — Université de Montpellier, France

Ernesto Serrano Guevara — Université de Neuchâtel, Switzerland

Estefania Yánez Cardoso — University of Southampton, UK

Esther Parra Mora — University of Queensland, Australia

Fabian Calero — University of Waterloo, Canada

Fabián Corral Carrera — Universidad Carlos III de Madrid, Spain

Fabián Cuzme — Universidad Técnica del Norte, Ecuador

Felipe Ebert — Universidade Federal de Pernambuco (UFPE), Brazil

Felipe Leonel Grijalva Arévalo — Escuela Politécnica Nacional EPN, Ecuador

Fernando Borja Moretta — University of Edinburgh, UK

Fernando Darío Almeida García — Universidade Estadual de Campinas (UNICAMP), Brazil

Firas Raheem — University of Technology, Iraq

Francisco Pérez — Universidad Politécnica de Valencia, Spain

Franklin Parrales Bravo — Universidad Complutense de Madrid, Spain

Gabriel López Fonseca — Sheffield Hallam University, UK

Gema Rodriguez-Perez — LibreSoft/Universidad Rey Juan Carlos, Spain

Jorge Rivadeneira Muñoz	University of Southampton, UK
Jorge-Luis Pérez-Medina	Universidad de las Américas, Ecuador
José Carrera Villacres	Université de Neuchâtel, Switzerland
José Quevedo Guerrero	Universidad Politécnica de Madrid, Spain
José Teodoro Mejía Viteri	Universidad Técnica de Babahoyo, Ecuador
Josue Flores de Valgas	Universitat Politécnica de València, Spain
Juan Carlos Santillán Lima	.
Juan Jiménez Lozano	Universidad de Palermo, Argentina
Juan Lasso Encalada	Universitat Politècnica de Catalunya, Spain
Juan Maestre Ávila	Iowa State University, USA
Juan Miguel Espinoza Soto	Universitat de València, Spain
Juan Romero Arguello	University of Manchester, UK
Juan Zaldumbide Proaño	University of Melbourne, Australia
Juliana Cotto Pulecio	Universidad de Palermo, Argentina
Julio Albuja Sánchez	James Cook University, Australia
Julio Balarezo	Universidad Técnica de Ambato, Ecuador
Karina Jimenes	Universidad de las Américas, Ecuador
Leopoldo Pauta Ayabaca	Universidad Católica de Cuenca, Ecuador
Lorena Guachi Guachi	Yachay Tech, Ecuador
Lorena León Quiñonez	Universidade Estadual de Campinas (UNICAMP), Brazil
Lorenzo Cevallos Torres	Universidad de Guayaquil, Ecuador
Lucia Rivadeneira Barreiro	Nanyang Technological University, Singapore
Luis Benavides	Universidad de Especialidades Espíritu Santo, Ecuador
Luis Carranco Medina	Kansas State University, USA
Luis Felipe Urquiza	Escuela Politécnica Nacional EPN, Ecuador
Luis Pérez Iturralde	Universidad de Sevilla, Spain
Luis Rodrigo Barba Guamán	Universidad Técnica Particular de Loja, Ecuador
Luis Torres Gallegos	Universitat Politècnica de València, Spain
Manuel Beltrán Prado	University of Queensland, Australia
Manuel Eduardo Sucunuta España	Universidad Técnica Particular de Loja, Ecuador
Manuel Sucunuta España	Universidad Politécnica de Madrid, Spain
Marcelo Alvarez V.	Universidad de las Fuerzas Armadas ESPE, Ecuador
Marcelo Rodrigo García Saquicela	Universidad Técnica Estatal de Quevedo, Ecuador
Marcia Bayas Sampedro	Vinnitsa National University, Ukraine
Marco Falconi Noriega	Universidad de Sevilla, Spain

Marco Molina Bustamante	Universidad Politécnica de Madrid, Spain
Marco Santórum Gaibor	Escuela Politécnica Nacional, Ecuador/Université Catholique de Louvain, Belgium
Marco Tello Guerrero	Rijksuniversiteit Groningen, The Netherlands
María Belén Mora Arciniegas	Universidad Técnica Particular de Loja, Ecuador
María del Carmen Cabrera Loayza	Universidad Técnica Particular de Loja, Ecuador
Maria Dueñas Romero	RMIT University, Australia
María Escalante Guevara	University of Michigan, USA
María Miranda Garcés	University of Leeds, UK
María Montoya Freire	Aalto University, Finland
María Ormaza Castro	University of Southampton, UK
Mariela Barzallo León	University of Edinburgh, UK
Mariela Tapia León	Universidad de Guayaquil, Ecuador
Mario González	Universidad de las Américas, Ecuador
Marius Giergiel	University of Science and Technology, Poland
Mauricio Domínguez	.
Mauricio Verano Merino	Eindhoven University of Technology, The Netherlands
Maykel Leiva Vázquez	Universidad de Guayaquil, Ecuador
Miguel Arcos Argudo	Universidad Politécnica de Madrid, Spain
Miguel Botto Tobar	Eindhoven University of Technology, The Netherlands
Miguel Fornell	Universidad Politécnica de Cataluña, Spain
Miguel Gonzalez Cagigal	Universidad de Sevilla, España
Milton Navas	Universidad de las Fuerzas Armadas ESPE, Ecuador
Milton Román-Cañizares	Universidad de Las Américas, Ecuador
Mohamed Kamel	Military Technical College, Egypt
Mónica Villavicencio Cabezas	Université du Quebec À Montréal, Canada
Nancy Jacho Guanoluisa	Universidad de las Fuerzas Armadas ESPE, Ecuador
Narcisa Crespo Torres	Universidad Técnica de Babahoyo, Ecuador
Nathaly Verónica Orozco Garzón	Universidad de las Américas, Ecuador
Nelson Oswaldo Piedra Pullaguari	Universidad Técnica Particular de Loja, Ecuador

Omar S. Gómez	Escuela Superior Politécnica del Chimborazo (ESPOCH), Ecuador
Orlando Erazo Moreta	Universidad de Chile, Chile/Universidad Técnica Estatal de Quevedo, Ecuador
Oswaldo López Santos	Universidad de Ibagué, Colombia
Pablo León Paliz	Université de Neuchâtel, Switzerland
Pablo Ordoñez Ordoñez	Universidad Politécnica de Madrid, Spain
Pablo Palacios Jativa	Universidad de Chile, Chile
Pablo Saá Portilla	University of Melbourne, Australia
Patricia Ludeña González	Politecnico di Milano, Italy
Paúl Hernán Mejía Campoverde	Universidad de las Fuerzas Armadas ESPE, Ecuador
Paul Rosero	Universidad Técnica del Norte, Ecuador
Paulina Tatiana Mayorga Soria	Universidad de las Fuerzas Armadas ESPE, Ecuador
Paulo Esteban Chiliguano Torres	Queen Mary University of London, UK
Pedro Neto	University of Coimbra, Portugal
Prasanta Ghosh	ICEEM, India
Praveen Damacharla	Purdue University Northwest, USA
Rafael Campuzano Ayala	Grenoble Institute of Technology, France
Rafael Jiménez	Escuela Politécnica del Litoral (ESPOL), Ecuador
Ramiro Santacruz Ochoa	Universidad Nacional de La Plata, Argentina
René Rolando Elizalde Solano	Universidad Técnica Particular de Loja, Ecuador
Ricardo Llugsi	Escuela Politécnica Nacional EPN, Ecuador
Ricardo Martins	University of Coimbra, Portugal
Roberto Gonzalo Sánchez Albán	University of Bern/Neuchatel/Fribourg, Switzerland
Roberto Larrea Luzuriaga	Universitat Politècnica de València, Spain
Roberto Murphy	.
Roberto Sánchez Albán	Université de Lausanne, Switzerland
Rodrigo Cueva Rueda	Universitat Politècnica de Catalunya, Spain
Rodrigo Saraguro Bravo	Escuela Superior Politécnica del Litoral (ESPOL), Ecuador
Rodrigo Tufiño Cárdenas	Universidad Politécnica Salesiana, Ecuador/Universidad Politécnica de Madrid, Spain
Samanta Cueva Carrión	Universidad Politécnica de Madrid, Spain
Santiago Leonardo Solórzano Lescano	Universidad de las Américas, Ecuador

Sergio Montes León — Universidad de las Fuerzas Armadas (ESPE), Ecuador

Shirley Coque — Universidad Politécnica Salesiana, Ecuador

Sixto Reinoso — Universidad de las Fuerzas Armadas ESPE, Ecuador

Tony Flores Pulgar — Université de Lyon, France

Vanessa Echeverría Barzola — Université Catholique de Louvain, Belgium

Vanessa Jurado Vite — Universidad Politécnica Salesiana, Ecuador

Verónica Fernanda Falconí Ausay — Universidad de las Américas, Ecuador

Veronica Segarra Faggioni — Universidad Técnica Particular de Loja UTPL, Ecuador

Verónica Yépez Reyes — South Danish University, Denmark

Victor Hugo Rea Sánchez — Universidad Estatal de Milagro, Ecuador

Vladimir Robles-Bykbaev — Universidad Politécnica Salesiana, Ecuador

Voltaire Bazurto Blacio — University of Victoria, Canada

Washington Velásquez Vargas — Universidad Politécnica de Madrid, Spain

Wayner Bustamante Granda — Universidad de Palermo, Argentina

Wellington Cabrera Arévalo — University of Houston, USA

William Venegas — Escuela Politécnica Nacional EPN, Ecuador

Wilmar Hernández — Universidad de las Américas, Ecuador

Xavier Merino Miño — Universidad Santa María, Chile

Yan Pacheco Mafla — Royal Institute of Technology, Sweden

Yessenia Cabrera Maldonado — Pontificia Universidad Católica de Chile, Chile

Yuliana Jiménez Gaona — Università di Bologna, Italy

Organizing Institutions

Sponsoring Institutions

Contents

e-Business

Communication

e-Learning

Security

Electronics

Low-Cost Embedded System for Shop Floor Communications and Control Based on OPC-UA

Hector F. Bano[1]($\boxtimes$) ⓘ, Carlos A. Garcia[1] ⓘ, Andres Cabrera A[1] ⓘ,
Esteban X. Castellanos[1] ⓘ, Jose E. Naranjo[1] ⓘ, and Marcelo V. Garcia[1,2]($\boxtimes$) ⓘ

[1] Universidad Tecnica de Ambato, UTA, 180103 Ambato, Ecuador
{hbano0123,ca.garcia,ag.cabrera,excastellanos,jnaranjo0463,
mv.garcia}@uta.edu.ec
[2] University of Basque Country, UPV/EHU, 48013 Bilbao, Spain
mgarcia294@ehu.eus

Abstract. The Cyber-Physical Systems (CPS) are an emerging paradigm based on exponential development within the design of complex industrial systems, which include a subset of control mechanisms and process monitoring through algorithms based on the development of software applications and included tightly within hardware devices of data acquisition adaptable and reconfigurable. These characteristics plays an important role in shop floor communications and control systems, because its main function is to provide flexibility in data management and easy integration into real physical systems. The current research work presents the design of a Data Acquisition System (DAS), based on the adaptability of software control skills integrated within a structural framework presented by the OPC UA (Unified Architecture), acting as a communication system, synchronization and data processing through a generic user interface where the control levels easily integrate the skills available in the system and synchronize their execution in real time in an optimal way.

Keywords: Industry 4.0 · OPC-UA · Embedded systems · CPS systems

1 Introduction

Currently, several researches have been presented about OPC-UA and its fundamental relationship within Industry 4.0 [1,2], thematic with which it is tried to face the challenges that arise around the implementation of interconnection systems in different production processes and large-scale data management. A central part of these definitions is their transparent link in all information processing instances in which a value chain is based on flexible control systems and is directly related to a connection to any type of integrated hardware device, which inevitably leads to operational synergy effects where the main goal is to

© Springer Nature Switzerland AG 2020
M. Botto-Tobar et al. (Eds.): ICAETT 2019, AISC 1067, pp. 3–12, 2020.
https://doi.org/10.1007/978-3-030-32033-1_1

exchange information efficiently, safely and quickly, as well as reflecting positive responses to the demand for re-engineering costs [3].

An industrial manufacturing environment is characterized by its extensive heterogeneity in relation to distributed processing resources, which include automation devices, running on different platforms. This heterogeneity requires the use of an adequate approach to support the interoperability between servers/clients OPC-UA and their physical resources [4,5]. In many of the cases, the deployment and implementation of new and different industrial control platforms is complex and is developed according to a specific industrial case to be solved with the use of control application, such as implementation of systems based on multiple agents (MAS) or within "Plug and Produce" systems (PnP) directly related to Cyber-Physical Systems (CPS) [6,7]. In both situations, the focus on the development of applications oriented to personalized controls within very elementary computational elements represents an important starting point in its technological scalability of research and implementation, where different challenges are included in the development of interfaces generic to simplify and to increase its application within more efficient control systems.

The contribution of this research highlights the integrated capabilities within the architecture presented by OPC-UA and focused on elementary systems of data acquisition. Where a fluid parameterization is integrated into an OPC-UA client application started as a controller and an OPC-UA server integrated into a Single-Board Computer (SBC). In addition, the reference architecture is validated with a combined implementation of OPC-UA software features and application development stacks under C++ and Python languages, demonstrate its performance capabilities in heterogeneous control without excessive hardware/software resource overload, presented as a solution approach in elementary specifications within the subset of mechanisms integrated in CPS systems.

This article lists the following sections: In Sect. 2, a general description of the system implementation is presented. In Sect. 3, the Case Study is shown and the results obtained are described in Sect. 4. The article ends with the corresponding conclusions based on the research carried out, as well as a perspective towards future work in Sect. 5.

2   System Implementation

The complete system of the SAD is designed based on a structure where four key elements interact:

1. Embedded Computing.
2. Power Control.
3. Network Communication Protocol.
4. Operational control.

Where (1), (2) and (3), interpret the operational synergy of the OPC-UA server with the iteration of elements that relate to a software extension and an electronic hardware control. While (3) and (4), they specify the operational synergy of the OPC-UA client, where real-time execution control is generated.

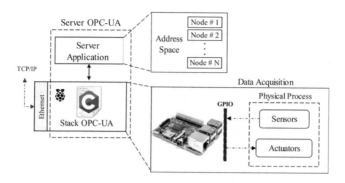

Fig. 1. Hardware architecture of Raspberry Pi Model B as OPC-UA server.

2.1 OPC-UA Implementation

OPC-UA Server Hardware Integration. An OPC-UA server is a software application that meets one or more specifications defined by the OPC Foundation according the case of implementation. The server performs the role of communication interface in a bidirectional manner, which means that customers can read and write to the embedded SAD devices through the OPC-UA protocol services. In a client/server architecture, the server is usually the slave, while the client is implemented as a master. Where the slave is oriented to work in the field levels of the pyramid automation so that it is structured as an independent device. And since one of the objectives of this research is the use of low-cost hardware platforms, an SBC is proposed as an "Embedded-UA" server (see Fig. 1). Its application reasons derive from its high integration ease within multiplatform system interfaces and its different functionalities, designed as communication platforms between plant levels [8].

In particular, the SBC Raspberry Pi Model B has been chosen as the platform for the creation of this system. Its computing performance and connectivity capabilities, compared to its low cost and size, make this platform an interesting solution for this type of industrial control systems [9].

OPC-UA Server Software Integration. Within the network communication protocol, the OPC-UA server application API is integrated based on the open62541 project (see Fig. 2), which implements the binary protocol stack OPC-UA written in a common subset of C99 and C++ 98 languages [10]. Its main goal is to interpret the main control and relate it to the management of feedback signals to the physical pins of the SBC.

The OPC-UA server and its main functionalities are integrated into a Server.c file. Extended functions thanks to a set of external libraries, whose purpose is to obtain the highest integrated performance in time of execution and control. Its management of attributes and OPC-UA characteristics within the server are executed within a structured programming initially proposed by the stack, where its strong point of applicability is based on its speed of communication

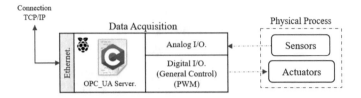

Fig. 2. OPC-UA Server using open62541 communication stack.

and its response in real time, in an intention to incorporate and present all the information processing capabilities of the SBC.

The subset of programming routines OPC-UA presents a scalable structure intended for the creation of its main function. Parameters such as:

1. The control of the life cycle.
2. Configuration of the communication network port.
3. Implementation of a physical control.
4. Creation of nodes.

In this way, the real-time control configuration of the server is structured in the most optimal way possible, where different control feedback systems with closed or open loop applications can also be integrated by the OPC-UA client, according to be the case of the implementation.

OPC-UA Client Software Integration. OPC-UA client is represented by a generic Application Programming Interface (API) software whose purpose is to visualize and present process control options based on the data consumed at the time of execution from the server. The FreeOpcUa/python-opcua [11] project provides the corresponding programming API under object-oriented programming, ideal for its integration and development (see Fig. 3).

In a similar way, several subsets of programming routines are integrated based on the OPC Foundation guidelines and based on a thread programming framework where a scalable structure is presented for the creation of the main GUI responsible for managing all visualization and control processes.

Where options like:

1. Explore the OPC-UA address space.
2. Selection of nodes for reading and writing.
3. Call subscription and access to server data from the desired node.
4. Visualization and control of specific processes in real time.

They are included within their lines of code in an intrinsic way in the Client.py file. In the same way as the server, these four basic parameters manage the characteristics of an OPC-UA client within a minimum configuration.

As already detailed, its structuring is based on the application of different stages Multithreading class, which allow to process and control in a better way

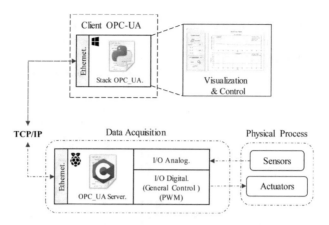

Fig. 3. OPC-UA client integration architecture using Python OPC-UA API.

the information processed from the server, when managing individual subprocesses and to share the information between the implemented classes. The benefits of this programming strategy derive from an adequate optimization in the management of information in real time and the use of PC resources. The client configuration is used to establish connection parameters and additional configurations used by the server. The configuration should not be modified after passing it to a client and currently only one client can use one configuration at a time.

3   Case Study

The proposed case study describes a laboratory system aiming at illustrating a scale factory automation application. In particular, the process plant is the system FESTO® MPS-PA Compact Workstation that was developed and designed with training purposes in the automation field [12]. This system combines 4 processes in closed loop, which can be operated individually. These are: level, flow, pressure and a temperature system. This industrial platform has sensors and actuators that allow easy manipulation of their analog electrical signals, which in turn are industrially regulated under the HART standard where their current values are limited between the ranges of 4 to 20 mA. Specifically, the implementation of control is applied in the Level and Temperature systems (see Fig. 4), where its general operation scheme is detailed according to the elements involved for each process. For which the main objective is to manage the values in real time of these processes with the server and control them by implementing a Graphical User Interface (GUI) within an OPC-UA client, as detailed below:

1. Level System, in this process the server manages the B101 level sensor signal, which is then reproduced by the client in its PID control stage, generating, a PWM feedback signal that controls the flow of water provided by the centrifugal pump P101.

2. Temperature System, in the same way in this process the server manages an OPC-UA object based on the temperature sensor incorporated in the model, to be later interpreted by the client. Its open loop control is performed by manipulating the On/Off controls of the heater implemented in the GUI.

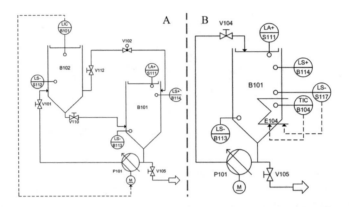

Fig. 4. Case study P&ID diagrams; (A) Controlled level system, (B) Controlled temperature system.

The structure showed in Fig. 5, details the application of the case study implemented. The SAD system manages the real-time data of the objects involved in the selected processes within their address space, to be subsequently consumed, controlled and visualized by the OPC-UA client through a bidirectional connection.

The OPC-UA server achieves effective access to control and feedback signals from the plant and makes them accessible by implementing OPC-UA nodes based on the real objects to be manipulated, thus allowing easy interpretation and real-time control of the processes selected by the OPC-UA client as detailed in Fig. 6. In this Figure we can observe two different sections in GUI:

1. Control panel: two sections are displayed for the control and manipulation of level and temperature processes by the end user. The first control function is responsible for generating the PID configurations necessary for closed loop level control. On the other hand, its second control function is responsible for manipulating the heater incorporated in the open loop control with a simple On/Off control.
2. Real time Charts: this section presents in a graphical way all the control processes carried out in the programming of the client in an intrinsic way and that is executed according to the main panel, with a first graph in the upper part showing the evolution of the PID control of Level as a function of time, while the second graph shows the temperature of the fluid in the reservoir tank of the module.

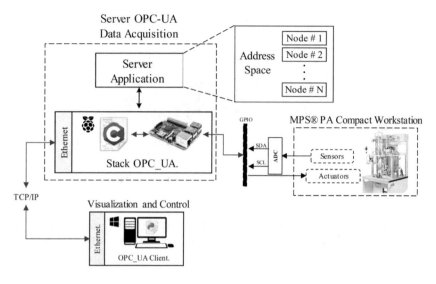

Fig. 5. Case study implementation.

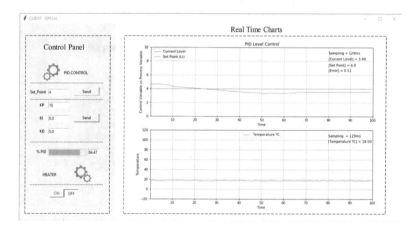

Fig. 6. GUI for OPC-UA client real-time execution.

4   Experimental Results

In the final stage of implementation, the use of Wireshark software is proposed, with which the performance of the OPC-UA protocol can be visualized within the system, depending on the bandwidth used and its capacity to transmit data packets.

Figure 7 shows the general network traffic when the communication between the OPC -UA server and client is implemented. Communication packets are usually measured in bytes/seconds. When applying the optic data filter, it shows a

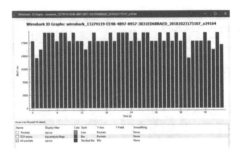

Fig. 7. OPC-UA server/client system bandwidth usage.

bandwidth that reaches a maximum of 18000 bits/s and a minimum of 13500 bits/s, all this with an average of 16000 bits/s for the whole packet group captures.

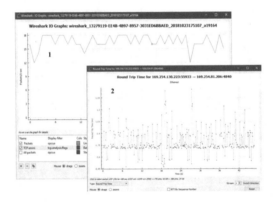

Fig. 8. OPC-UA packages processing.

Reference (1) in Fig. 8, shows the number of packages processed in a request/response transaction, with a maximum of 18 Packets/s, and a minimum of 12 Packets/s, which in turn generate an average of 16 Packets/s for each time interval mentioned. Likewise, Reference (2) in Fig. 8 measures the time delta between the capture of an OPC-UA under TCP packet and the corresponding delayed ACK for each packet. In conclusion, the proposed system has a delay in the request/response process that does not exceed 1.4 ms. This time measured, has an important relationship with the Reference (1) because, if this time is long, it could show some kind of delay in the network and generate a packet loss, collision or other negative aspects of communication, but that in our case study is not a major problem.

In the same way, the amount of internal memory used to run the server application is evaluated (see Fig. 9), if we remember that the Raspberry Pi 3

Model B has a 1 GB RAM theoretically in its factory specifications, but that in reality only has 927.32 MB of capacity and that for this case study only 18.35% (±10%) of its total capacity is necessary, that is to say that around 170 MB of internal memory is used to work optimally.

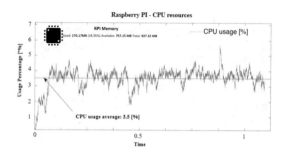

Fig. 9. Use of CPU resources of the Raspberry Pi.

5   Conclusions and Future Work

The implemented research presents a practical and exponentially developing app-roach to a basic data acquisition system based on the architecture presented by OPC-UA, whose main objective is the manipulation and direct management of data at field level and in real time execution, where its feasibility is based on the synergy of integration in automation systems under low cost CPS architectures, which helps to introduce new concepts related to SAD within Industry 4.0.

By introducing OPC-UA technology, an open and flexible communication model has been developed, as a substitute for complex and traditional systems dependent on private software or hardware suppliers. In the implementation of the system, horizontal and vertical interconnections between application levels and field devices are achieved, where two industrial field networks are easily included: Ethernet TCP/IP and Wireless LAN for monitoring the system, which confirms the interoperability of heterogeneous field networks at the management level.

In future research, it is intended to evaluate the performance of the system and its response algorithm against similar application protocols in industrial automation, such as the MQTT protocol. In addition to inquiring into consid-erations about the efficient binary coding of OPC-UA based on the UA services proposed and their security of information encryption within embedded devices.

Acknowledgment. This work was financed in part by Universidad Tecnica de Ambato (UTA) and their Research and Development Department (DIDE) under project CONIN-P-0167-2017.

References

1. Schwarz, M.H., Borcsok, J.: A survey on OPC and OPC-UA: about the standard, developments and investigations. In: XXIV International Conference on Information, Communication and Automation Technologies (ICAT), pp. 1–6 (2013)
2. Schlechtendahl, J., Keinert, M., Kretschmer, F., Lechler, A., Verl, A.: Making existing production systems. Industry 4.0-Ready **9**(1), 143–148 (2015)
3. Garcia, M.V., Irisarri, E., Perez, F., Estevez, E., Marcos, M.: OPC-UA communications integration using a CPPS architecture. In: IEEE Ecuador Technical Chapters Meeting (ETCM), pp. 1–6 (2016)
4. García, M.V., Irisarri, E., Pérez, F., Estévez, E., Marcos, M.: An open CPPS automation architecture based on IEC-61499 over OPC-UA for flexible manufacturing in oil & gas industry. IFAC-PapersOnLine **50**(1), 1231–1238 (2017)
5. Hensel, S., Graube, M., Urbas, L., Heinzerling, T., Oppelt, M.: Co-simulation with OPC UA. In: IEEE 14th International Conference on Industrial Informatics (INDIN), pp. 20–25 (2016)
6. Dias, J., Barbosa, J., Leitao, P.: Deployment of industrial agents in heterogeneous automation environments. In: IEEE 13th International Conference on Industrial Informatics (INDIN), pp. 1330–1335 (2015)
7. Koziolek, H., Burger, A., Doppelhamer, J.: Self-commissioning industrial IoT-systems in process automation: a reference architecture. In: IEEE International Conference on Software Architecture (ICSA), pp. 196–19609 (2018)
8. Palm, F., Gruner, S., Pfrommer, J., Graube, M., Urbas, L.: Open source as enabler for OPC UA in industrial automation. In: IEEE 20th Conference on Emerging Technologies & Factory Automation, pp. 1–6 (2015)
9. Raspberrypi.org: Raspberry Pi - Github (2018). https://github.com/raspberrypi. Accessed 27 Sept 2018
10. Open62541.org: open62541: an open source implementation of OPC UA. https://open62541.org/. Accessed 20 Dec 2018
11. Gruner, S., Pfrommer, J., Palm, F.: RESTful industrial communication with OPC UA. IEEE Trans. Ind. Inform. **12**(5), 1832–1841 (2016)
12. Festo: MPS® PA compact workstation (2018). https://www.festo-didactic.com. Accessed 02 Nov 2018

Study of Feature Extraction Methods for BCI Applications

Milton León[1](✉) ⓘ, Diego Orellana[1] ⓘ, Luis Chuquimarca[2] ⓘ,
and Ximena Acaro[3] ⓘ

[1] Universidad Nacional de Loja, 110103 Loja, Ecuador
moltilion@live.com, diego.orellana@unl.edu.ec
[2] Universidad Estatal Península de Santa Elena, 240204 La Libertad, Ecuador
lchuquimarca@upse.edu.ec
[3] Universidad de Guayaquil, 090514 Guayaquil, Ecuador
ximena.acaroc@ug.edu.ec

Abstract. In this work, different types of feature extraction methods for the detection of alpha waves and sensorimotor rhythms were analyzed. These signals were acquired through EEG. For the detection of alpha waves, the discrete wavelet transform and a neural network were used as feature extraction and classification methods respectively, resulting in an average detection accuracy of 89,1%. A BCI for the recognition of alpha waves was implemented through this method. Additionally, three feature extraction techniques for the identification of sensorimotor rhythms were proposed and studied. Discrete wavelet transform, autoregressive components and spatial filtering were the analyzed techniques; the classification method used was a neural network. An average accuracy of 60,81%, 61,78% and 64,59% was obtained for each method, respectively.

Keywords: Brain-computer interface · BCI · Digital signal processing · EEG · Brain-machine interface · Biopotentials

1 Introduction

A brain-computer interface is a human-machine interaction mechanism that translates the brain activity of a person into commands for physical and software systems [1]. The operation of a BCI is based in the recording and processing of a person's brain activity in order to identify the action that they wish to perform.

The detection of the brain signals is achieved in two stages: the feature extraction stage and the feature classification stage. In the first stage, a series of features useful for the identification of the user's intentions are extracted from the recorded brain signal. In the second stage, translation algorithms are applied to these features in order to recognize the type of mental activity generated and execute the appropriate control action [2].

Initially, the research of brain-computer interfaces was not in the main interest of the scientific community, as the goal of trying to identify the mental state or intentions of an individual was regarded as strange or rare. Nevertheless, with the advance of computer science, semiconductor technology and signal processing techniques, BCI technology became an actively researched field, currently with more than 100 active

© Springer Nature Switzerland AG 2020
M. Botto-Tobar et al. (Eds.): ICAETT 2019, AISC 1067, pp. 13–23, 2020.
https://doi.org/10.1007/978-3-030-32033-1_2

research groups around the world [3]. Initial successful results in the processing of neural signals have also contributed to direct attention to this field. In the same manner, the potential applications of this technology as aid for severely disabled people has stressed the necessity of encouraging the development of this subject.

Still, BCI technology is a young area whose applications can be mostly found inside the laboratory. Mayor challenges for its growth include the development of a common language and a general framework for BCI design. Another challenge is the correct detection and labeling of the brain signals, because they can be hidden by various other mental activities occurring at the same time. Also, they can be corrupted by electrophysiological artifacts such as electrooculography (EOG) and electromyography (EMG). For overcoming this problem, various methods of feature extraction have been proposed but the advantages and disadvantages of these techniques remain unclear.

2   Materials and Methods

2.1   Alpha Waves Detection and Implementation of a BCI

Brain Signals Recording. The brain signals were acquired by means of electroencephalography (EEG). The equipment used for this was a 32 bits OpenBCI biopotential's amplifier, which has 8 channels with a sampling frequency of 250 Hz and 24 bits resolution. Two electrodes placed at locations O1 and O2 of the international 10/20 system were used for the EEG recording. These locations correspond to the occipital area of the scalp, which was chosen because alpha rhythms appear with great intensity on it [3, 4].

Training Paradigm. A training paradigm was developed in order to record EEG activity that can be used as learning examples for the classifier. Both alpha activity and regular brain activity were recorded.

In this paradigm the user sits in front of a white screen with a cross in its center. At the beginning of the test, an acoustic indicator is produced for telling the user that the procedure has started; at that moment the user must fix its sight on the cross. Later, at t = 10 s, a second acoustic stimulus is produced for telling the person to close its eyes until the end of the test. At t = 25 s the last acoustic stimulus is produced for signaling the end of the test. Figure 1 depicts this process.

Alpha waves appear with great intensity when the eyes are closed [3, 4]. EEG examples of regular mental activity are recorded when the subject has their sight on the cross; on the other hand, samples of alpha activity are acquired when the subject has their eyes closed. The training sessions were run for five days, with five sessions per day. This amounts to a total of 25 sessions.

Signals Pre-processing. During the pre-processing stage, the portions of interest of the signals were extracted from the recordings. The recordings have two parts, regular brain activity and alpha waves. The regular brain activity portions were extracted without any changes, but the portions with alpha rhythms were extracted from three

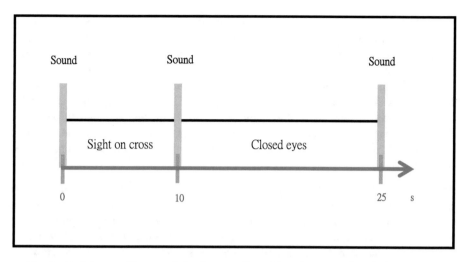

Fig. 1. Training paradigm used for the recording of learning examples of alpha waves.

seconds after the appearance of the second acoustic stimulus until the end of the sessions. This was made in order to avoid ambiguous signals produced during the subject's reaction period when changing from mental activity. The extracted portions were saved in a matrix.

The signals in the matrix were divided in windows of 250 samples. Every window is equal to a 1 s period since the sampling frequency of the EEG is 250 Hz. Each window was filtered with a 4th order bandpass Butterworth filter with a 1–35 Hz passband.

The amplitude of the brain waves was normalized in order to obtain a common reference framework for comparing them; Min-Max normalization was used for this. From the windows total, 80% was used for training the classification algorithm and 20% was reserved for testing its classification accuracy.

Feature Extraction. The feature extraction method applied was the discrete wavelet transform (DWT). With this approach, an analysis window of 250 samples is divided in 4 decomposition levels. The mother wavelet used was a 10th order Daubechies. The decomposition levels are described in Table 1.

Table 1. Decomposition levels obtained with the DWT.

Decomposition level	Group of coefficients	Frequency range
1	D1	62,5–125 Hz
2	D2	31,25–62,5 Hz
3	D3	15,625–31,25 Hz
4	D4	7,813–15,625 Hz
4	A4	0–7,813 Hz

As alpha waves operate in the frequency band of 8 to 12 Hz [5], the D4 level was the only decomposition level analyzed, as it contains this band. Next, the mean absolute value, standard deviation and average power of the coefficients of this level were computed.

These parameters represent the frequency distribution and the amount of changes in the frequency distribution of the D4 level [6]; for this reason, they were selected as features for the classification algorithm. Since the EEG was acquired through two channels (two electrodes) and we computed three features from each channel, a total of 6 features was obtained. This feature extraction method is the same as the one used by Xu et al. [6].

Feature Classification. Artificial neural networks (ANN) had been used for EEG pattern classification with excellent results [7]. Because of that, an ANN consisting of one hidden layer and one output layer was used for the classification process. The hidden layer has 20 neurons and uses the tangent-sigmoid function as activation function. The output layer has 2 neurons, each of them acting as one of the two possible outputs of the BCI (presence of alpha waves or regular neural activity) and uses the softmax function as activation function.

The training algorithm used was the scaled conjugate gradient algorithm. The extracted analysis windows were used for the ANN learning process; 80% of these windows were used for the training and the remaining 20% was saved in order to test the trained network.

It is necessary to point out that the numerical software used for implementing the ANN normalizes the input features automatically, so that the data falls in the range between 1 and −1. This property was kept for this project.

BCI Implementation. A BCI for the detection of alpha waves was implemented using the aforementioned process. The system activates a set of outputs depending on the nature of the neural activity recorded; if the brain signals are alpha rhythms the BCI activates a luminous indicator, an acoustic indicator and a LCD screen. When there are no alpha waves the indicators and the screen remain off. Here are the steps that the feature extraction and classification algorithm follows for detecting the alpha activity (See Fig. 2):

1. One second of EEG recording is acquired through O1 and O2 electrodes.
2. The acquired signals are filtered with a 4th order Butterworth filter with a passband of 1–35 Hz.
3. Each channel is normalized with Min-Max normalization.
4. The time windows from both channels are decomposed in 4 levels with the DWT.
5. The absolute value, standard deviation and average power of the D4 level coefficients is computed.
6. The wavelet features are combined in one vector, which will have a total of 6 elements because is composed from 3 features obtained from 2 channels.
7. The feature vector is feed to the ANN in order to identify the kind of mental activity recorded.
8. The ANN output is presented to the output system, which is driven by a microcontroller.

2.2   Feature Extraction and Classification of Sensorimotor Rhythms

Three feature extraction techniques were proposed and analyzed with the purpose of discriminating between left hand motor imagery and right hand motor imagery. The classification process was made with the help of an ANN.

Brain Signals Acquisition. Sensorimotor rhythms are brain oscillations related with the movement of the body, and they can be modified with the intention of movement, without any actual muscular action. Two types of sensorimotor rhythms have been identified: Mu (μ) rhythms which operate in the 8–12 Hz band and Beta (β) rhythms which operate in the 18–25 Hz band [8].

The brain signals used were obtained from the IIIA and 2A datasets given by the BCI competitions III and IV, respectively. The IIIA dataset contains the EEG recordings of three subjects who were asked to perform 4 kinds of motor imagery: left

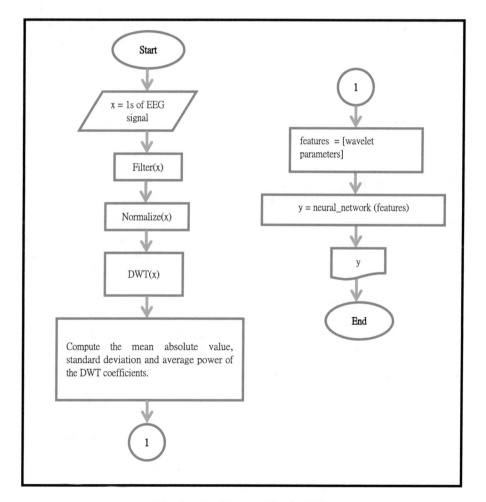

Fig. 2. Algorithm used by the BCI.

hand, right hand, foot and tongue movement. It was recorded using 60 channels with a sampling frequency of 250 Hz. The signals were filtered with a 1–50 Hz passband filter.

The 2A dataset contains the information of 9 subjects who were also asked to perform 4 kinds of motor imagery: left hand, right hand, both feet and tongue. The signals were recorded with 22 electrodes at a 250 Hz sampling frequency and later filtered with a 0.5–100 Hz bandpass filter.

Signals Pre-processing. Both datasets were acquired through several channels, nevertheless previous studies show that information given by various EEG channels is redundant [7]. Also, it has been shown that the neural activity that is mostly correlated with the fists movement is contained almost exclusively in the C3, Cz and C4 electrodes of the international 10/20 system [9]. Only those channels were used in this work.

Next, the interest information was extracted from each EEG recording. It is important to say that the datasets were obtained using training paradigms with procedures similar to the one used for the alpha waves classification. The extraction criterion was to take the portions with the relevant motor imagery information 1 s after the appearance of the cue that informs the subject of the kind of activity that they should generate. This was done in order to avoid ambiguous signals produced when the subject reacts and changes their mental activity.

Two matrices were constructed for each person with the selected data. One contains the left hand motor imagery information and the other has the right hand motor imagery information. Tests corrupted by the presence of visual artifacts were discarded. Both matrices were divided in 250 samples time windows. Since the two datasets used a 250 Hz sampling frequency, the windows are equivalent to a 1 s recording. The feature extraction and classification process were made with these windows.

First Feature Extraction Method. Firstly, the analysis windows were normalized with Min-Max normalization. Later, the DWT was used in order to decompose the signals in 4 levels as shown in Table 1. Levels D3 and D4 have the same frequency bandwidth as the μ and β rhythms respectively, thus only the coefficients of these levels were used.

The mean absolute value, standard deviation and average power of the coefficients were computed. These magnitudes are the classification features. In summary we compute three values for the two levels, resulting in 6 features for one channel. As the signal was acquired through three channels, a total of 18 features is computed.

Second Feature Extraction Method. Firstly, the amplitude of the analysis windows was normalized. Next, two kinds of features were extracted: wavelet features and AR coefficients. The wavelet features are the three values described in method one, giving a total of 18 features.

AR parameters were also extracted from the normalized channels. A 6th-order AR model was used and the AR coefficients were obtained using the Burg method, as was proposed by Xu et al. [6]. The model gives 6 coefficients, giving a total of 18 AR features when the three channels are considered.

Third Feature Extraction Method. As first step, the normalized windows were spatially filtered using Common Spatial Patterns (CSP) in order to enhance the discrimination between the motor imagery classes.

Next, three types of features were extracted: wavelet features, AR coefficients and the variance of the signals. For computing the two first kinds of features the same methods as before were applied.

The decision of using the variance of the signals as a classification feature, was made because CSP maximizes the variance of one class while minimizing the variance of the other class [10]. In order to use the variance of the windows as features, first the signals were filtered with two passband filters; one of them had a passband of 8–12 Hz and the other had a passband of 13–30 Hz. Please note that these passbands are equal to the μ and β bands. After this, the variance of each filter output was computed. These two values are the ones that will be used as classification features. As two variances were computed for each channel, a total of 6 features was obtained.

Finally, the features extracted by the two aforementioned methods and the variance of the signals are combined into one vector with a total length of 42 elements.

Feature Classification. The structure of the ANN used and the learning algorithm implemented are the same as those used for the alpha rhythms classification. The learning process of the neural network was done as follows: two matrices containing the motor imagery of each hand were constructed for each person. The signals in both matrices were divided in 250 samples windows. Then, 80% of the first matrix windows are used as learning examples; the same is done for the second matrix. After the training, the 20% remaining windows from each matrix are used for testing purposes.

3 Results

3.1 Performance of the Alpha Waves Detection Method

The neural network for the alpha waves detection was trained ten different times and each of them was tested with the testing windows group for ensuring the best generalization capabilities of the network. The obtained results are shown in Table 2. The network with the best identification accuracy was used for the BCI. Please note that for this case, two training states show the best performance. One of them is chosen without any specific criterion.

Table 2. Performance of the different training states of the ANN used for alpha waves detection.

User	Percentage of successfully classified windows for every training state (%)										Best result (%)
	1	2	3	4	5	6	7	8	9	10	
Woman	88,2	88,2	82,7	88,2	86,4	88,2	89,1	86,4	89,1	88,2	89,1

3.2   Performance of the Sensorimotor Rhythms Detection Methods

First Method Performance. As before, the testing of the first method used for sensorimotor rhythms classification was made with the aid of the testing windows. A neural network was created for each person of the two datasets and trained ten different times in order to ensure the best generalization capability. Table 3 shows the testing results for the IIIA dataset and Table 4 does the same for the 2A dataset.

Table 3. Performance of the different training states of the ANN on the IIIA dataset, applying the first method proposed.

User	Percentage of successfully classified windows for every training state. First method. IIIA dataset (%)										Best result (%)
	1	2	3	4	5	6	7	8	9	10	
1	60	51,1	64,4	64,4	64,4	53,3	46,7	48,9	57,8	64,4	**64,4**
2	48,3	55,2	55,2	51,7	58,6	58,6	65,5	58,6	51,7	41,4	**65,5**
3	48	52	52	68	44	60	52	72	64	48	**72**

The best result for each subject is chosen. Later, in order to have a more global vision of the performance of the method, the best results of every network were averaged. For the IIIA dataset an average accuracy of 67,30% was achieved, for the 2A dataset, 58,64%. Table 9 shows this result.

Table 4. Performance of the different training states of the ANN on the 2A dataset, applying the first method proposed

User	Percentage of successfully classified windows for every training state. First method. 2A dataset (%)										Best result (%)
	1	2	3	4	5	6	7	8	9	10	
1	48,8	59,5	46,4	50	59,5	50	50	53,6	41,7	45,2	**59,5**
2	34,9	42,2	36,1	53	44,6	51,8	51,8	51,8	41	47	**53**
3	66,3	60,2	69,9	62,7	62,7	63,9	56,6	67,5	66,3	65,1	**69,9**
4	50,6	46,8	50,6	51,9	53,2	45,6	50,6	54,4	46,8	45,6	**54,4**
5	51,3	42,3	42,3	37,2	44,9	53,8	50	38,5	53,8	47,4	**53,8**
6	52,2	55,1	49,3	47,8	52,2	56,5	49,3	47,8	49,3	46,4	**56,5**
7	55,6	50,6	56,8	50,6	44,4	46,9	43,2	42	53,1	49,4	**56,8**
8	62,5	56,3	58,8	61,3	62,5	58,8	57,5	57,5	56,3	48,8	**62,5**
9	58,6	58,6	54,3	57,1	38,6	61,4	54,3	54,3	54,3	50	**61,4**

Second Method Performance. The performance of the second method was tested in the same way as the first method. Tables 5 and 6 shows the testing results for the IIIA and 2A datasets respectively. The average accuracy for the IIIA and 2A datasets is shown in Table 9.

Table 5. Performance of the different training states of the ANN on the IIIA dataset, applying the second method proposed.

User	Percentage of successfully classified windows for every training state. Second method. IIIA dataset (%)										Best result (%)
	1	2	3	4	5	6	7	8	9	10	
1	51,1	71,1	57,8	64,4	64,4	53,3	51,1	46,7	62,2	66,7	**71,1**
2	51,7	48,3	62,1	51,7	48,3	51,7	58,6	51,7	58,6	55,2	**62,1**
3	48	36	68	60	56	36	56	64	52	60	**68**

Table 6. Performance of the different training states of the ANN on the 2A dataset, applying the second method proposed

User	Percentage of successfully classified windows for every training state. Second method. 2A dataset (%)										Best result (%)
	1	2	3	4	5	6	7	8	9	10	
1	53,6	46,4	56	44	50	47,6	41,7	47,6	51,2	53,6	**56**
2	42,2	56,6	48,2	48,2	51,8	49,4	50,6	45,8	44,6	49,4	**56,6**
3	68,7	67,5	56,6	60,2	65,1	65,1	62,7	66,3	69,9	63,9	**69,9**
4	48,1	51,9	60,8	45,6	49,4	54,4	50,6	51,9	45,6	40,5	**60,8**
5	51,3	44,9	46,2	44,9	59	41	51,3	53,8	47,4	46,2	**59**
6	52,2	44,9	47,8	52,2	44,9	52,2	47,8	56,5	43,5	52,2	**56,5**
7	49,4	49,4	56,8	38,3	54,3	50,6	56,8	48,1	42	56,8	**56,8**
8	48,8	56,3	53,8	51,2	52,5	48,8	52,5	47,5	57,5	51,2	**57,5**
9	62,9	67,1	54,3	57,1	61,4	55,7	54,3	67,1	60	61,4	**67,1**

Third Method Performance. The third method performance was tested just as the other methods. Tables 7 and 8 shows the classification accuracy for the IIIA and 2A dataset. Table 9 illustrates the average accuracy of the third method.

Table 7. Performance of the different training states of the ANN on the IIIA dataset, applying the third method proposed.

User	Percentage of successfully classified windows for every training state. Third method. IIIA dataset (%)										Best result (%)
	1	2	3	4	5	6	7	8	9	10	
1	48,9	51,1	51,1	48,9	53,3	55,6	46,7	46,7	57,8	53,3	**57,8**
2	65,5	31	48,3	55,2	44,8	51,7	51,7	72,4	55,2	51,7	**72.4**
3	68	44	52	48	76	64	72	56	64	72	**76**

Table 8. Performance of the different training states of the ANN on the 2A dataset, applying the third method proposed.

User	Percentage of successfully classified windows for every training state. Third method. 2A dataset (%)										Best result (%)
	1	2	3	4	5	6	7	8	9	10	
1	51,2	52,4	58,3	35,7	58,3	53,6	52,4	53,6	53,6	52,4	**58,3**
2	51,8	51,8	49,4	49,4	60,2	47	51,8	47	51,8	43,4	**60,2**
3	73,5	72,3	69,9	74,7	75,9	74,7	71,1	69,9	73,5	75,9	**75,9**
4	50,6	55,7	44,3	49,4	45,6	54,4	49,4	54,4	58,2	49,4	**58,2**
5	48,7	47,4	44,9	38,5	60,3	37,2	44,9	57,7	46,2	47,4	**60,3**
6	46,4	47,8	40,6	52,2	46,4	50,7	50,7	58	46,4	47,8	**58**
7	53,1	61,7	58	54,3	60,5	51,9	55,6	51,9	46,9	54,3	**61,7**
8	55	56,3	66,3	60	61,3	55	61,3	47,5	48,8	57,5	**66,3**
9	60	64,3	70	58,6	62,9	65,7	61,4	44,3	60	67,1	**70**

Table 9. Average accuracy of the feature extraction methods.

	Rhythm's detection methods		
	IIIA (%)	2A (%)	Total
First	67,30	58,64	60,81
Second	37,07	60,02	61,78
Third	68,73	63,21	64,59

4   Conclusions

The detection of alpha waves with high levels of accuracy was obtained through the proposed method based on wavelet features and classification by neural networks. This result allowed the implementation of a simple BCI which activates a set of outputs every time alpha rhythms are perceived. As future work, we propose the development of a training paradigm that will allow the user to generate alpha waves without the necessity of closing their eyes, so that the BCI can be handled without the involvement of any muscular activity. Also, since BCIs are real-time systems, the time window of 1 s used in this work needs to be reduced in order to allow a more fluent use of the BCI.

As for the discrimination of left hand and right hand motor imagery, the highest accuracy obtained was of 64,59% when using the third proposed method. Although a better result that the ones obtained with the other two methods, it still has plenty of room for improvement. As part of our future work, the refining of these methods will be pursued until a more acceptable identification accuracy can be reached, features extraction and classification methods.

Finally, it is important to note that even though EEG is a method that rely in several channels of recordings, this work shows, as it has been demonstrated several times before, that a neural potential can be detected using only the channels over the areas of the scalp associated with that potential.

References

1. Graimann, B., Allison, B., Pfurtscheller, G.: Brain-computer interfaces: a gentle introduction. In: Brain-Computer Interfaces, pp. 1–27. Springer, Heidelberg (2009)
2. Schalk, G., Mellinger, J.: A Practical Guide To Brain-Computer Interfacing with BCI2000. Springer (2010)
3. Nicolas-Alonso, L.F., Gomez-Gil, J.: Brain computer interfaces, a review. Sensors 12(2), 1211–1279 (2012)
4. Rao, R.P.: Brain-computer interfacing: an introduction. Cambridge University Press, Cambridge (2013)
5. Cohen, M.X.: Analyzing Neural Time Series Data: Theory and Practice. MIT Press, Cambridge (2014)
6. Xu, B., Song, A.: Pattern recognition of motor imagery EEG using wavelet transform. J. Biomed. Sci. Eng. 1(01), 64 (2008)
7. Alomari, M., Awada, E., Samaha, A., Alkamha, K.: Wavelet-based feature extraction for the analysis of EEG signals associated with imagined fists and feet movements. Comput. Inf. Sci. 7(2), 17 (2014)
8. McFarland, D.J., Miner, L.A., Vaughan, T.M., Wolpaw, J.R.: Mu and beta rhythm topographies during motor imagery and actual movements. Brain Topogr. 12(3), 177–186 (2000)
9. Neuper, C., Pfurtscheller, G.: Evidence for distinct beta resonance frequencies in human EEG related to specific sensorimotor cortical areas. Clin. Neurophysiol. 112(11), 2084–2097 (2001)
10. Lotte, F., Guan, C.: Regularizing common spatial patterns to improve BCI designs: unified theory and new algorithms. IEEE Trans. Biomed. Eng. 58(2), 355–362 (2011)

Dynamics of a Unicycle-Type Wheeled Mobile Manipulator Robot

José Varela-Aldás[1]([✉]) [ID], Fernando A. Chicaiza[2] [ID],
and Víctor H. Andaluz[3] [ID]

[1] Facultad de Ingeniería y Tecnologías de la Información y la Comunicación,
Grupo de Investigación en Sistemas Industriales, Software y Automatización,
Universidad Tecnológica Indoamérica, 180103 Ambato, Ecuador
josevarela@uti.edu.ec
[2] Instituto de Automática, Universidad Nacional de San Juan,
5400 San Juan, Argentina
fachicaiza@inaut.unsj.edu.ar
[3] Departamento de Eléctrica y Electrónica,
Universidad de las Fuerzas Armadas - ESPE, 171103 Sangolquí, Ecuador
vhandaluz1@espe.edu.ec

Abstract. Mathematical modeling is widely used in robotics to display behaviors and design control algorithms; the dynamic model of a robot describes the movements considering the action of internal and external forces of the system. This paper presents the dynamics of a unicycle mobile manipulator robot using the Euler-Lagrange proposal and conditioning the model to obtain the output velocities in the robot from the desired velocities. Additionally, the dynamic model is complemented by the identification of parameters using experimental data obtained from the AKASHA robot. The results present the signals applied in the identification, the values of the dynamic constants are found and the validation data which indicate a correct performance of the proposal are graphically shown.

Keywords: Mobile manipulator · Dynamic modeling · Euler-Lagrange · Parameters identification

1 Introduction

Mobile manipulators are capable robots of combining locomotion and manipulability, eliminating some limitations which exist in the manipulator and the mobile robot when they work independently [1]. Due to the complexity of these robots, it is common to analyze their operation by simulation [2], allowing observing its behavior and designing control algorithms [3]. The current proposals make variations which improve the performance of these robots [4] or make the most of their advantages [5].

It is common to use mathematical models to understand the physical behavior of a robot, allowing the analysis of the system variables [6]. The modeling of mobile manipulative robots has allowed to optimize their characteristics [7], as well as, implement autonomous tasks, where it is common to use the kinematic model of the

© Springer Nature Switzerland AG 2020
M. Botto-Tobar et al. (Eds.): ICAETT 2019, AISC 1067, pp. 24–33, 2020.
https://doi.org/10.1007/978-3-030-32033-1_3

system, because it allows to know its basic behavior and design autonomous movements [8, 9]. On the other hand, the static model allows to analyze its physical behavior only using the fixed positions of the joints [10], and the most complex model is one which defines its dynamic behavior.

Dynamic modeling is used in robotics to analyze the behavior of the system in various conditions [11], even when a human operator interacts with the system [12]. The complexity of these models makes it hard to have all the parameters of the mathematical expression, for this reason, it is common to identify experimentally the dynamic parameters which guarantee the accuracy of the obtained model [13], where these models will allow a mobile manipulator to perform controlled movements with greater accuracy [14].

Regarding similar jobs, [15] uses a dynamic model which decouples the robotic platform and the robotic arm to control the operative end of a mobile manipulator; the experiments shown a correct functioning in the tracking of trajectories; [16] performs kinematic modeling through mathematical transformations and dynamic modeling using the Lagrange formula, where the results are validated using the displacements in the joints of the robot; and [17] determines the dynamic model of a mobile manipulator assisted by a multibody system; the errors of the tests indicate a maximum error of 15%. The main contribution of the proposed model in this paper is to obtain the dynamics using robot velocities as analysis variables.

This work presents the dynamic modeling of a unicycle mobile manipulator robot, considering the arm-mobile system as a single, the model has as input variables the desired velocities for the robot actuators, and the real velocities of the robot as output variables. The results are obtained using the AKASHA robot (a research robot built using Dynamixel Pro smart motors). The document is organized as follows: *(i) Modeling* details the kinematic and dynamic model by the Euler-Lagrange method and other equations; *(ii) Identification of parameters* presents the method for identifying the dynamic parameters; *(iii) Results* shows the determined constants and the validation of the model; and, *(iv) Conclusions* indicates the main interpretations of this work.

2 Modeling

2.1 Kinematic

All the components of the modeling are established, as shown in Fig. 1, establishing three points of interest ($\mathbf{h_1}$, $\mathbf{h_2}$, $\mathbf{h_3}$). Nomenclature of the mathematical components is described in Table 1. By means of geometry, the position of $\mathbf{h_1}$ is determined as a function of the elements, and by applying the temporal derivative, Eq. (1) is obtained, which describes the velocities in the mobile base of the robot.

$$\dot{\mathbf{h}}_1 = \begin{bmatrix} \dot{h}_{1x} \\ \dot{h}_{1y} \\ \dot{h}_\theta \end{bmatrix} = \begin{bmatrix} \dot{O}_x - l_1 \dot{\theta} S_\theta \\ \dot{O}_y + l_1 \dot{\theta} C_\theta \\ \dot{\theta} \end{bmatrix} \tag{1}$$

Remark: To reduce the presentation of mathematical expressions, consider that S_θ, S_i, and S_{ij} are equivalent to $sin(\theta)$, $sin(q_i)$, and $sin(q_i + q_j)$, respectively, and similar equivalency with cosine.

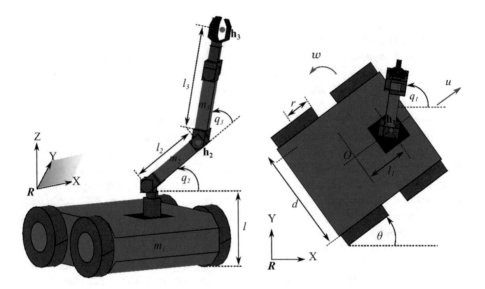

Fig. 1. Elements of the mathematical modeling of the mobile manipulator.

Table 1. Nomenclature of the components of the mathematical model.

Nomenclature	Definition
r	Wheel radius
d	Distance between wheels
l	Height from ground to the first link
$l1$	Distance between the center of the mobile and the arm
$l2$	Length of the first link of the arm
$l3$	Length of the second link of the arm
m_1	Mass of the mobile robot
m_2	Mass of the first link of the arm
m_3	Mass of the second link of the arm
q_1	Angle of rotation in Z of the arm
q_2	Angle of rotation in Y of the first link of the arm
q_3	Angle of rotation in Y of the second link of the arm
θ	Angle of rotation in Z of the mobile robot
u	Linear velocity of the mobile robot
ω	Rotating velocity of the mobile robot
$\mathbf{h_3}$	Position XYZ of the end effector
$\mathbf{h_2}$	XYZ position of the end of the first link of the arm
$\mathbf{h_1}$	XY position of the base of the arm in the mobile robot
$\mathbf{R}$	Reference system

Similarly, the geometry of $\mathbf{h_2}$ is obtained, and the velocities at the end of the first link of the manipulator is found by derivation according to Eq. (2), complemented with Eqs. (3), and (4).

$$\dot{\mathbf{h}}_2 = \begin{bmatrix} \dot{h}_{2x} \\ \dot{h}_{2y} \\ \dot{h}_{2z} \end{bmatrix} = \begin{bmatrix} \dot{O}_x - (l_1 + l_2 C_2 C_1)\dot{\theta}S_\theta - l_2 C_2 S_1 \dot{\theta}C_\theta + aC_\theta - bS_\theta \\ \dot{O}_y + (l_1 + l_2 C_2 C_1)\dot{\theta}C_\theta - l_2 C_2 S_1 \dot{\theta}S_\theta + aS_\theta + bC_\theta \\ l_2 C_2 \dot{q}_2 \end{bmatrix} \tag{2}$$

$$a = (-l_2 C_2 S_1 \dot{q}_1 - l_2 C_1 S_2 \dot{q}_2) \tag{3}$$

$$b = (l_2 C_1 C_2 \dot{q}_1 - l_2 S_1 S_2 \dot{q}_2) \tag{4}$$

Finally, is obtain the kinematics of the end effector of the robot, the Eq. (5) define the velocities at the operative end, complemented with Eqs. (6), (7), (8), and (9).

$$\dot{\mathbf{h}}_3 = \begin{bmatrix} \dot{h}_{3x} \\ \dot{h}_{3y} \\ \dot{h}_{3z} \end{bmatrix} = \begin{bmatrix} \dot{O}_x - (l_1 + c)\dot{\theta}S_\theta - e\dot{\theta}C_\theta + mC_\theta - nS_\theta \\ \dot{O}_y + (l_1 + c)\dot{\theta}C_\theta - e\dot{\theta}S_\theta + mS_\theta + nC_\theta \\ (l_3 C_{23} + l_2 C_2)\dot{q}_2 + l_3 C_{23}\dot{q}_3 \end{bmatrix} \tag{5}$$

$$c = l_2 C_2 C_1 + l_3 C_{23} C_1 \tag{6}$$

$$e = l_2 C_2 S_1 + l_3 C_{23} S_1 \tag{7}$$

$$m = -(l_2 C_2 S_1 + l_3 C_{23} S_1)\dot{q}_1 - (l_2 C_1 S_2 + l_3 S_{23} C_1)\dot{q}_2 - l_3 S_{23} C_1 \dot{q}_3 \tag{8}$$

$$n = (l_3 C_{23} C_1 + l_2 C_1 C_2)\dot{q}_1 - (l_3 S_{23} S_1 + l_2 S_1 S_2)\dot{q}_2 - l_3 S_{23} S_1 \dot{q}_3 \tag{9}$$

2.2 Dynamic

The mobile manipulator dynamics is analyzed with the Euler-Lagrange formulation $L = K - G$, where the kinetic energy K is defined in Eq. (10) and the potential energy G in Eq. (11).

$$K = \frac{1}{2}\dot{\mathbf{h}}_1^{\mathrm{T}} m_1 \dot{\mathbf{h}}_1 + \frac{1}{2}\dot{\mathbf{h}}_2^{\mathrm{T}} m_2 \dot{\mathbf{h}}_2 + \frac{1}{2}\dot{\mathbf{h}}_3^{\mathrm{T}} m_3 \dot{\mathbf{h}}_3 \tag{10}$$

$$G = (m_2 + m_3)gh + m_2 gl_2 S_2 + m_3 gl_2 S_2 + m_3 gl_3 S_{23} \tag{11}$$

Applying the Lagrangian it is obtain the dynamic Eq. (12), where $\mathbf{M(q)}$ is the inertial matrix, $\mathbf{C(q, \dot{q})}$ the centrifugal and centripetal force matrix, $\mathbf{g(q)}$ is the vector of gravity, and the vector $\mathbf{q}$ is composed of O_x, O_y, θ, q_1, q_2, and q_3.

$$\mathbf{f} = \mathbf{M(q)}\ddot{\mathbf{q}} + C(\mathbf{q}, \dot{\mathbf{q}})\dot{\mathbf{q}} + \mathbf{g(q)} \tag{12}$$

Next, the model is conditioned to express it in function of the robot velocities $\mathbf{v} = [\mu \;\; \omega \;\; \dot{q}_1 \;\; \dot{q}_2 \;\; \dot{q}_3]^{\mathrm{T}}$, taking advantage of the unicycle-type characteristics of the robot.

Starting from the transformation (13), its corresponding derivative, and replacing them in (12), the model (14) is obtained.

$$\dot{\mathbf{q}} = \mathbf{S}(\theta)\mathbf{v} \tag{13}$$

$$\mathbf{f} = \mathbf{M}(\mathbf{q})\mathbf{S}(\theta)\dot{\mathbf{v}} + M(\mathbf{q})\dot{\mathbf{S}}(\theta,\dot{\theta})\mathbf{v} + \mathbf{C}(\mathbf{q},\dot{\mathbf{q}})\mathbf{S}(\theta)\mathbf{v} + \mathbf{g}(\mathbf{q}) \tag{14}$$

Additionally, the transformation $\mathbf{B}(\theta)$ of forces towards torques of the actuators is use, and the model of the motors, where $\mathbf{D}$ and $\mathbf{E}$ are constant matrices and V, are the voltages applied to the motors. Also, a PD controller of the Eq. (15) is added in order to supply the desired velocities to the robot $\mathbf{v}_{\mathbf{ref}}$, where $\mathbf{L}$ and $\mathbf{J}$ are the gains of the control; then is obtained the expression (16).

$$\upsilon = \mathbf{L}[\mathbf{v}_{\mathbf{ref}} - \mathbf{v}] - \mathbf{J}\dot{\mathbf{v}} \tag{15}$$

$$\mathbf{f} = \mathbf{B}(\theta)[\mathbf{D}[\mathbf{L}\mathbf{v}_{\mathbf{ref}} - \mathbf{L}\mathbf{v} - \mathbf{J}\dot{\mathbf{v}}] - \mathbf{E}\mathbf{v}] \tag{16}$$

Finally, the mathematical expressions (14) and (16) are matched by finding the dynamic model of Eq. (17). The model found is a function of the velocity of the robot, it is common to control the commercial robots by their velocities (robot AKASHA). The problem of the model is to determine the constants which make up the dynamic matrices, for which the identification of parameters is proposed.

$$\mathbf{v}_{\mathbf{ref}} = \bar{\mathbf{M}}(\mathbf{q})\dot{\mathbf{v}} + \bar{\mathbf{C}}(\mathbf{q},\dot{\mathbf{q}})\mathbf{v} + \bar{\mathbf{g}}(\mathbf{q}) \tag{17}$$

3   Identification of Parameters

Figure 2 presents the methodology used to identify the dynamic parameters, this proposal uses the known values such as the lengths and masses in the robot and experimental data to determine the missing components of the dynamic model. The variables measured are the positions and velocities of the robot, and the accelerations are deduced to apply the identification algorithm. The mobile manipulator used is the AKASHA robot and the identification is carried out offline.

Reordering the model (17) determines Eq. (18), which groups the unknown parameters $\chi_1, \chi_2, \chi_3, \ldots, \chi_{16}$ for the identification.

$$\Gamma(\mathbf{q}, \mathbf{v}, \dot{\mathbf{v}})\chi = \mathbf{v}_{ref} \tag{18}$$

Considering the j-th interaction in the robot data acquisition, the identification proposal is presented in (19).

$$\bar{\Gamma}(\mathbf{q}, \mathbf{v}, \dot{\mathbf{v}})\chi = \bar{\mathbf{v}}_{ref} \tag{19}$$

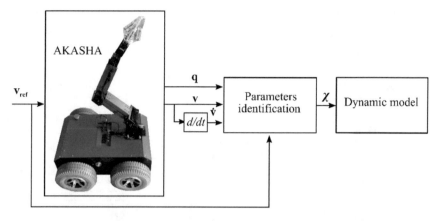

Fig. 2. Diagram of identification of dynamic parameters.

To solve the Eq. (19), the mathematical artifice (20) is used, which allows to solve the unknowns by the problem of the inverse matrix.

$$\bar{\Gamma}(\mathbf{q}, \mathbf{v}, \dot{\mathbf{v}})^T \bar{\Gamma}(\mathbf{q}, \mathbf{v}, \dot{\mathbf{v}}) \chi = \bar{\Gamma}(\mathbf{q}, \mathbf{v}, \dot{\mathbf{v}})^T \bar{\mathbf{v}}_{ref} \tag{20}$$

Equation (21) presents the definitive identification proposal, which determines the 16 unknown parameters using experimental data. In the next section, the results obtained are presented.

$$\chi = \left[\bar{\Gamma}(\mathbf{q}, \mathbf{v}, \dot{\mathbf{v}})^T \bar{\Gamma}(\mathbf{q}, \mathbf{v}, \dot{\mathbf{v}}) \right]^{-1} \bar{\Gamma}(\mathbf{q}, \mathbf{v}, \dot{\mathbf{v}})^T \bar{\mathbf{v}}_{ref} \tag{21}$$

4 Results

4.1 Identification

AKASHA robot has intelligent motors that allow the changing of velocities from the Matlab environment directly, where the experiments are designed with a duration of 10 s. In order to obtain a reliable model it is important to determine the dynamic parameters using input signals which excite the robot to the maximum. In this identification, step type reference velocities with varied amplitudes are used in all system inputs.

Figure 3 shows the identification velocities in the manipulator, showing differences with respect to the actual velocities of the robot. Due to the physical components of the manipulator, the angular velocities in the articulations show a dynamic response (shown in the lag of the read velocities with respect to the reference ones).

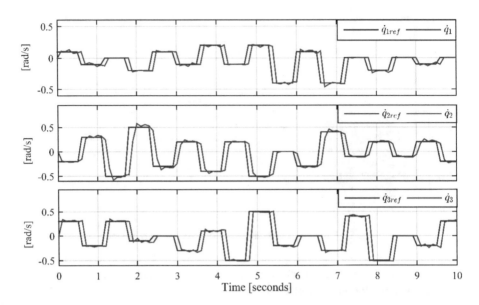

Fig. 3. Identification velocities in the manipulator robot.

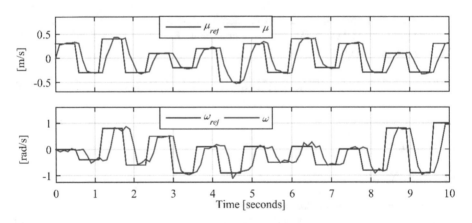

Fig. 4. Identification velocities in the mobile robot.

Table 2. Dynamic parameters.

Parameter	Value	Parameter	Value	Parameter	Value
χ_1	0.1575	χ_7	0.1400	χ_{13}	0.9917
χ_2	0.0055	χ_8	0.0437	χ_{14}	0.9625
χ_3	4.4307e–04	χ_9	0.0059	χ_{15}	0.0580
χ_4	5.9075e–04	χ_{10}	0.0079	χ_{16}	0.0773
χ_5	0.0035	χ_{11}	0.1575		
χ_6	0.0047	χ_{12}	1.0333		

Figure 4 shows the identification velocities in the mobile robot, observing the dynamic response of the system. The mobile robot velocities indicate a lag with respect to the reference velocities and damping in the response of the system, thus, the angular velocity of the mobile robot presents greater dynamic effects.

The values obtained in the identification of the 16 dynamic parameters are presented in Table 2. It is important to note that the units of the parameters are not of interest for the purpose of this work.

4.2 Validation

Once the dynamic constants are included in the model, validation is carried out using other input signals, in this case is use sinusoidal velocities with different amplitude and frequency values.

Figure 5 shows the validation velocities in the manipulator, showing a correct correspondence in the velocities calculated by the model $(\dot{q}_{im})$. The velocity obtained by the model in the second articulation of the arm presents greater differences with respect to the real velocity of the robot, this due to the weight of all the links. This is how errors in the response of the dynamic model have a maximum value of 3%.

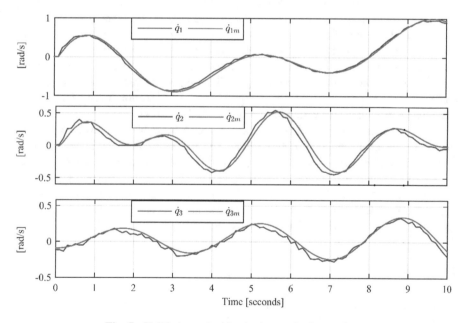

Fig. 5. Validation velocities in the manipulator robot.

Figure 6 presents the validation velocities in the mobile robot, comparing the velocities generated by the dynamic model with the actual velocities of motions, denoting a minimum difference. The velocities given by the model present an accurate tracking of the real velocities.

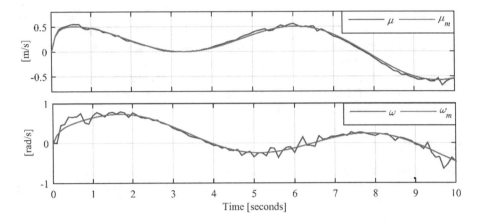

Fig. 6. Identification velocities in the mobile robot.

5 Conclusions

In this work, the dynamics of the mobile manipulator robot is analyzed with the Euler-Lagrange proposal, determining the dynamic equations of the robot from the kinetic and potential energy of the robotic system. To obtain the model in function of the velocities of the robot, different physical models are applied, where the unicycle-like characteristics are used. The model found allows to interpret the dynamic response of the system with the control variables (v_{ref}). These equations require several unknown constants that are grouped for off-line experimental identification, taking into account that angular positions and velocities are inputs of the system.

The experiments are performed in the AKASHA robot and the results are strictly related to the characteristics of this robot. The results of the identification present differences between the desired velocities (inputs) and the actual velocities of the mobile manipulator (outputs) due to the dynamics of the robot, indicating the importance of the model. The 16 dynamic currents are determined and incorporated into the dynamic model for validation, the final results indicate an adequate dynamic response, guaranteeing a model with 3% errors in the validation test.

References

1. Carius, J., Wermelinger, M., Holtmann, K., Hutter, M.: Deployment of an autonomous mobile manipulator at MBZIRC. J. Field Robot. 1–16 (2018)
2. Avilés Sánchez, O.F., Rubiano M, O.G., Mauledoux M, M.F., Valencia, A., Jiménez Moreno, R.: Simulation of a mobile manipulator on webots. Int. J. Online Eng. **14**, 90–102 (2018)
3. Velasco, P.M.: Control based on linear algebra for mobile manipulators. Mech. Mach. Sci. **50**, 79–86 (2018)

4. Ortiz, J.S., Molina, F., Andaluz, V.H., Varela, J., Morales, V.: Coordinated control of a omnidirectional double mobile manipulator. In: Lecture Notes in Electrical Engineering, pp. 278–286 (2017)
5. Varma Kalidindi, V., Vick, A., Krüger, J.: Maximization of operational workspace of a mobile manipulator system. In: 4th International Conference on Control, Automation and Robotics Maximization, pp. 431–436 (2018)
6. Bostelman, R., Foufou, S., Hong, T.: Modeling performance measurement of mobile manipulators. In: IEEE 7th Annual International Conference on CYBER Technology in Automation, Control, and Intelligent Systems, pp. 300–305 (2017)
7. Ancona, R.: Redundancy modelling and resolution for robotic mobile manipulators: a general approach approach. Adv. Robot. **1864**, 1–10 (2017)
8. Galicki, M.: Optimal kinematic finite-time control of mobile manipulators. In: 11th International Workshop on Robot Motion and Control, pp. 129–134 (2017)
9. Giftthaler, M., Farshidian, F., Sandy, T., Stadelmann, L., Buchli, J., Motivation, A.: Efficient kinematic planning for mobile manipulators with non-holonomic constraints using optimal control. In: International Conference on Robotics and Automation (ICRA), pp. 3411–3417 (2017)
10. Han, Y., Wu, J., Liu, C., Xiong, Z.: Static model analysis and identification for serial articulated manipulators. Robot. Comput. Integr. Manuf. **57**, 155–165 (2019)
11. Giordano, A.M., Garofalo, G., De Stefano, M., Ott, C., Albu-Schffer, A.: Dynamics and control of a free-floating space robot in presence of nonzero linear and angular momenta. In: 55th Conference on Decision and Control (CDC), pp. 7527–7534 (2016)
12. Haninger, K., Surdilovic, D.: Identification of human dynamics in user-led physical human robot environment interaction. In: 27th IEEE International Symposium on Robot and Human Interactive Communication (RO-MAN), pp. 509–514. IEEE (2018)
13. Jin, H., Liu, Z., Zhang, H., Liu, Y., Zhao, J.: A dynamic parameter identification method for flexible joints based on adaptive control. IEEE/ASME Trans. Mechatronics **23**, 2896–2908 (2018)
14. Ratajczak, J., Tchon, K.: On dynamically consistent Jacobian inverse for non-holonomic robotic systems. Arch. Control Sci. **27**, 555–573 (2017)
15. Sandy, T., Buchli, J.: Dynamically decoupling base and end-effector motion for mobile manipulation using visual-inertial sensing. In: International Conference on Intelligent Robots and Systems (IROS), pp. 6299–6306 (2017)
16. My, C.A., Bien, D.X., Hieu, C., Packianather, M.: An efficient finite element formulation of dynamics for a flexible robot with different type of joints. Mech. Mach. Theory **134**, 267–288 (2019)
17. Fuentes, A.T., Kipfmueller, M., Prieto, M.A.J.: 6 DOF articulated-arm robot and mobile platform : dynamic modelling as multibody system and its validation via experimental modal analysis. In: IOP Conference Series: Materials Science and Engineering (2017)

Development of a Simulator with Two Degrees of Freedom of the Direction System of Massey-Ferguson's 3640 Agricultural Tractors

André Mixán⬤, Andy Mamani⬤, Leonardo Vinces,
and Christian Del Carpio$^{(\boxtimes)}$⬤

Universidad Peruana de Ciencias Aplicadas (UPC), Lima, Peru
{u201510236, u201319389, pcelcdel}@upc.edu.pe,
leonardo.vinces@upc.pe

Abstract. The present work proposes a simulator of the direction system of an agricultural tractor with the purpose of detecting and correcting operating errors that frequently carry economic and operative consequences in different companies. Therefore, the target of this work is to reduce the incidence rate of such errors by using the proposed simulator to give the operators an adequate training. This simulator is made up of a mechatronic structure that simulates the mechanical direction system, a bank of sensors of positioning of axis, strain gauges to detect forces of acceleration and deacceleration, a controller based on a reduced board processor and a visual interface that virtually shows the movements realized and the operation errors report. The operation guide made by the manufacturer was used as a reference for the validation of the simulator in order to verify if the system identifies clearly the errors typified in the documentation.

Keywords: Simulator · Agricultural tractor direction system · Mechanical design · Unity · Stewart platform

1 Introduction

In general, the companies of the agricultural industry have reported that a great number of their clients declare to have problems with the training of the operators of their tractors due to the lack of training centers or programs with such purpose. In order to face this problem, the present work intends to develop a simulator of the direction system of agricultural tractors, specifically in the Massey Ferguson 3640 tractor, since it is a commercial model in great demand and on which many later models are based. In the literature on simulators, you can find multiple works on these mechanisms dedicated to flying vehicles, formula one, heavy duty mining vehicles, etc. For example, Ortega and Sigut in [1] propose developing a scaled moving platform for high realism flight simulation which they called "Albatros". In this document, the authors base their work on a Stewart platform with pneumatic actuators and magnetic sensors and present the use of a low-cost controller as their contribution. The later must be noted since it's not a robust dispositive or with proven support and therefore might be harmful under certain circumstances. On the other hand, from a point closer to the present work,

M. Botto-Tobar et al. (Eds.): ICAETT 2019, AISC 1067, pp. 34–43, 2020.
https://doi.org/10.1007/978-3-030-32033-1_4

Ojados develops in [2] a tractor conduction simulator with immersive virtual reality for training with the purpose of avoiding work related risks. In his work, the author saw convenient the use of virtual reality environment using Oculus Rift glasses in the videogame development engine Unifty, and the construction of a moving platform with three degrees of freedom. In consequence, with the mentioned tools the author focuses his work in the prevention of over turnings of agricultural tractors, given that his research presents as a fact that the 40% of deaths and injuries are provoked by these events. As a remarkable point, it can be mentioned that the use of an immersive virtual reality environment is sometimes counterproductive since the mechanical actuators are not visible as the glasses block the sight of them.

2 Description of the Proposed Simulator

The block diagram of the proposed simulator is shown in the Fig. 1. It is composed by three stages that will be continued below.

Fig. 1. Block diagram of the proposed simulator

2.1 Mechanical Design

In order to simulate the operation of a Massey Ferguson model 3640 agricultural tractor, a simulator has been designed consisting of a platform resting on a steel structure driven by two pneumatic pistons and two electrovalves with closed centers. The base of the simulator consists of a steel frame and has dimensions of 1.2 m × 1 m, while the moving platform has dimensions of 1 m × 0.8 m. An open cabin similar to the one present in the model tractor has been placed on top of the moving frame. Steel tubes have been used in order to lighten the weight of the frames as much as possible and the mentioned dimensions where obtained from measurements taken from the model in question.

The mechanical design was primarily focused on the development of the moving platform, the base on which this platform rests and the universal joints that allow both frames to freely articulate as the correct functioning of these components of the simulator are critical for the safety of the occupier.

2.1.1 Design of the Moving Frame
Due to the fact that the moving platform will be subject to different levels of acceleration during its operation and that it must safely support the weight of a person and all the equipment required for the operation of the simulator, it is important that the material of which this frame is built of can support these loads.

To determine the material from which the base and the moving platform will be built of, a local search of different metallic tube providers was conducted. Three different materials were considered which were structural Steel ASTM A500 of both grade A as in grade B and the ASTM A513 Steel. Finally, it was decided to use square ASTM A500 grade A structural steel tubes to construct both frames of the simulator. This material was chosen because out of the ones considered it was the one that offered the necessary stiffness at an accessible cost. Table 1 shows the most relevant mechanical properties for the design of the mentioned materials along with their approximate costs.

Table 1. Mechanical properties

Material	Elastic limit (MPa)	Rupture tension (MPa)	Cost of a 6 m long bar ($)
ASTM A500 grade A	270	310	45
ASTM A500 grade B	315	400	50
ASTM A513	500	600	70

In order to consider the acceleration of the moving platform due to its movements in the design, a drive test was carried out using a real tractor of the mentioned model in which the inclination angles of the cabin as the tractor was crossing an irregular terrain were measured using an IMU (Inertial Measuring Unit). With the information of the inclination angles measured during the drive test and using the geometry of the design of simulator, it was possible to calculate the angular velocity, angular acceleration and finally the linear acceleration of the corners of the moving platform, which are the points that would develop the greatest linear accelerations during the operation of the simulator. The graphs of the inclinations measured during the drive test are presented in the Figs. 2 and 3. In the Fig. 2, the Y axis represents an axis parallel to the ground and with positive direction towards the front of the tractor, while in the Fig. 3, the Z axis represents an axis parallel to the ground and with positive direction towards the right side of the tractor. The values of the axes of the ordinates in the graphs represent the angles measured in sexagesimal angles, while the values on the axes of the abscissas represent the elapsed time of the test in minutes. The sampling frequency of the IMU was of 10 Hz.

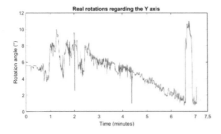

Fig. 2. Inclinations measured with respect to the Y axis

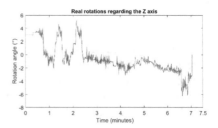

Fig. 3. Inclinations measured with respect to the Z axis

As a result of these tests, it was determined that the highest linear velocity developed by any point of the tractor was of 0.2 m/s. Having determined this, and as the linear acceleration and linear velocity of the corner of the simulator are the same as the ones of the rod of the pneumatic cylinder attached to that corner, an extension test of one the cylinders was carried out in order to determine its linear acceleration. In this test, the pneumatic cylinder had a linear potentiometer attached to its rod in order to determine the extension and velocity of the rod at any given time of the test. The result of this test was that the rod reaches a velocity of 0.2 m/s in approximately 0.1 m. Taking into account that the rod had an initial velocity of 0 m/s and assuming that it performs a uniformly varied rectilinear movement, it is possible to calculate the acceleration of the rod and of the point of the platform attached to it using (1).

$$a = \frac{V_f^2}{2 \times d} = 0.2 \, \mathrm{m/s^2} \tag{1}$$

Where a is the linear acceleration of the rod in m/s^2, V_f is the final velocity of 0.2 m/s and d is the distance in which the rod reaches the velocity V_f, 0.1 m in this case. As a result of these tests, it was determined that the highest linear acceleration that the simulator can develop on its corners due to the pneumatic cylinders drive is 0.2 m/s^2, which when added to the acceleration of the gravity gives a total acceleration of approximately 10 m/s^2.

Once the material was chosen and having determined the maximum linear acceleration that the mobile platform will experience, we proceeded with the selection of the appropriate profile for the tubes with which the structure will be built. For this purpose, a generic design of the structure with the geometry of the mobile frame mentioned above was made. In this design the only variables are the length of the sides of the tube and its thickness, while the frame of the mobile platform itself has a squared shape and its sides have a length of 1 m and 0.8 m. Once the generic design was done, stress and deformation tests were made for each tube profile using the Auto-desk Inventor 2017 software.

Considering the results obtained previously, the acceleration of the gravity was set at 10 m/s^2 in the simulations to consider the combined effect of the acceleration produced by the real gravity and that produced by the pneumatic cylinders drive. For these simulations the information of the selected material, structural steel ASTM A500 grade A, available in the native library of materials available in the program, was used. Additionally, since the simulator must be able to support the load of a person weighing approximately 100 kg plus another 20 kg of instruments and devices, a load of 120 kg was configured in the form of a force of 1200 N (120 kg with a gravity of 10 m/s^2) uniformly distributed over the upper face of the mobile frame. Finally, the holes through which the pins that secure the mobile platform to the frame of the base and to the cylinders which are located in the rear corners and the center of the frame were set as fixed points.

As a result of these simulations, the square profile with a side length of 76.2 mm and a thickness of 3 mm was selected. Table 2 shows the results of the simulations on the selected profile, along with those of the immediate smaller and those of the immediate larger.

Table 2. Results of the simulations

Dimensions of the squared tube (mm)	Maximum deformation (mm)	Safety factor
Side length: 63.5/Thickness: 2.5	0.22	0.95
Side length: 76.2/Thickness: 3	0.14	1.37
Side length: 101.6/Thickness: 4	0.08	1.55

The results of the simulations on the frame designed using the selected tube are presented in the Fig. 4. As it can be seen in this figure, the maximum deformation obtained is given in the front part of the frame and is of 0.14 mm, a rather small value considering the dimensions of the frame. On the other hand, as can be seen in Fig. 5, the minimum safety factor is 1.37 and although the ideal would be to have a safety factor as high as possible, in this case the areas that present this value are at the edges of the holes of the bolts that hold the frame and are very small. If it is desired to increase the minimum safety factor, this could be achieved by increasing the number of bolts per joint point from 4 to 6 or 8 bolts, with which the load between the points of union would be better distributed and therefore the stress in those points would be reduced.

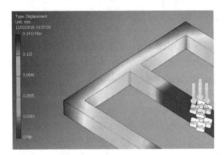

Fig. 4. Heat diagram of the maximum deformations obtained

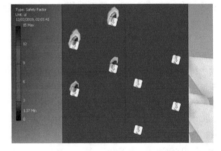

Fig. 5. Heat diagram of the minimum safety factor obtained

For the design of the base frame, the same procedure was followed using the same tube, obtaining slightly better results in terms of safety factor and maximum deformation.

2.2   Control System

To carry out the control of the simulator, one must know the extensions that the actuators must have in order for the simulator to perform the movements. Therefore, these values were determined through the inverse kinematics of the system in a similar way to that expounded by Eftekhari and Liu in [3] and [4] respectively. This process begins by defining an inertial and a non-inertial coordinate system, considering the first to be the plane of the base frame of the simulator located on the ground and the second, in the mobile frame of the simulator. Once the described coordinate systems were defined, a group of vectors was defined in order to be able to perform the calculations of the inverse kinematics of the simulator. Both the distribution of the coordinate systems and the vectors defined for that purpose are shown in Fig. 8.

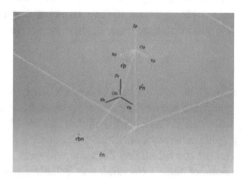

Fig. 8. Diagram of the simulator's vector analysis

In Fig. 8, the vector r_p is defined once, while the other vectors must be defined for each of the actuators that the simulator has, being n the sub index of the actuator. When these vectors are added, (2) is obtained for each actuator of the simulator.

$$r_p + P_n - r_n - r_{bn} = 0 \qquad (2)$$

As the vector P_n is defined in the non-inertial coordinate system and the other vectors are defined in the inertial system, it is not possible to calculate the sum shown in (2) as it is expressed. For this it is necessary to define the vector P'_n, which allows the vector P_n to be expressed in the inertial reference system, for which it is necessary to define the rotation matrix R^l_p as indicated in (3).

$$R^l_P = \begin{bmatrix} cos\theta cos\psi sin\theta sin\varphi sin\psi & sin\theta sin\varphi cos\psi - sin\psi cos\theta & cos\varphi sin\theta \\ cos\varphi sin\psi & cos\varphi cos\psi & -sin\varphi \\ cos\theta sin\varphi sin\psi - cos\psi sin\theta & sin\theta sin\psi + cos\theta sin\varphi cos\psi & cos\varphi cos\theta \end{bmatrix} \qquad (3)$$

Where:

R^l_p: Transformation matrix used to express a vector defined in the non-inertial coordinate system in the inertial one.
θ: Euler angle regarding the x axis.
φ: Euler angle regarding the y axis.
ψ: Euler angle regarding the z axis.

Once calculated the transformation matrix R_P^l, it's possible to calculate the vector P_n' using (4).

$$P_n' = P_n \times R_P^l \tag{4}$$

Since the objective of performing the inverse kinematics is to determine the extension that each of the actuators must have to reach a particular position and orientation of the mobile platform, the vector r_n must be calculated using (5).

$$r_n = r_{bn} - r_p - P_n' \tag{5}$$

Once the vector r_n has been calculated, the required length of any given actuator can be determined by obtaining the module of the vector r_n associated to that actuator.

In order to control the inclinations of the mobile frame, (5) will be used in the programming of the graphic environment, which will determine the inclinations of the virtual tractor when crossing the simulation field and will enter said inclinations to (4) along with the dimensions of the mobile frame to calculate the extension that each cylinder should have. Once these extensions are calculated, they will be sent to the microcontroller that controls the cylinders.

The microcontroller used was the TMDS570LS31 from Texas Instruments. This embed is based on an ARM-Cortex-R4F. As indicated by the manufacturer in [5], it has been specifically designed to be used in industrial applications that require high safety in terms of reliability of the handling of its inputs and outputs, and supports adverse operating conditions such as work environments with low or high temperatures (−40 to 125 °C), so it is ideal to perform the control of the actuators, a task that is vital for the safety of the occupant of the simulator. Given that the proposed system requires 6 digital inputs, 6 analog inputs and 2 digital outputs, the TMDS570LS31 meets the desired requirements. This microcontroller has a native USB port that will facilitate the transmission of data between itself and the computer.

2.3   Graphic Simulation

The virtual environment has been developed in the UNITY software. The developed environment has a dimension of 1024 × 1024 m long by wide, a maximum height of 100 m and a minimum unit for distance of one meter. As for the type of terrain, it was considered that it was similar to the land of the cultivated areas of the Peruvian coast, so the texture Albedo (RGB) Smoothness (A) whose appearance is similar to this land was used. Likewise, elevations have been created in the terrain with various shapes as shown in points 1 to 5 in Fig. 9 and in Fig. 10, with a maximum elevation of six meters located in point 1. In the Fig. 9 and in Fig. 10 there is shown a panoramic view and a top view of the graphic environment respectively.

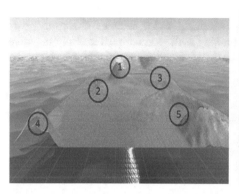

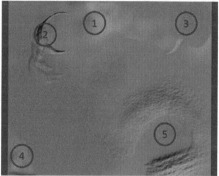

Fig. 9. Panoramic view of the graphic environment

Fig. 10. Top view of the graphic environment

The communication between the PC that contains the graphical environment and the microcontroller is serial at a baud rate of 9600 with 8 bits, without parity bit and with two stop bits. The data that is transmitted corresponds to the information that will be necessary to determine the errors commited the operator. The number of data that the graphic environment program receives is seven and these are shown in Fig. 11.

4 bits	2 bits	2 bits	8 bits	8 bits	2 bits	8 bits
Current gear shift	Curent type of gear shift	Sense of travel	Level of pressure on the brake pedal	Level of pressure on the acceler-ator pedal	Level of pressure on the clutch pedal	Position of the steering wheel

Fig. 11. Reception frame of the graphic environment

The number of data that the graphical program sends is two. These are the extensions that the rods of the cylinders must have for the simulator to reach the position that the virtual tractor has. These extensions are obtained by means of Eqs. (2), (3), (4) and (5), from the angles of Euler with respect to the axes x, and z obtained from the mobile in the virtual environment.

8 bits	8 bits
Extension of cylinder 1	Extension of cylinder 2

Fig. 12. Sending frame of the graphic environment

2.3.1   Error Detection Method

The error detection algorithm contemplates critical cases such as changing gears without using the clutch, accelerating and braking at the same time, etc. Only negative results are shown, as these are displayed in the message panel. Figure 13 shows the flow diagram for error detection.

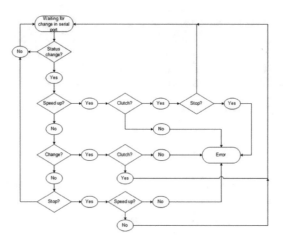

Fig. 13. Flow diagram for the error detection

3   Results Obtained

Having checked the strength of the frames and universal joints designed, they were assembled and added to the general assembly of the simulator. Below are two views of the overall assembly of the complete designed simulator.

Fig. 14. Rear left view of the general assembly of the simulator

Fig. 15. Rear right view of the general assembly of the simulator

A 5-min test session was carried out, in which intentional maneuvers were made wrongly. In this way, it was visually checked that the system detects the faults of the operator and shows them in its report as shown in Fig. 16. Likewise, as can be seen in the right part of the observer's view, it indicates the type and the number of the gear.

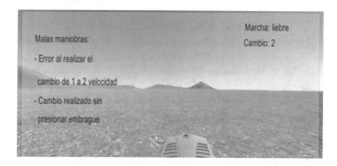

Fig. 16. Operator view

References

1. Ortega, J.: A low-cost mobile prototype for high-realism flight simulation. Revista Iberoamericana de Automatica e Informatica Industrial **13**(3), 293–303 (2016). https://doi.org/10.1016/j.riai.2016.05.002
2. Ojados, D.: Development and assessment of a tractor driving simulator with immersive virtual reality for training to avoid occupational hazards. Comput. Electron. Agric. **143**, 11–118 (2017). https://doi.org/10.1016/j.compag.2017.10.008
3. Eftekhari, M.: Emulation of pilot control behavior across a Stewart platform simulator. Robotica **36**(4), 588–606 (2018). https://doi.org/10.1017/s0263574717000662
4. Liu, K.: Kinematic analysis of a stewart platform manipulator. IEEE Trans. Ind. Electron. **40** (2) (1993). https://doi.org/10.1109/41.222651
5. Texas Instruments: Hercules Safety Microcontrollers. http://www.ti.com/en/download/mcu/SPRB204.pdf?DCMP=hercules&HQS=hercules-mc. Accessed 29 Jan 2019

Database Proposal for Correlation of Glucose and Photoplethysmography Signals

Christian Salamea[1,2], Erick Narvaez[1(✉)], and Melisa Montalvo[1]

[1] Grupo de Interacción, Robótica Automática (GIIRA),
Universidad Politécnica Salesiana Sede Cuenca, Cuenca, Ecuador
enarvaez@ups.edu.ec
[2] Speech Technology Group, Information and Telecommunications Center,
Universidad Politécnica de Madrid, Ciudad Universitaria Av.
Complutense, 40, 28040 Madrid, Spain

Abstract. This work presents a Database that contains Photoplethysmography signals, glucose levels, weight, height and age of 217 patients. The information of biologic activity was obtained using the handle Empatica E4 Wristband, the glucose level using laboratory blood chemistry analyzers (Cobas 6000), and the physical parameters using standardized instruments. The database comprises a forward training a total of 5576 samples and another segment of validation to a total of 2164 samples. The Database has been used to evaluate different prediction techniques based on Machine Learning (Random Forest, Artificial Neural Network, Support Vector Machine, Gradient Boosting Machine). The implementation of these algorithms provides up to 90% average accuracy, a correlation of 0.88 and a satisfactory evaluation in the Error Diagram of Clarke. According to the results obtained, the proposed database is appropriate for training and verification of existing correlation between photoplethysmography signals and blood glucose level.

Keywords: PPG · Glucose · Database · MFCCs · Machine Learning

1 Introduction

Photoplethysmography is a technique used in the measurement of volumetric changes in the blood, directly associated with the light variation intensity applied in the Blood Flow [1]. The instruments that have won the photoplethysmography signal (PPG) are composed of a light emitting source and a photodetector, with these devices, are perceived the difference between the amount of the light reflected and absorbed by the volumetric flow [2]. The PPG signals are used in different fields of biomedicine. The uses of these signals are diverse, such as the detection of cardiovascular function status [3], heart rate meters [4], heart rate monitors [5], stress detection devices [6], and glucose level estimation [7, 8].

A considerable variability has been identified between the different databases (DB) of PPG signals and that each DB is oriented to a specific application. For example, in [8] it can be seen the database has been obtained from 100 patients specifying only the signals PPG and the glucose level. In [9] the DB generated from 30

© Springer Nature Switzerland AG 2020
M. Botto-Tobar et al. (Eds.): ICAETT 2019, AISC 1067, pp. 44–53, 2020.
https://doi.org/10.1007/978-3-030-32033-1_5

patients has been used to develop a non-invasive glucose measurement system based on PPG signals. On the other hand, some DBs are oriented to a specific segment, as in [7] where patients diagnosed with diabetes are included. In other cases, although DB is related to glucose levels, the use of PPG signals as the main component is not observable. In [10] it is observed that the database is constituted by information of people diagnosed with diabetes, and their glucose level is measured at different times of the day. In [7–9] the metric most used for the evaluation of results is the Clarke Diagram, which is accepted as a standard metric for determinate both, accuracy and errors in electronic glucose measurement devices [5]. In addition, metrics are used for the evaluation of the proposed system, such as mean square error, absolute mean error correlation and standard deviation [7, 8, 14].

This article presents a DB composed of PPG signals from 217 participants between the ages of 18 to 65 years, including information about their corresponding physical characteristics (weight-height), the value of glucose and data related to diabetes. This has been implemented in the generation of a methodological proposal for the obtaining of the correlation between these characteristics and the blood glucose level through artificial intelligence systems of supervised learning, this proposal has been approved by The Ethics Committee of Research on Human Beings of the San Francisco University of Quito. The systems obtained have been evaluated by means of the Clarke Error Diagram and system performance metrics proposed in [7–9].

2   Description of the Technique

The DB proposed contains information obtained, on the one part, using the bracelet Empatica E4 Wristband and, on the other, using the clinical-chemical and immunology analysis unit (Cobas 6000). The information obtained with the bracelet has been used as entry parameter vectors in a supervised machine learning system and the data of the unit of analysis as the objective values (labels) used in the training phase.

To obtain the labels, which in essence is the information that the system expected to "learn" from the information obtained from the bracelet Empatica E4, the traditional invasive blood glucose measurement method has been used. The blood sample is obtained fasting to avoid possible alterations in blood chemistry tests [11]. The information of the referential value is measured by means of the Cobas 6000 clinical-chemical analysis unit that has been designed as a fully automated random access analyzer for the clinical chemistry laboratory with the capacity to determinate concentrations or activities of various substances in bodily fluids such as: enzymes, substrates, electrolytes, proteins, drugs, cardiac markers, tumor markers and hormones [15]. The results obtained in the Cobas 6000 are corroborated with a traditional glucose measurement instrument, the "Accu-Check Preforma Nano" that performs quantitative measurement of blood glucose in venous, arterial, neonatal and fresh capillary blood [16].

On the other hand, the vectors of characteristics that are going to be the inputs of the proposed system correspond to the measurement of the PPG signals obtained from the Empatica E4 Wristband bracelet (E4) (Fig. 1). The bracelet is composed of a series of sensors designed for the continuous and real-time acquisition of biometric data [12]. Among the sensors are:

1. Photoplethysmography sensor (PPG)
2. Electro dermal activity sensor
3. 3-axis accelerometer
4. Optical thermometer

The signals produced by each sensor are the following:

– Pulse of blood volume (BVP), sampling at 64 Hz
– Interval between pulsations (IBI)
– Electro dermal activity (EDA), sampling at 4 Hz. Acceleration XYZ with sampling at 32 Hz
– Skin temperature, sampling at 4 Hz

Fig. 1. E4 wristband Empatica bracelet [13].

During the procedure to obtain the signals, the patient must be at rest and seat, should not perform movements with the arm where the bracelet is placed since this may generate unwanted signals within the PPG data set. This procedure is carried out in a period of 3 to 5 min after the first data acquired by the bracelet, which is considered sufficient for the system training. Taking into account the sampling rate of the bracelet of 64 samples per second, 9600 samples were obtained representing an average of information acquisition of 2.5 min. The signals obtained from the E4 bracelet are transferred via a USB 2.0 connection or a Bluetooth wireless connection to a personal computer. The software used to manage the signals from the bracelet is known as "E4 manager", which allows import all of the data collected in an "Excel" type format. The graphs corresponding to the BVP, IBI, EDA and temperature signals can be observed in a conventional computer using the software "E4 connect".

From the totality of the PPG signals, segments of 5 s are extracted from each patient, after ruling out corrupt information. Thus, an output vector of 320 samples is obtained, which is used as an input vector for the pre-processing of the information. This procedure increases the DB to a size of 7740 samples. From the total information obtained, the DB has been divided into 5576 training samples and 2164 validation samples.

On the other hand, in addition to the data obtained from the bracelet (PPG signals), the DB includes information on physical parameters such as the weight and height of the participants. These factors are useful in obtaining the body mass index (1), which has a direct relationship with the blood glucose level [12].

$$ICM = \frac{Height\ (m^2)}{Weight\ (Kg)} \tag{1}$$

The information contained in the DB was obtained from 217 participants of the Public Regional Hospital "Jose Carrasco Arteaga" in Cuenca-Ecuador. The information has been collected considering the following conditions of fitness:

- Patients between 18 and 65 years old. The age limits were adjusted to the restrictions on the dimensions of the bracelet, which was designed for this age range (summary of the characteristics in Tables 1 and 2).
- Pregnant women diagnosed with diabetes were not allowed to participate. The participation of patients in this condition within the age range indicated above was allowed just after signing the consent form. The study doesn't generate any direct risk for the state, and welfare will be ensured throughout the process.
- Patients diagnosed or not with diabetes, in the age indicated in previous paragraphs, the percentage indices of the patients diagnosed are seen in Fig. 2.

Persons under 18 years of age and individuals deprived of their liberty are not included.

Table 1. Physical characteristics of the participants

Measure	Max	Min	Average	Deviation
Age (years)	65	22	48.98	11.34
Weight (Kg)	140	38	75.87	16
Height (m)	1.9	1.41	1.6	0.09

Table 2. Characteristics of the samples collected

Measure	Max	Min	Average	Deviation
Glucose (mg/dL) laboratory	390.7	58.6	114.05	50.13
Glucose (mg/dL) immediate	363	67	115.66	44.7

The reference glucose values of the DB are between 50 to 400 mg/dl, which means that the DB contains information on extreme cases of hyperglycemia and hypoglycemia. The number of patients within the acceptable range of blood glucose is within 70 to 120 mg/dl, as shown in Table 3.

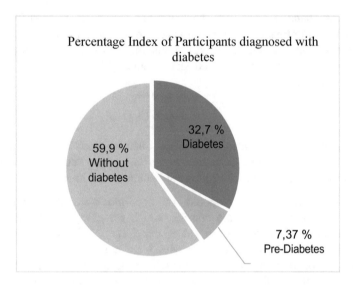

Fig. 2. Percentage index of participants diagnosed with diabetes.

Table 3. Glucose level data obtained.

Range	Glucose value established	Data
1	300–400	5
2	200–300	7
3	100–200	78
4	50–100	124

The information of the DB has been used in the application of various techniques of digital signal processing and "Machine Learning" [7–9], whose results have been made visible using the following metrics: the quadratic error mean (MSE), the absolute mean error (MAE), the standard deviation and the correlation (R), the latter is carried out by relating the reference values and the estimated values of the evaluation data set of the system. The MSE is used in the measurement of the average loss of the system, the standard deviation allows us to observe the average distance between the estimated data compared to the actual values of validation. The MAE indicates an average of the magnitude of the error, without considering the direction of the error produced by the previous metrics, and finally, the correlation indicates the direction of the error defining the level of association between the estimated and reference values of the system used. The mathematical formulations used to calculate the statistical values used are presented below:

$$MSE = \frac{1}{n}\sum_{i=1}^{n}(p_i - r_i)^2 \qquad (2)$$

$$MAE = \frac{1}{n}\sum_{i=1}^{n}|p_i - r_i|^2 \tag{3}$$

$$Standar\ Dev. = \sqrt{\frac{\sum_{i=1}^{n}(p_i - r_i)^2}{n}} \tag{4}$$

$$R = \frac{\sum_{i=1}^{n}[(r_i - \bar{r})(p_i - \bar{p})]}{\sum_{i=1}^{n}(r_i - \bar{r})^2 \sum_{i=1}^{n}(p_i - \bar{p})^2} \tag{5}$$

In Eqs. (2), (3), (4) and (5), n represents the total of the validation samples used, pi the prediction values obtained from the systems and the actual values obtained from the clinical analysis. Together with the evaluation data, the use of the Clarke Error Diagram used in [7–9] is proposed. The data located within Zones A and B represent accurate or acceptable glucose results for use, the remaining zones of the diagram indicate erroneous correlations that lead to negative results for zone C, while significant system failures occur in zones D and E correspondingly. In the latter, the estimation could lead to erroneous diagnoses of hypoglycemia and hyperglycemia [14] (Fig. 3).

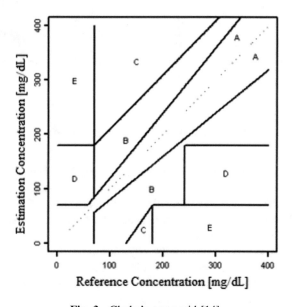

Fig. 3. Clarke's error grid [14].

3   Results and Discussion

The DB is composed of 327 characteristics of approximately 9600 expanded records of 217 patients. The characteristics are divided into three sections. The first one represents the physical characteristics of the participants, the weigh with a maximum of 140 and a minimum of 22 kg, the size in range of 1.9 to 1.41 m, the age in range of 65 to 22 years and the index of body mass (kg/m²) considered as normal if the values are among the indices of 18.5 to 24.9., The percentage ranges are presented in Fig. 4.

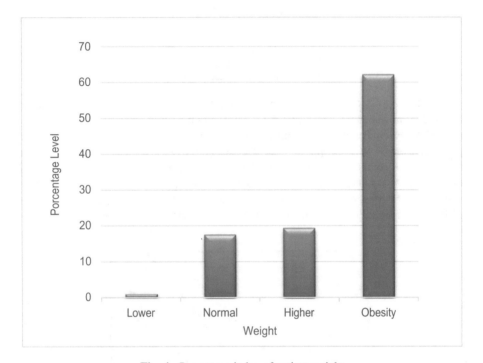

Fig. 4. Percentage index of patient weight

The second section is represented by the characteristics of the PPG signals, these are obtained from the MFCCs of the PPG, the feature vector made up of a vector of 319 characteristics. Which has been normalized between values of 0 to 1 in standard MFCCs. This information in conjunction which the personal information makes the input vector for the proposed system.

The last section is represented by the objective information of fasting blood glucose. The data are divided between a level lower than 70 mg/dl is considered Hypoglycemia, the normal range of 80 to 115 mg/dl, a high range of 116 to 180 mg/dl with cases of prediabetes or poorly controlled diabetes are considered, and levels higher than 215 mg/dl with are considered critical levels, the percentage indices are presented in Fig. 5.

The results of the Glucose characteristics and body mass index presented show that DB has a range of parameters that cover the majority of cases related to diabetes and blood glucose measurement in patients.

The technique of Cepstral Coefficients in Mel Frequency (MFCCs) was implemented for the extraction of the hemodynamic characteristics of PPG signals. The MFCCs coefficients obtained from the PPG signals together with the physical

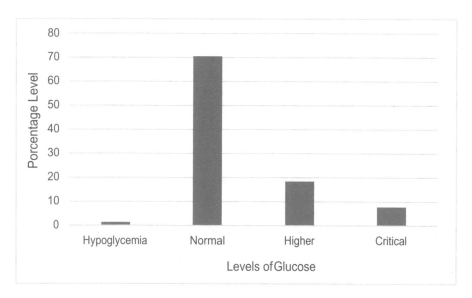

Fig. 5. Percentage index of patient weight.

parameters of patients are implemented as input vector in supervised learning methods with an output vector with information fasting glucose levels. The supervised learning systems used for the evaluation of the implementation of the proposed Database are Artificial Neural Networks (NN), Random Forest (RF), Gradient Boosting Machine (GBM) and Vector Support Machines (SVM). For each method, the results offered by the applied topologies were evaluated in different ways, to then modify the configurations of each method, strengthening the prediction system. The topologies that best presented acceptable accuracy according to the evaluation metrics used are: for RNA a multilayer Perceptron topology of 3 internal layers of 5 neurons each with tanh activation function. For the vector support machine method, a topology with a Kernel function of the polynomial type of grade 3. For the GBM system consisting of 50 internal estimators and a learning rate of 0.1. Finally, for RF, a configuration of 20 decision trees was implemented. The evaluation of the performance of the systems by

Table 4. Results of the system performance evaluation.

System	MSE	MAE	Standard Dev.	R
RF	202.6	8.59	14.23	0.88
NN	386.3	15.6	19.65	0.65
GBM	264.2	9.26	16.27	0.84
SVM	396.5	16.3	19.91	0.55

means of the metrics of (2), (3), (4) and (5) obtained are presented in Table 4.

The technique that presented the best result through the evaluation tools is that of Random Forest, which achieves levels of correlation comparable to other more complex technologies that require higher costs. When making the observations based on the Clarke Error, it is concluded that the system has high precision and the possibility of replacing traditional glucose measurement systems, the information of the precision of the system is specified in the MAE of 8.59 which represents a precision of 91.4%.

4 Conclusions

The DB presented in this article presents a broad universe of data, because we found information of 50 mg/dL up to values of 400 mg/dL, which represent the extreme cases of hypoglycemia and hyperglycemia and a concentration of data within the range of 90–150 mg/dL. Offering a broad universe for the applicability of the DB raised. The use of physical parameters within the agglomerate of data has allowed differentiation between those PPG signals that may be similar in some patients allowing an adequate implementation of the DB.

The results suggest that the PPG signals indicate that the characteristics proposed by the DB promise to be an alternative to the current glucose measurement technology.

The proposed time interval allows obtaining an average feature vector of 2.5 min which, when divided into segments of 5 s, allows a considerable increase in the size of the resulting DB, improving the characteristics that can be used for the systems. Due to the confidentiality of patient data required by the Ecuadorian Ethics Committee and the hospital's internal policies, this database is not freely accessible but can be considered as a metric for the collection of information in various fields for PPG signals and studies related to the subject of glucose.

For future work it will be necessary to collect information in different cities of the country increasing the universe of information, because the height of the city with reference to sea level is a parameter within the PPG signals that should be considered.

To obtain more information about the database, contact csalamea@ups.edu.ec.

References

1. Yamakoshi, T., Lee, J., Matsumura, K., Yamakoshi, Y., Rolfe, P., Kiyohara, D., Yamakoshi, K.: Integrating sphere finger-photoplethysmography: preliminary investigation towards practical non-invasive measurement of blood constituents. PLoS ONE (2015)
2. Allen, J.: Photo plethysmography and this application in clinical physiological measurement. Physiol. Measur. **28**, R1–R39 (2007)
3. Pastoriza Beltrán, J.J.: Desarrollo de un módulo para la teledetección automática del estado de las funciones cardiovasculares. Universidad Estatal Península de Santa Elena, La Libertad-Ecuador (2017)
4. Molina, C., Bladimir, Á.: Módulo Didáctico de un Medidor de la Frecuencia Cardiaca Mediante Fotopletismografía, para el Laboratorio de Instrumentación Biomédica de la Universidad de las Fuerzas Armadas. Universidad De Las Fuerzas Armadas, ESPE, Latacunga-Ecuador (2015)

5. Claudia, A., Duato, A.: Diseño e Implementación de un Pulsómetro Digital Basado en la Fotopletismografía. Universidad politécnica de Valencia, Valencia-España (2014)
6. Martín Sánchez, D.: Diseño de un Dispositivo para la Detección del Estrés a Partir de la Señal de Fotopletismografía. Universidad de Sevilla, Sevilla-España (2015)
7. Monte-Moreno, E.: Non-Invasive estimate of blood glucose and blood pressure from a photoplethysmograph by means of machine learning techniques. Artif. Intell. Med. **53**, 127–138 (2011)
8. Ramasahayam, S., Arora, L., Chowdhury, S.R.: FPGA based smart photoplethysmograph and online correction of motion artificial. In: Sensor of Everyday Life. Springer (2017)
9. Pai, P.P., Sanki, P.K., De, A., Banerjee, S.: NIR photoacoustic spectroscopy for non-invasive glucose measurement. Department of E & ECE, IIT Kharagpur (2015)
10. Diabetes-Database, Machine Learning Repository. http://archive.ics.uci.edu/ml/datasets/Diabetes. Accessed 19 Dec 2018
11. Quevedo, M.C.C., Seringe, S.E.: Preparación del paciente y colección de muestras para análisis de Laboratorio Clínico, Hospital Oncológico Docente "Conrado Benítez" (1999)
12. Empatica: "Utilizing the PPG/BVP signal," Empatica Support (2016). https://support.empatica.com/hc/en-us/articles/204954639-Utilizing-the-PPG-BVP-signal. Accessed 21 Dec 2018
13. Alvarado Torres, L., Gonzales Torres, A., Hernández Reséndiz, M.V., Mercado Hurtado, D.C., Morales García, D., Anaya Loya, M.A.: Relación del Índice de Masa Corporal y las Concentraciones de Glucosa Sérica en Jóvenes Adultos Queretanos. Facultad de Ciencias Naturales. Universidad Autónoma de Querétano, México (2010)
14. Yadav, J., Rani, A., Singh, V., Murari, B.M.: Prospects and limitations of non-invasive blood glucose monitoring using near-infrared spectroscopy. Biomed. Signal Process. Control **18**, 214–227 (2015)
15. Harman-Boehm, I., Avner Gal, A.M., Raykhman, J.D., Zahan, E.N., Mayze, Y.: Non-invasive glucose monitoring: a novel approach. J. Diabetes Sci. Technol. (2009)
16. Roche: cobas 6000 analyzer series: The success story continues. Roche Diagnostics International Ltd., pp. 1–36 (2013)
17. Roche Diagnostics: Accu-Chek Performa Nano (2015). https://www.accuchek.cl/medidores-deglucosa/performa-nano#productDetails. Accessed 21 Dec 2018

Speed Estimator for a Hydraulic System

Erick Narvaez[1]([⊠]), Pablo Sáenz[2], and Walter Orozco[1]

[1] Grupo de Interacción, Robótica Automática (GIIRA),
Universidad Politécnica Salesiana Sede Cuenca, Cuenca, Ecuador
enarvaez@ups.edu.ec
[2] Universidad Politécnica Salesiana Sede Cuenca, Cuenca, Ecuador

Abstract. The estimation of the speed of rotating hydraulic systems (motors) is an important task in processes (control) that merit an exact measurement of the rotating speed. The implementation of a filtering technique based on adaptive filtering algorithms offers a new proposal in the treatment of feedback signals from hydraulic systems (Speed). The implementation of adaptive filtering in obtaining training parameters of Machine Learning algorithms for the estimation of the speed of a variable speed hydraulic system proposes a novel and highly applicable technique. In this article, a speed estimator system for a rotating hydraulic system is proposed using the adaptive filtering technique based on the noise canker topology in conjunction with multilayer neural networks, evaluated by: the mean square error, the absolute average error, the standard deviation and the correlation obtaining values of 0.59, 0.19, 0.23 and 0.99 in comparison with its counterpart of conventional census (transducer). According to the results, the system is appropriate for a speed estimation of the proposed rotary hydraulic system.

Keywords: Neural networks · Hydraulic · Transductor · Signal · Noise · Filter · Speed estimator

1 Introduction

Hydraulic systems are currently a large sector of the industry due to its reliability in these systems are currently a large sector of the industry due to its reliability in torque against heavy loads, its main feature is its low speed and high mechanical noise [1]. The transducers that record the speed of this type of systems are constituted by a mechanical-electrical coupling which can be susceptible to faults leading to erroneous information readings due to high torque and forces driven by the hydraulic actuators compared to the electric actuators [2]. The implementation of these systems generates stochastic electrical noise and difficult filtering for system feedback signals. The identification of a mathematical model that effectively defines a hydraulic system is a laborious task, these models involve specific parameters depending on physical and constructive parameters of the elements such as valves, hoses, engines, etc. [10–12]. Regarding the application of filtering algorithms in applied in which the output signal is required to be as "pure" as possible, it results in the application of analog or digital

© Springer Nature Switzerland AG 2020
M. Botto-Tobar et al. (Eds.): ICAETT 2019, AISC 1067, pp. 54–62, 2020.
https://doi.org/10.1007/978-3-030-32033-1_6

filters in cascade, causing the system to be robust creating a load computational high [3]. Due to these problems, the implementation of adaptive filtering systems is an attractive field in the implementation of various fields such as the improvement of radiation in wireless antennas [4], filtering of biomedical signals [5], elimination of noise in ECG signals [6], image noise filtering [7], active power filters [8]. This type of filtering algorithms has shown a better convergence time compared to conventional analog and digital filters [9]. So the implementation of adaptive filters for the elimination of mechanical noise produced in hydraulic systems is an attractive technique.

It has been identified in relation to the implementation of artificial neural networks in robust power systems such as induction motor failure detection [13], induction motor controls [14], emission prediction in diesel engines [15], other fields. These have demonstrated the efficiency and great utility of intelligent algorithms in the prediction of information. In the field of hydraulics, the implementation of intelligent systems includes systems such as an identification and control recognition system for a system based on hydraulic pistons [16], fuzzy controls for pistons and motors [17, 18] correspondingly. Each one of the proposed systems has shown satisfactory results, in turn, the implementations of this type of algorithms in hydraulic rotary systems have not been exploited in their entirety.

This paper presents a proposed methodology for the estimation of the speed of a hydraulic motor by means of the adaptive filtering of the feedback signal (speed) as objective parameters of prediction through artificial neural networks. Which are compared with a conventional speed measurement transducer.

2 Description of the Technique

The proposed system is composed in the first instance by an adaptive filter (noise canceller) that conditions the feedback signal of the hydraulic system (rotating speed of the engine), this information obtained are the objective values (labels) of a speed estimator based on artificial neural networks that use as input vectors the current and voltage supplied to the proportional valve of the hydraulic system.

To obtain information from the labels, which in essence is the information that the system is expected to "learn" from the information of an adaptive filtering system with noise canceller topology (Fig. 1). This type of topology is far from the other adaptive filtering systems due to its speed of convergence when implemented in computer systems [9]. The signal coming from the hydraulic system is acquired through the DAQ card, which allows a signal sampling at different frequencies of analog and digital signals. This signal is filtered to be stored in a database in conjunction with the input vectors (current voltage of the valve).

d(n)

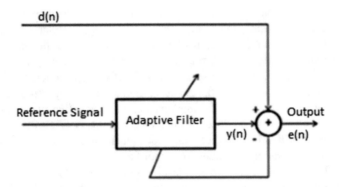

Fig. 1. Adaptive filter noise canceling configuration [9].

On the other hand, the vectors of characteristics that are going to be the inputs of the proposed system correspond to the measurement of the current and voltage signals of the hydraulic proportional valve which is part of the hydraulic circuit. Which is made up of various elements which are essential to transmit and direct the hydraulic energy to the active element of the system (hydraulic motor), these elements are the pump, hoses, connectors, maintenance unit, proportional valve 4/3, the tank of oil, hydraulic motor and manometers (Fig. 2).

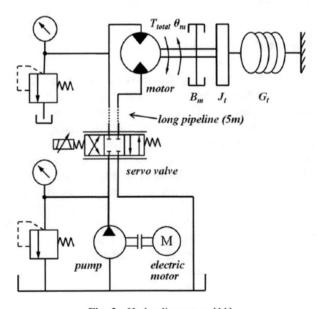

Fig. 2. Hydraulic system [11].

During the process of obtaining the signals at a rate of acquisition of 1 kHz, in which the hydraulic system was subjected to various states (variable speed). Of which a database of approximately 10,000 samples was obtained, which were subdivided into a percentage of 80% for training and 20% for validation. The reference values are in a 2–10 V operating range in a ratio of 10 to 1 rpm.

The statistical metrics used for the evaluation of the performance of the prediction methods, these metrics are the following: the mean square error (MSE), the absolute mean error (MAE), the standard deviation and the correlation (R), this test is it performs between the reference values and the values estimated by the prediction methods. The MSE is used to measure the average of the squares of the losses generated by the system, the standard deviation is the parameter that measures the estimated average distance of the data in comparison with those obtained by means of the prediction method. The MAE indicates an average of the magnitude of the error, without considering the direction of the error produced by the previous metrics, finally, the correlation indicates the direction of the error defining the level of association between the estimated and reference values of the system used. The mathematical formulations used to calculate the statistical values used are presented below:

$$MSE = \frac{1}{n}\sum\nolimits_{i=1}^{n}(p_i - r_i)^2 \tag{1}$$

$$MAE = \frac{1}{n}\sum\nolimits_{i=1}^{n}|p_i - r_i|^2 \tag{2}$$

$$Standar\,Dev = \sqrt{\frac{\sum_{i=1}^{n}(p_i - r_i)^2}{n}} \tag{3}$$

$$R = \frac{\sum_{i=1}^{n}[(r_i - \bar{r})(p_i - \bar{p})]}{\sum_{i=1}^{n}(r_i - \bar{r})^2 \sum_{i=1}^{n}(p_i - \bar{p})^2} \tag{4}$$

In Eqs. (1), (2), (3) and (4), n represents the total of the validation samples used, p_i the prediction values obtained from the systems and r_i the values of samples that are extracted from the validation database of the filtered feedback signal.

In addition, several system operation tests were carried out in order to obtain as much information as possible to corroborate the operation of the system.

3   Results and Discussion

The adaptive filtering technique with noise eliminator topology was implemented with an LMS convergence algorithm (Least-Mean-Square) for the extraction of the feedback signal from the transducer of the hydraulic system. The signal obtained is implemented as an entry vector in a supervised learning system based on artificial neural networks. The various configurations proposed were evaluated to later modify them, strengthening the estimation system. The topology that presents an acceptable precision according to the metrics presented in (1), (2), (3) and (4), is a multilayer Perceptron of an internal layer of 65 neurons with an activation function of the tansig type and the output layer corresponds to a linear function as seen in Fig. 3 and the performance obtained from the training in Fig. 4.

The adaptive filtering technique with noise eliminator topology was implemented with an LMS convergence algorithm (Least-Mean-Square) for the extraction of the feedback signal from the transducer of the hydraulic system. The signal obtained is implemented as an entry vector in a supervised learning system based on artificial neural networks. The various configurations proposed were evaluated to later modify them, strengthening the estimation system. The topology that presents an acceptable precision according to the metrics presented in (1), (2), (3) and (4), is a multilayer Perceptron of an internal layer of 65 neurons with an activation function of the tansig type and the output layer corresponds to a linear function as seen in Fig. 3 and the performance obtained from the training in the Fig. 5.

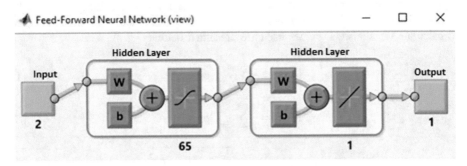

Fig. 3. Topology of the neural network used for training

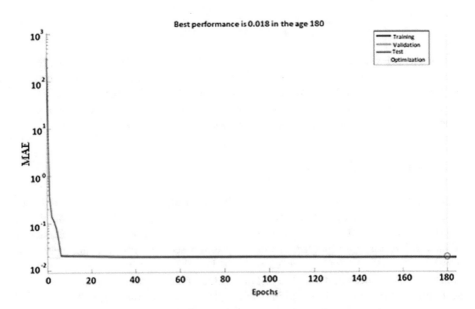

Fig. 4. Performance obtained from neural network training

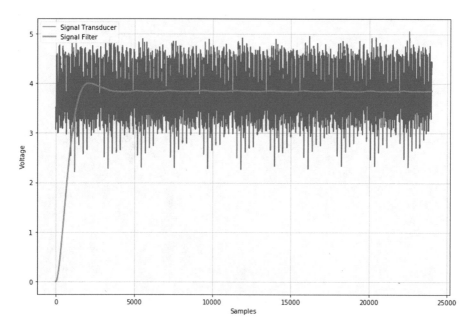

Fig. 5. Signal filter vs Signal transducer

The most reliable database obtained is presented with data of increase of speed, because, when making data capture of increase and decrease of speed, causes errors in the training of the neural network due to the return of the spool of the proportional valve. This problem is known as reel displacement hysteresis due to several factors between this friction and spring return force of the valve, this generates erroneous or corrupted data that directly affect the training of the neural network. The proposed solution is a rapid restart of the power system or in turn, the implementation of an analog or digital flow sensor, the first solution was implemented due to costs and market availability of this type of sensor.

The evaluation of the performance of the speed estimator measured by the metrics of (1), (2), (3) and (4) is shown in Table 1 and visualized in Fig. 6 in the operation in front of the conventional system of the signal filtered transducer.

Table 1. Results of the system performance evaluation.

System	MSE	MAE	Standard Dev.	R
NN	0.59	0.19	0.23	0.998

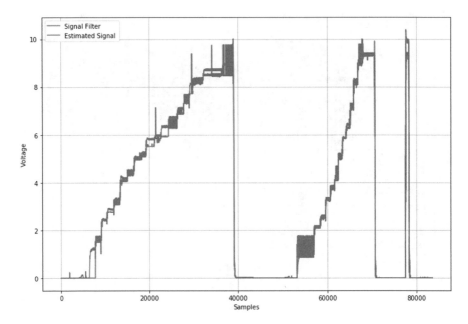

Fig. 6. Graphic result of the neural network, behavior of the plant vs virtual sensor.

In addition to the evaluation results of the neural network performance, a test of 11 possible scenarios was carried out, among which we have variable speed, power cuts, hydraulic power cuts. These scenarios were evaluated by (2) obtaining the results shown in Fig. 7.

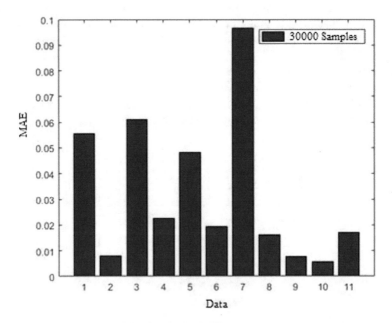

Fig. 7. MAE of different tests.

The technique presented offers comparable levels of evaluation with conventional census technologies that require constant maintenance and higher cost. When carrying out the observations of the proposed system, it is concluded that the system has high pressure and the possibility of replacing conventional speed census systems for hydraulic motors with variable speed configuration.

4 Conclusions

The proposed system presents applicability due to the results exposed in the implementation of the same, also presents clear advantages over conventional transducer systems as these do not suffer mechanical fatigue or a defective electrical mechanical coupling.

The speed estimation system is superior in comparison with its physical counterpart because the exposed results are compared against a filtered signal to obtain a reliable reference and their training patterns proposed.

The neuronal network has limitations due to the mechanical behavior of the proportional valve, such as problems of friction and return of the internal reel of the device. The erroneous data observed are: obtain different values of current and input voltage for the same speed output. This leads to erroneous data which cause problems in the training and implementation of the neural network.

The response of the implementation of adaptive filtering in noise canceller topology with an LMS convergence algorithm presents a satisfactory response to a conventional filtering system, which means a low computational load and a much faster convergence. This system does not require a calibration depending on the noise generated by the hydraulic plant at different speeds. When implementing this type of adaptive filtering algorithm, it represents for our neural network reliable feedback data for its training and implementation.

It is advisable to use a filter for the measurement of the valve current since oscillations in this parameter can generate an erroneous database for the training and implementation of the neural network.

For future work, the implementation of a third input parameter to the neural network is proposed, which is the flow of incoming oil to the hydraulic motor in order to solve the limitations of the proposed neural network.

References

1. Enrique, N.M.J.: Análisis y modelado del control de una válvula Direccional Hidráulica, Instituto Politécnico Nacional de México (2011)
2. Czekaj, D.: Aplicaciones de la Ingeniería: 3. Maquinaria Hidràulica en Enbarcaciones Pesqueras Pequeñas (1988)
3. Ortiz, D., Quintero, O.: Una aproximación al filtrado adaptativo para la cancelación de ruidos en señales de voz monofónicas (2014)

4. Orozco-Tupacyupanqui, W., Nakano-Miyatake, M., Pérez-Meana, H.: Filtro Híbrido Adaptativo en Cascada para Arreglos Lineales Uniformes de Antenas. Inf. Tecnol. **25**(4), 103–122 (2014)
5. Taş, İ.C., Arica, S.: Partition of medical signals with a two band adaptive filter bank. In: 2018 26th Signal Processing and Communications Applications Conference (SIU), Izmir, pp. 1–4 (2018)
6. Gilani, S.O., Ilyas, Y., Jamil, M.: Power line noise removal from ECG signal using notch, band stop and adaptive filters. In: 2018 International Conference on Electronics, Information, and Communication (ICEIC), Honolulu, HI, pp. 1–4 (2018)
7. Qinlan, X., Hong, C.: Image denoising based on adaptive filtering and multi-frame averaging filtering. In: 2009 International Conference on Artificial Intelligence and Computational Intelligence, Shanghai, pp. 523–526 (2009)
8. Pereira, R.R., da Silva, C.H., da Silva, L.E.B., Lambert-Torres, G.: Application of adaptive filters in active power filters. In: 2009 Brazilian Power Electronics Conference, Bonito-Mato Grosso do Sul, pp. 770–774 (2009)
9. Shiogai, K., Sasaoka, N., Itoh, Y., Kinugasa, Y., Kobayashi, M.: Fast convergence adaptive IIR notch filter using allpass filter and adaptive line enhancer. In: 2013 International Symposium on Intelligent Signal Processing and Communication Systems, Naha, pp. 279–284 (2013)
10. Stefański, T., Zawarczyński, Ł.: Analysis of inverter-fed drive of hydraulic pump in volumetric control system. In: 2016 21st International Conference on Methods and Models in Automation and Robotics (MMAR), Miedzyzdroje, pp. 146–151 (2016)
11. Long, Z., Guan, B., Chen, S., Chen, G., Guo, S.: Modeling and simulation of hydraulic motor tracking servo motor driving load. In: 2018 International Conference on Smart Grid and Electrical Automation (ICSGEA), Changsha, pp. 118–121 (2018)
12. Dong, C., Lu, J., Meng, Q.: Position control of an electro-hydraulic servo system based on improved Smith predictor. In: Proceedings of 2011 International Conference on Electronic & Mechanical Engineering and Information Technology, Harbin, pp. 2818–2821 (2011)
13. Villada, F., Cadavid, D.R.: Diagnostico de fallas en motores de inducción mediante la aplicación de redes neuronales artificiales. Inf. Tecnol. **18**(2), 105–112 (2007)
14. Sowilam, G.: Aplicación de las redes neuronales en los sistemas de control vectorial de los motores de inducción. Tesis doctoral, UPC, Departament d'Enginyeria Elèctrica (2000)
15. Narváez, F., Sierra Vargas, F.E., Montenegro Mier, M.A.: Modelo basado en redes neuronales para predecir las emisiones en un motor diésel que opera con mezclas de biodiésel de higuerilla. Informador Técnico [S.l.], vol. 76, p. 46, December 2012
16. Espinosaa, R.D.C., Cárdenas, R.G.: Identificación y control digital con redes neuronales para un sistema hidráulico. Vector **12**, 32–39 (2017)
17. Rairán, A.D.: Implementación de un controlador Difuso para la regulación de posición de un cilíndrico hidráulico lineal, Bogotá D.C. Tecnura, no. 19, pp. 18–28 (2007)
18. Daniel, C., Pablo, G.: Diseño e implementación de un control difuso de velocidad para el motor hidráulico HRE HIDRAULIC, Universidad Politecnica Salesiana, Cuenca-Ecuador. http://dspace.ups.edu.ec/handle/123456789/16369. Accessed 21 Mar 2019

Comparative Study of Sensors Applied to a Three-Dimensional Scanner

Alex Pérez, Javier Rojas$^{(\boxtimes)}$, and Gustavo Caiza

Salesian Polytechnic University, Quito, Ecuador
aperezch@est.ups.edu.ec, {jrojasu,gcaiza}@ups.edu.ec

Abstract. A three-dimensional scanner obtains a digital model of any object in a detailed and fast way, its operation is based on the acquisition of distance measurements from a reference to each point of the object's contour, which, when organized according to the reference system, form a cloud of three-dimensional points. In the present work, a comparative study between infrared and ultrasonic distance sensors is carried out to define its functionality as an active part of a three-dimensional scanner. It uses a mechanical structure with fixed base and ability to move the sensor in the x, y and z axes, the control is done with an arduino card and a visual interface in MatLab that allows the variation of the acquisition parameters to obtain quantitative data and visualize the three-dimensional point cloud. The infrared sensor obtains better results when acquiring the cloud of points of the contour of an object and using a fixed base structure with circular sweep.

Keywords: Three-dimensional scanner · Digital model · Distance sensor

1 Introduction

The three-dimensional printing of objects has made great advances in recent years, becoming a tool for continuous use in industries like mechanics, auto parts, reverse engineering, in which parts such as pinions, shafts, bearings, among others, are made [1, 2]. The ease to make the prototype and construction of parts and pieces in different materials and from a digital model has aroused great interest in industries that require tailor-made and customized parts such as the industry of physical and medical reha-bilitation for the construction of customized splints or prostheses for the human body [3] and the possibility of quickly and easily making modifications or adjustments using a computer. The huge amount of data to process requires reliable automatic data processing [4].

Commonly the digital model is carried out by specialized personnel and equipment, being a complicated process to obtain models in applications of replacement or replication of a specific piece or in the elaboration of customized items [5].

A three-dimensional scanner obtains the digital model of objects quickly; its active part is a sensor that measures the distance between a reference and each point of the contour of the object, these values are ordered and digitally processed to obtain a three-dimensional image [6]. The accuracy and detail of the model depends directly on the data acquired by the sensor, as well as on its processing. Much research about sensors

© Springer Nature Switzerland AG 2020
M. Botto-Tobar et al. (Eds.): ICAETT 2019, AISC 1067, pp. 63–73, 2020.
https://doi.org/10.1007/978-3-030-32033-1_7

in a 3D scanner have been done like [1] where the performance of time-of-flight sensors (ToF) is explored, it uses a rotating base structure and make a comparison with an infrared distance sensor however doesn't explore the effect of angular velocity or time between each measurement. In [7] a fixed base structure is used with linear sweep of the sensors, focusing on the processing of acquired data to eliminate false or redundant values however it doesn't stablish any comparison between sensors, in [8] the combination of a distance sensor and a digital camera is evaluated, the combination obtain a better perception of the object depth and the environment to obtain the digital model by reconstructing images. In [9] explores the application of foot scanning based on laser triangulation, it proposes a fixed architecture with rotating and linear type 3D laser sensors and cameras to obtain 360° level. Other research like [10] y [11] uses Photogrammetry, a scanning technique based on cameras and image processing with good results but doesn't establish any comparison with other scanning techniques. In this article, we propose a comparison between non-contact sensors as active part in a fixed base structure 3D scanner, sensors have rotational movement on the X-Y axis and linear displacement on the Z axis so that it surrounds the object to be scanned, this configuration is very useful in applications like scanning of limbs of the human body for the construction of splints or custom sports and protection equipment.

To control the acquisition of data an arduino card is used in conjunction with a Human Machine Interface (HMI) implemented in Matlab, used to vary the data acquisition parameters and processing them to visualize a cloud of points of the object's contour. The acquisition of data is done with different sampling times, speed, direction of sweep and separation with the object to obtain comparative values between the different sensors.

The article is organized as follows, Sect. 2 describes the basic concepts of the study and the used methodology, Sect. 3 shows the design of the mechanical structure and its control and Sect. 4 details the tests performed and the obtained results.

2   Materials and Method

2.1   Three-Dimensional Scanner

A three-dimensional scanner is an electromechanical device that allows you to quickly and accurately acquire the contour of an object in the form of a digital model, its operation is based on measuring the distance from the location of the sensor element to a point on the object from various angles of reference to cover the entire surface of the object.

Types of Three-Dimensional Scanners
Movable base structure, the object is on a mobile base so that it rotates on its own axis while the sensors remain static acquiring different contour measurements [1, 12].

Fixed base structure, the object remains static on a fixed base while the sensors rotate around it acquiring contour measurements [7].

Structure without base, it does not have a mechanical structure, the instrumentation moves freely, this group can include technologies based on image acquisition from different angles and the combination of cameras and sensors.

2.2 Hardware

Distance Sensor. It is an electronic device that allows to determine the distance, is the active and fundamental part of a three-dimensional scanner, the accuracy of this and the way in which the object is sampled depends on the accuracy of the digital model.

Contact Distance Sensor. It is a sensor that measures distance as a function of the time it takes to move from a fixed point to make contact with the object. It requires a structure that moves the sensor at a certain speed [7, 13].

Non-contact Distance Sensor. It is a sensor that uses radiation means such as light or ultrasound, the distance is determined as a function of the time of flight it takes to return the signal emitted when hitting the object, it requires a transmitter and a receiver [13].

Infrared Type Distance Sensor. The sensor sends a beam of infrared light to the object and its reflection is concentrated by a receiver lens to determine the distance by triangulation between the emitter, object and receiver. The sensor delivers analog voltage that is interpreted as distance with the help of the distance vs. voltage graph provided by the manufacturer in the datasheet [13, 14].

Ultrasonic Type Distance Sensor. The ultrasonic sensor cyclically emits an acoustic pulse of high frequency and short duration towards the object and detects the echo produced by the return of the signal by measuring the time it takes. The distance is determined by Eq. (1) [13, 15].

$$d = (t * vs)/2 \tag{1}$$

Where: d (distance); vs (speed of sound in the air, 340 m/s); and t (elapsed time)

Stereoscopic Methods. They are based on the acquisition of digital images that allow to determine the depth of an object, by means of the analysis of the pixels or triangulation, in combination with a sensor of distance without contact allows to obtain measurements of more precise distances [12].

3 Design of the Sampling and Testing System

This section presents the design of the fixed base structure to acquire sensor's data in different tests scenarios to compare its behavior as tridimensional scanner active part, a block diagram of the structure is shown in Fig. 1.

Fig. 1. Sampling and testing system's block diagram.

3.1 Computer System

It is a HMI that was developed with the Matlab computer software. Its function is to allow the visualization of the data obtained by the sensors while the sampling of the contour of the object to be scanned is made, it also allows the entry of experimental parameters such as: the separation distance of the sensors, the rotational and vertical displacement, sampling rate, among others. The visualization of the point cloud of the object's contour is obtained by forming a matrix of distances.

3.2 Electromechanical System

It is the mechanical and electronic structure that guides the movement of the sensor around the object and at different heights, until taking samples of the entire contour, the sensor is located on an aluminum arm whose rotational movement is made by a system of pulleys coupled to a stepper motor while the vertical movement is carried out by means of a bolt coupled to a second stepper motor. The system is controlled by an Arduino development card that is responsible for giving the necessary signals to command the motors, as well as the acquisition of data from the sensors according to the configuration made in the HMI.

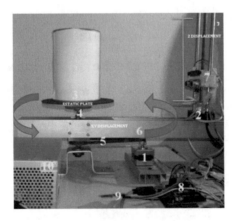

Fig. 2. Electromechanical system.

The basic components of the electromechanical system can be observed in Fig. 2, and listed in Table 1.

Table 1. Electromechanical system components

ID number	Component
1	Stepper motor for XY axis displacement
2	Stepper motor for Z axis displacement
3	Object to be scanned
4	Static plate
5	Coupling band for aluminum arm displacement
6	Aluminum arm
7	Sensor support
8	Arduino development card
9	SD card socket
10	Power source

3.3 Sensors

According to a previous investigation of the state of the art, a three-dimensional Scanner with fixed base structure require non-contact distance sensor as active component, so in this study infrared and ultrasonic distance sensors are evaluated. To select the sensors factors like availability in local market, measurement range, theoretical accuracy and price are considered.

Sensor SHARP GP2Y0A41SK0F and HC-SR04 are used because they satisfy the measuring range from 5 cm to 20 cm considered in medical applications, its accuracy reported in the datasheet are 0.4 cm and 0.3 cm respectively and fulfill the initial requirement for the scanner, additionally these sensors are easy to acquire in the local market and have low cost. Other important factor to use them is its output signal because is easy to connect and process with an Arduino [16].

4 Tests and Results

In order to obtain the qualitative data of each of the sensors in the implementation of a three-dimensional scanner, several tests are performed in which the acquisition conditions of distance measurements are varied and subsequently the point cloud is contrasted.

4.1   Accuracy Test

The objective of this test is to obtain the accuracy of each sensor in the distance measurement in order to establish its margin of error; considering the limitations of the mechanical structure, the sensors are located between 10 cm and 30 cm away from the object, at intervals of 5 cm to obtain sufficient data and cover the entire measurement range; 5000 measurements are made without altering the position of the sensor, the obtained data is analyzed to determine the accuracy and precision of each sensor. The results of this analysis for different separation distances between the object and the sensor are shown in Tables 2 and 3.

Table 2. Distance measurements with SHARP laser sensor GP2Y0A41SK0F.

Real distance (x [cm])	Average distance ($\bar{x}$)	Relative error (e) $\|x - \bar{x}\|/x$	Standard Deviation (σ)
10	10,635	0.0635	0,55943
15	14,695	0.0203	0,82769
20	20,687	0.03435	1,9319
25	25,037	0.00148	2,6826
30	31,05	0.035	3,7467

Table 3. Distance measurements with the HC-SR04 ultrasonic sensor.

Real distance (x [cm])	Average distance ($\bar{x}$)	Relative error (e) $\|x - \bar{x}\|/x$	Standard Deviation (σ)
10	10,885	0.0885	0,0576
15	15,078	0.0052	0,12762
20	18,943	0.05285	0,18963
25	24,941	0.00236	0,23628
30	30,529	0.01763	0,41499

According to statistics theory, sensor's precision is defined as the mean standard deviation, so considering the results we obtain a precision of 1.95 for the infrared sensor and 0.21 for the ultrasonic sensor. The accuracy of each sensor is obtained from the relative error, we obtain 3.1% for the infrared sensor is and 3.3% for the ultrasonic sensor is. In addition for both sensors, it can be seen that 10 cm of separation has the best repeatability, while the smallest error occurs with a separation of 25 cm.

4.2   Sweep Test

To obtain measurements of the contour of an object it can be surrounded with circular or linear movement, the objective of this test is to determine the best movement technique to obtain the digital model; we experimented with objects with circular contour and flat contours with the presence of angles, the measurements were taken for the two types of sweeping with a separation distance of 25 cm, measured from the center of the fixed base, and at intervals of 3° for the circular sweep and 0.5 cm for the linear sweep, smaller intervals generate many redundant measurements, the point cloud obtained with the HMI can be seen in Fig. 3.

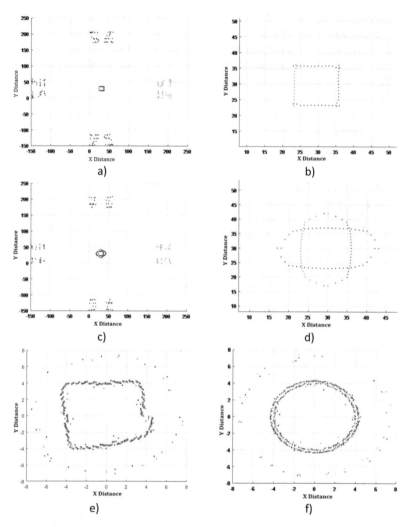

Fig. 3. Scanning test results for a cubic and cylindric object. (a) Cube linear (b) Cube's contour zoom. (c) Cylinder linear (d) Cylinder's contour zoom. (e) Cube circular. (f) Cylinder circular.

According to Fig. 3, with the linear sweep for circular objects an elliptical model is obtained while for objects with flat surfaces the model is very close to the real one, in addition many unnecessary measurements obtained outside the object need to be eliminated.

4.3   Resolution

In order to obtain an object's detailed point cloud with minimum blank spaces is logical to assume that a lot of measurements are needed however it will produce some false and redundant data and require more processing. The objective of this test is to determine the maximum resolution that can be obtained with each sensor, so that the acquired data sample shows the entire contour of the object with the least amount of false and redundant data; in this test the contour of a cylindrical object is acquired using circular sweeping, the measurements are taken at different displacements, from 0.6° according to the limitations of the mechanical structure.

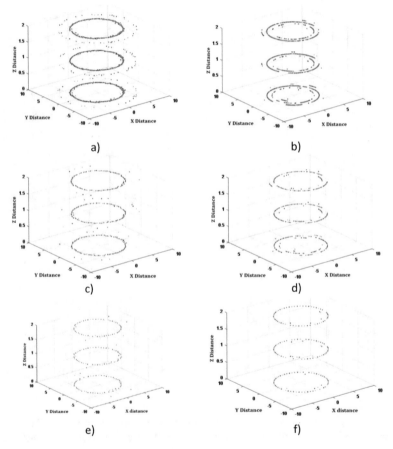

Fig. 4. Scanning contours for a cylindrical object. (a) Infrared sensor, 0.6° intervals (b) Ultrasonic sensor, 0.6° intervals (e) Infrared sensor, 3° intervals. (f) Ultrasonic sensor, 3° intervals. (g) Infrared sensor, 6° intervals. (h) Ultrasonic sensor, 6° intervals.

Figure 4 shows some of the obtained results. For both sensors a better model is obtained when measurements are made at 0.6° intervals, larger intervals generate empty spaces in the point cloud. It is also observed that the laser type sensor acquires a more real model since with the ultrasonic sensor a double contour is appreciated, this was consistent in each of the carried-out tests.

4.4   Distance Test

The aim is to analyze the influence of the separation distance between the object and the sensor in the measurements while a circular sweep is carried out and to obtain recommendable limits to obtain a good digital model. For this test we get the model of the object maintaining the displacement in 0.6°, according to the results of the previous test, and varying the separation of the sensor from 10 cm to 30 cm measured from the center of the base, according to the limitations of the mechanical structure.

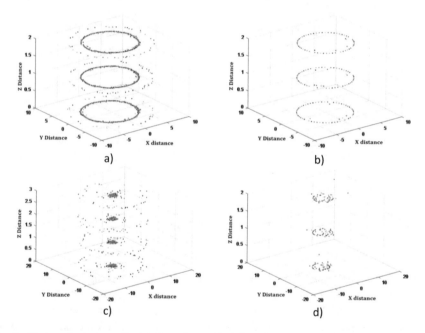

Fig. 5. Distance's test results (a) Scanning with infrared sensor, 13 cm. (b) Scanning with ultrasonic sensor, 13 cm. (c) Scanning with infrared sensor, 28 cm. (d) Scanning with ultrasonic sensor, 13 cm.

Figure 5 shows some results, for the two sensors it can be seen that the closer the sensor of the object is, the better digital model is, starting from 15 cm apart the model begins to present voids and at a greater distance it completely distorts.

4.5   Environmental Influence

Through this test we wanted to determine the influence of the experimental environment when scanning an object. Taking into account the different accuracy, sweep, resolution and distance tests, and the contour of the object shown in Fig. 6a was acquired on three different days, at the same time, in an internal and external environment, the results are shown in Fig. 6.

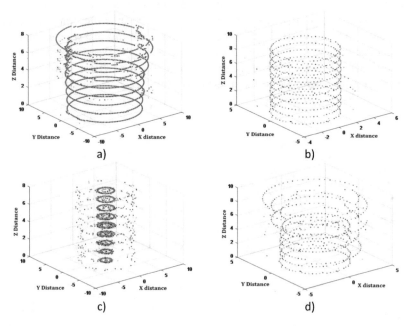

Fig. 6. Environment's influence test. (a) Infrared sensor in interior. (b) Ultrasonic sensor in interior. (c) Infrared sensor in exterior. (d) Ultrasonic sensor in exterior.

Observing the point cloud it can be seen that, when working in a closed room, called the internal environment, a more accurate shape is acquired, while in the open air the conical shape of the object is not detected and there are several false points, which may be due to the presence of dust, light and the influence of wind on the measurements, this effect has a greater influence on the ultrasonic sensor.

5   Conclusions

According to the different tests carried out, the infrared sensor obtains better results than the ultrasonic sensor when acquiring the point cloud of the contour of an object, by using a fixed base structure with a circular sweep. This structure is suitable for objects with circular contours and without the presence of angles, the best results are obtained when working in a closed environment, with a separation distance of 10 cm to

15 cm between the object and the sensor and with a circular displacement of 0.6° between each measurement to reduce the amount of empty spaces and false points. For objects with flat surfaces and presence of angles, it is convenient to use a linear sweep, adjusting the displacement of the sensor to the limits of the object. For the digital model to be complete and can be used, the acquired point cloud must undergo a digital processing that eliminates the false points and forms the contour of the object filling empty spaces.

References

1. Gutierrez-Villalobos, J.M., Dimas, T.-V., Mora-Vazquez, J.C.: Simple and low cost scanner 3D system based on a Time-of-Flight ranging sensor. In: 2017 XIII International Engineering Congress, pp. 1–5 (2017)
2. Herman, K., Fernandez, W.: A comparative study of selected methods for ultrasonic signals energy estimation for target strength and distance evaluation, pp. 1–3 (2017)
3. Suarez, O.D., Castro, D.A.: Diseño de un sistema de perfilación de férulas mediante el escaneo 3d de muñeca, brazo y tobillo para fines de, vol. 1(1), pp. 94–113 (2019)
4. Guo, X., Wang, F., Ma, Y., Du, D.: Research on three-dimensional point clouds processing for standing tree volume based on laser scanner. In: 2009 Second International Symposium on Knowledge Acquisition and Modeling, vol. 3(4), pp. 206–208 (2009)
5. Zhang, H., Wen, Y., Lv, G., Shen, J., Wang, X.: "drawings of buildings I: I". In: 2010 2nd Conference on Environmental Science and Information Application Technology, vol. 2, pp. 511–514 (2010)
6. Yoshida, K.: Self-localization method for three-dimensional handy scanner using multi spot laser. In: 2014 IEEE Symposium on Computational Intelligence for Multimedia, Signal and Vision Processing, pp. 1–5 (2014)
7. Zacatenco, U.: Escaner 3D de Alta Precision. Instituto Politécnico Nacional, Mexico
8. Julio, A.: Implementacion de un sistema para digitalización de objetos (2015)
9. Lee, M., Baek, S., Park, S.: 3D foot scanner based on 360 degree rotating-type laser triangulation sensor (2017)
10. Kirsten, E., et al.: 3d data acquisition using stereo camera, pp. 9214–9217 (2018)
11. Abd-Raheem, A., Aldeiri, F.: Design of an automated 3D scanner. In: 2018 the International Arab Conference on Information Technology, pp. 1–5 (2018)
12. Romero, A., Orozco, J.: Modelado 3d de objetos usando matlab mediante sensor, October 2014
13. Germán Corona Ramírez, L., Abarca Jiménez, G.S., Mares Carreño, J.: Sensores y actuadores: aplicaciones con Arduino. Distrito Federal. Grupo Editorial Patria, MÉXICO (2014). ProQuest ebrary. Web. 12 October 2016. Copyright © 2014. Grupo Editorial Patria. All rights reserved, October 2016
14. D. Measuring and S. Unit, GP2Y0A41SK0F, pp. 1–9 (2002)
15. E. Parameter, Ultrasonic Ranging Module HC - SR04, pp. 3–6
16. Cajas, J.Y., Toapanta, J.M., Oñate, L.: Diseño y construcción de un escáner bifocal para la obtención de una nube de puntos a través de filtrado y adquisición de imágenes, pp. 18–24 (2014)

Diagnosis of Incipient Faults in Induction Motors Using MCSA and Thermal Analysis

William Oñate$^{(\boxtimes)}$, Ramon Perez, and Gustavo Caiza

Universidad Politécnica Salesiana, Quito, Ecuador
{wonate, rperezp, gcaiza}@ups.edu.ec

Abstract. Induction motors are widely used at industrial level and many researches have been conducted to predict faults and ensure their continuous operation. The present study shows a fault detection system based on the Motor Current Signature Analysis (MCSA) technique which is a non-invasive method, and the Fast Fourier Transform (FFT) algorithm to perform spectral analysis on the current to detect specific components that characterize faults in different conditions such as damaged bearings, shorted winding in turns, and broken bars. The tests were conducted using a motor test bench and the results show a spectral value obtained by FFT in the frequency of 300 Hz for faults in bearings of the motors 1 and 2. Results also reveal coincidence of 96.9% and 99.46% compared to the results obtained through the theoretical equations included in the MCSA technique, representing values that are within the fault frequency bands on 1x and 2x respectively.

Keywords: Three-phase motor · MSCA · Incipient faults · FFT

1 Introduction

Induction motors are widely used at industrial level due to its robust structure, low cost, efficiency, quick setting-up, and reliability. For these reasons, there are more than 300 million electric motors in the world which are used for different applications [1, 2]. This means that almost 80% of all electric motors in operation are of the induction type [3].

Due to the extended use of these motors at industrial level, many researches have been carried out to predict faults and guarantee their continuous operation, avoiding interruptions that may cause undesired cuts and economic loss. Several parameters should be considered such as time of service, setting-up errors, environmental factors, and improper maintenance which, together, may accelerate the motor fault rates [4].

The monitoring of the machine plays an important role in the detection of incipient faults since it allows to detect the beginning of a fault in the electric motor by means of a predictive maintenance before it develops to the point of stopping motor operation [3]. During the last two decades, many studies have been carried out to develop efficient techniques for detecting faults in induction motors. The MCSA is a well-established technique and a widely used non-invasive method [6] that permits to detect problems such as rotor broken bars, eccentricity of air gap, and mechanical imbalances [5]. This method consists in obtaining the amplitude of the spectral components of the current by

© Springer Nature Switzerland AG 2020
M. Botto-Tobar et al. (Eds.): ICAETT 2019, AISC 1067, pp. 74–84, 2020.
https://doi.org/10.1007/978-3-030-32033-1_8

applying the FFT algorithm and comparing specific components related to faults where a significant difference between the actual value and the previous one is taken as a fault indicator [7].

The researches have also studied the power produced by an electrical machine and is limited by its maximum operation temperature where an efficient cooling system allows the same machine to operate above its rated power without reducing the expected useful life [8]. For these reasons, researchers use methods such as the thermographic image analysis which has allowed to detect faults and know the motor condition. On the other hand, the spectral analysis of the FFT, used at the same time, has proven to be a good method to extract the characteristics of the signals for further processing [9]. The reliable assessment of the induction motor condition is an interesting topic for many companies due to the serious consequences that faults can generate.

The present research describes a fault detection system based on the Motor Current Signature Analysis (MCSA) technique, using the Fast Fourier Transform (FFT) algorithm to detect the specific components that characterize faults in different conditions such as damaged bearings, shorted winding in turns, and broken bars.

The paper is organized as follows: Sect. 2 describes the concepts and methodology used; Sect. 3 shows the system design and the analysis of results; and finally, Sect. 4 presents the conclusions.

2 Materials and Methods

2.1 Motor Current Signature Analysis

MCSA is a classic method to assess the rotor condition in induction motors, based on capturing the waveform of the current demanded by the motor in a steady state operation, analyzing it with the FFT and evaluating the amplitude of the harmonics that are amplified by the rotor fault [2]. According to the MCSA, the frequency of the motor current is scanned to detect specific components related to faults. On the basis of the relative amplitude of these components, a fault of the current spectrum can be inferred [10]. Fault can be then classified of electric or mechanic nature according to the following principle [11].

2.2 Major Faults in Induction Motors

Fault Caused by Broken Rotor Bars (BRB). The signal of current of a defective induction motor stator contains the harmonics of the fundamental and the characteristics of fault frequency components along with the noise. Most of the BRB faults begin with small cracks in a bar of the rotor or in the end ring and, subsequently, become major faults with multiple broken or cut bars [12]. The fault caused by the existence of broken bars in the rotor begins in those parts of the rotor which are not supported by its body, i.e. they concentrate at the union of the short-circuit ring or close to it [13]. A BRB causes electrical and magnetic asymmetry in the rotor, which introduces the components of lower side band in the stator current [12].

If the motor has broken bars, it creates unbalanced or asymmetric conditions that generate an additional rotating magnetic field in delay, that turns at slip rate and produces shorts in the stator windings, inducing a voltage and a current with the same frequency of the rotary field, whose location is given by [11]:

$$f_1(1 \pm 2ks) \tag{1}$$

Where:

f_{sb}: Frequency of side bands due to broken bars.
f_1: Frequency of the motor supply network.
k: Integer value (1, 2, 3...) depends on the frequency band to be obtained.
s: Slipping

Faults Caused by Damaged Bearings. The faults produced by damaged bearings are those that occur with greater frequency [13]. Faults in the bearings contribute to 40–50% of all the faults in the machine [14]. The equations corresponding to the frequency of each element of the bearing are shown below [15]. These equations depend on the geometry of the bearing, the number of balls or rollers, and the bearing rotation speed.
Fault in ball:

$$F_b = \frac{d_r}{d_b} f_{rm} \left[1 - \left(\frac{d_b \cos(\beta)}{d_r} \right)^2 \right] \tag{2}$$

Fault in outer raceway:

$$F_e = \frac{n}{2} f_{rm} \left[1 - \left(\frac{d_b \cos(\beta)}{d_r} \right) \right] \tag{3}$$

Fault in inner raceway:

$$F_i = \frac{n}{2} f_{rm} \left[1 + \left(\frac{d_b \cos(\beta)}{d_r} \right) \right] \tag{4}$$

Fault in cage:

$$F_j = \frac{n}{2} \left[1 - \left(\frac{d_b \cos(\beta)}{d_r} \right) \right] \tag{5}$$

Where:

F_b: Characteristic frequency of fault in balls.
F_e: Characteristic frequency of fault in outer raceway.
F_i: Characteristic frequency of fault in inner raceway.
F_j: Characteristic frequency of fault in cage.
d_r: Diameter of the bearing.

d_b: Diameter of the balls.

f_{rm}: Rotation frequency of the motor.

β: Contact angle between raceway and bearing balls.

n: Number of balls.

Equation (6) was used to determine failures caused by damaged bearings (F_{fr}) as a function of the failure diagnosis frequency (f_f).

$$F_{fr} = f_1 \pm k f_f \tag{6}$$

Faults Caused by Short-Circuited Turns. The fault caused by short-circuited turns of the same phase is one of the most critical faults because it starts with a small imperceptible short that becomes more severe over time and can cause permanent damage to the turns of the motor. The short circuit fault between turns causes a modification of the electrical circuit that produces a change in the flux density in the machine's air gap. The harmonics of interest in the stator current for this type of faults are listed below [16].

Frequencies below 400 Hz:

$$f_{stl} = f_1 \left[\frac{m}{p} (1 - s) \pm k \right] \tag{7}$$

Where:

f_1: Frequency of the motor supply network.

m: Integer value (1, 2, 3…) depends on the frequency to be obtained.

p: Number of pairs of poles in the motor.

s: Slipping

k: Integer value (1, 2, 3…) depends on the frequency band to be obtained.

2.3   Implementation

A test bench is used for applying and analyzing the tests of incipient faults, as shown in Fig. 1.

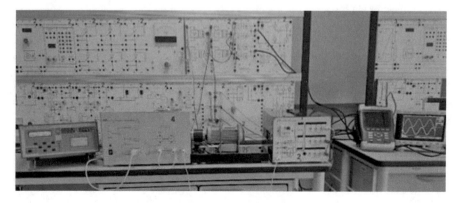

Fig. 1. Test bench for tested motors.

It consists of a magnetic power brake unit and a controller to set the required torque for testing motors up to 10 Nm, 3 current transducers up to 20 A, 3 V differential probes up to 500 V, an oscilloscope for the acquisition of signals of current of each phase in relation to the time, Matlab software to develop the transformation to frequency of the signals of current using the FFT, a thermographic camera to measure the temperature reached by the tested motor, and finally, two asynchronous squirrel cage motors with characteristics that are shown in Table 1.

Table 1. Characteristic data of tested motors

1 Motor		2 Motor	
Brand	FELISATTI	Brand	ABB
Type	10/C	Speed	1052 rpm
Voltage	220 Vac Δ	Voltage	220 Vac Δ
Current	0.85 A	Current	0.4 A
HP	0.1	Power	0.3 kW

Before being submitted to the fault test in this study, these motors were tested for vibration spectrum, impedance, and temperature control - Isolation in all three stator phases, concluding that the motors are in good operation under ISO 2372, IEEE 95-2002, and IEEE 43-2000; also reported, that these motors have approximately 1500 h of service. To develop the bearing fault, ball radial bearings were placed, which were discarded by the use and implemented for fault testing in the two engines mentioned above. Table 2 shows the characteristics of the bearings for the fault diagnosis frequency equations.

Table 2. Characteristics of the bearings.

Motor 1	Motor 2		
Series	602-2RS	6204-2RS	629-2RS
Diameter of ball	5.95 mm	7.565 mm	5.025
Average diameter between raceways	24.55 mm	33.025 mm	17.575
Number of balls	8	8	8

For developing the broken bars fault, drill holes were made in two bars of the squirrel cage in motor 1, leaving 2 mm of material before reaching the axis. To develop the short circuit fault in the turns of the same phase, it was necessary to identify the poles that make up each phase of motor 2, whereby the insulation of the winding is removed, and the turns are short-circuited with tin.

3 Analysis of Results

The stator currents obtained from motors with nominal load are in good working order and will be compared with the signals of current obtained in each fault to later be analyzed and interpret their spectral characteristics of incipient faults as in the case of bearing wears, broken bars, and short circuits between turns of the same phase. Finally, the errors are obtained between the theoretical and experimental data.

3.1 Bearing Faults

Figure 2 presents the spectral amplitude variation in the three phases of the system in approximately a range of frequencies from 68 to 110 Hz, in which a decrease in the spectral amplitude is less than −40 dBv. In addition, there is a small increase of amplitude in the frequency of 300 Hz with respect to the signals without faults, data that refers to Eqs. (3) and (6) indicating the presence of fault by bearings in the external raceway.

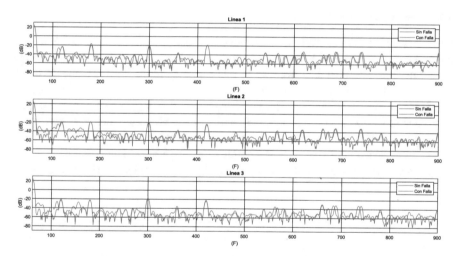

Fig. 2. Presence of bearing fault in all three phases of motor 1

In the same way, the motor 2 indicates the existence of spectral amplitude variations below −40 dBv approximately in the frequency of 80 Hz, as shown in Fig. 3, as a characteristic of the three phases. Additionally, it is observed that in the frequency of 300 Hz there is a slight increase in the spectral amplitude with respect to the signals without faults, making reference to the Eqs. (2) and (6) for ball fault in the bearing 629-2RS, and the Eqs. (5) and (6) for fault in cage in the bearing 6204-2RS.

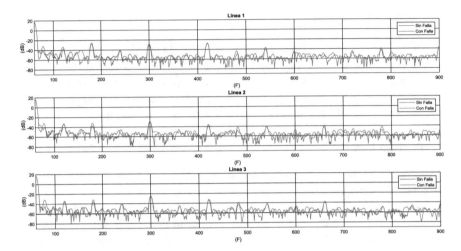

Fig. 3. Presence of bearing fault in all three phases of motor 2

3.2 Broken Bar Fault

Later, in motor 1, the analysis was developed for distorted three-phase currents when the motor presents broken bars operated at nominal load. In Fig. 4, it is observed that, during the range frequencies from 75 to 175 Hz, the amplitude decreases drastically below −40 dBv with respect to the signals without fault except for the spectrum amplitude in the frequency of 120 Hz, besides an existing increase in the spectral amplitude at the frequencies of 360, 600, and 720 Hz, characteristics that are repeated in the three phases. It is worth mentioning that with this type of fault the motor increased its speed with respect to the nominal one by 5%, so decreasing the slippage. For a better analysis on broken bar faults, the signal was broken at 100 Hz and it was

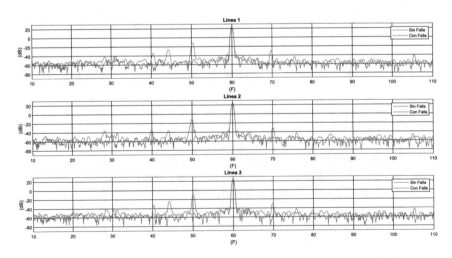

Fig. 4. Presence of broken bars, diagnosis by upper and lower lateral peaks

centered on the fundamental frequency. It can be considered that for defining such broken bar fault, [17] states that this kind of fault occurs when the spectra are greater than −20 dBv of the noise level.

The previous figure shows the existence of a lateral peak greater than 70 Hz and two lower lateral peaks next to the fundamental one at 40 and 50 Hz, the latter having an amplitude of approximately −10 dBv in the three phases that make up the system, which means that the motor has a fault inside caused by broken bars in the squirrel cage.

3.3   Short Circuit Fault

Figure 5 shows the distortion effects of the stator currents generated by a short circuit in 28 turns of phase 3 with respect to the signals without fault of the motor 2.

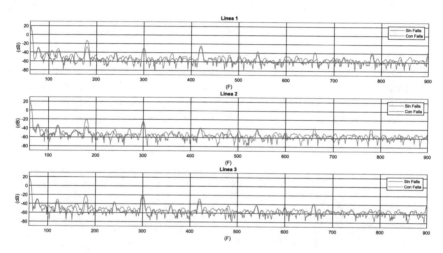

Fig. 5. Distortion of stator currents due to short circuit in phase 3.

There is an increase in the spectral amplitude over the noise level for the three phases that is found in the frequency of 180 Hz. By means of this diagnosis, the existence of stator faults was also determined. Phase 3 is the only one that presents a slight increase in amplitude in the frequency of 300 Hz, approximately 0.53 dBv over the noise level, meaning that there is a short circuit in the turns of this phase. As a result, the increase of temperature $T_{máx}$ in the stator windings is 20 °C with respect to the motor without fault. Continuing with the same test, the fault was increased by shorting 28 more turns of the phase 1 with which, all the spectral signal moved about 11 dBv over the fundamental without fault, showing an increase and distortion of the three-phase currents. Additionally, the spectrum in the frequency of 180 Hz significantly grew around 23.44 dBv over the noise level in all the phases, with which the diagnosis predicts the existence of stator fault. Besides, Fig. 6 shows a small growth in the spectrum over the noise level in the frequency of 300 Hz for phase 1 and 3, meaning that there is short circuit turns fault in these phases.

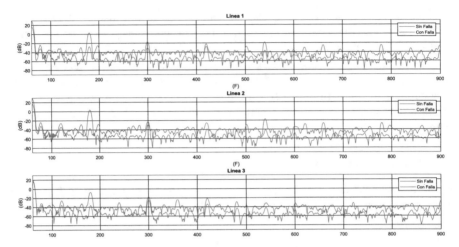

Fig. 6. Distortion of the stator currents by short circuit in phase 1 and 3.

Due to the increase in current and the distortion of the system caused by short-circuit fault in turns of the same phase, the temperature in the stator windings $T_{máx}$ considerably increased to 60 °C compared to the $T_{máx}$ windings without faults, as shown in Fig. 7. It is also mentioned that the motor was operating during the same time period and with the same nominal load.

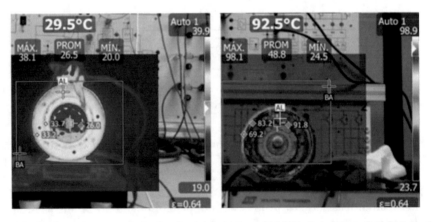

Fig. 7. Temperature increase due to short circuits in turns of the same phase 3 and phase 1.

The internal resistance of each coil of the three-phase system before the fault test was approximately 205 Ω, and, after the short-circuit fault test in the turns of phase 1 and 3, its internal resistance decreased to 71.9 Ω and 148.4 Ω respectively. Therefore, the system entropy is reflected with the increase of current and dissipated in the form of heat by the joule effect. Besides, the signals of current in the phases are distorted due to the loss of resistivity of the copper, changing the winding impedance and thus displacing the power factor by consumption of reagents.

4 Conclusions

The spectral value in the frequency of 300 Hz for the fault of bearings in the motors 1 and 2 obtained by the FFT from the analysis of the results reveal a coincidence of 96.9% and 98.9% compared to the results obtained through the theoretical equations implied in the MCSA technique, representing values that are within the fault frequency bands on 1x and 2x respectively. Furthermore, the spectrum at the frequency of 180 Hz for short circuit faults in turns of motor 2 showed a coincidence of 98.57% compared to the theoretical value which is in the band 2x of fault frequency for a phase, and with a coincidence of 99.14% when two phases present fault, being in 4x of the frequency band. Also, for the broken bar fault tests, the spectral value of the lower side band at 70 Hz has a similarity with the theoretical values of 99.85% and is in the band 1x of fault frequencies. In the same way, the spectral amplitude of the lower side band at 50 Hz is within the same band of fault frequency and has a similarity with the theoretical values in 99.8%. Finally, the spectral value of the upper side band at 40 Hz has a coincidence with the theoretical values in a 99.5% and is in the band 2x of frequency for this kind of fault. The analytical data compared with the experimental data do not present a total coincidence, due to the fluctuations of ±4% in the dynamometer control system, affecting the slip in the motors.

References

1. Antonino-Daviu, J., Rubbiolo, M., López, A.Q.: Diagnosis of the rotor condition in electric motors operating in mining facilities through the analysis of motor currents, pp. 1–8 (2017)
2. Antonino-Daviu, J.A., Member, S., Quijano-l, A., Rubbiolo, M.: Advanced analysis of motor currents for the diagnosis of the rotor condition in electric motors operating in mining facilities, vol. 54(4), pp. 3934–3942 (2018)
3. Fontes, A.S., Cardoso, C.A.V., Oliveira, L.P.B.: Comparison of techniques based on current signature analysis to fault detection and diagnosis in induction electrical motors, pp. 74–79 (2016)
4. Abhinandan, A.C.: Fault diagnosis of an induction motor through motor current signature analysis, FFT & DWT Analysis
5. Cox, R.W.: Algorithm for tracking the health of multiple induction motors using bus-level current. In: IECON 2018 - 44th Annual Conference of the IEEE Industrial Electronics Society, vol. 1, pp. 373–378 (2018)
6. Pineda-Sanchez, M., Antonino-Daviu, J.A.: Fractional fourier domain. IEEE Trans. Instrum. Meas. 59(8), 2065–2075 (2010)
7. Stopa, M.M., Filho, B.J.C., Lage, B.L.O.: An evaluation of the MCSA method when applied to detect faults in motor driven loads. In: IECON 2010 - 36th Annual Conference of the IEEE Industrial Electronics Society, vol. 2(1), pp. 760–765 (2010)

8. Climente-Alarcon, V., Nair, D., Sundaria, R., Antonino-Daviu, J.A., Member, S., Arkkio, A.: Combined model for simulating the effect of transients on a damaged rotor cage. IEEE Trans. Ind. Appl. **53**(4), 3528–3537 (2017)

9. Morales-Perez, C., Grande-Barreto, J., Rangel-Magdaleno, J., Peregrina-Barreto, H.: Bearing fault detection in induction motors using MCSA and statistical analysis. In: 2018 IEEE International Instrumentation and Measurement Technology Conference, pp. 1–5 (2018)

10. Gazzana, S.: An automated system for incipient fault detection and diagnosis in induction motors based on MCSA, pp. 1227–1232 (2010)

11. Rafael, J., Náthali, M.M.: Barras rotas y espiras cortocircuitadas en motores de inducción utilizando algoritmos de análisis

12. Naha, A., Member, S., Samanta, A.K.: A method for detecting half-broken rotor bar in lightly loaded induction motors using current. IEEE Trans. Instrum. Meas. **65**(7), 1614–1625 (2016)

13. Oswaldo, O., Contero, R.: UPS, QUITO (2016)

14. Kompella, G.R., Srinivasa, R., Sreenivasu, R.N.: Estimation of bearing faults in induction motor by MCSA using Daubechies wavelet analysis. In: 2014 International Conference on Smart Electric Grid, pp. 1–6 (2014)

15. Castellino, A.M., Donolo, P.D., Bossio, G.R., De Angelo, C.H.: Variables eléctricas Diagnóstico de fallas en los rodamientos de motores eléctricos empleando variables eléctricas, January 2007

16. Poncelas, O.: Diagnóstico de motores de inducción mediante la adquisición de corrientes de estator con sonda Rogowski Tesis Máster Ingeniería Electrónica Autor: Director, February 2016

17. El, N.: Universidad veracruzana (2012)

Three-Lead Electrocardiograph with Display and Printing System

Millard Escalona$^{(\boxtimes)}$ ⓘ, René Cortijo ⓘ, and Carlos Redrovan

Universidad Israel, Francisco Pizarro E4-142, Quito, Ecuador
{mescalona, recortijo, credrovan}@uisrael.edu.ec

Abstract. This document is the product of the work developed during 18 months framed in the line of research of the university, oriented to production and society. The motivation of the research team was to develop a prototype of an electrocardiograph of three derivations of bio-potential signals, which is a graphic representation of the bioelectric activity of the cells that make up the heart muscle. Allowing specialists in the area to evaluate the state of conduction of this organ and the appearance of pathologies caused by damage to the conduction tissues of the electrical signals of the heart safely, through a portable, compact equipment with visualization of these signals and wireless communication for remote visualization, between different mobile devices with Android operating system. This paper will indicate the design stages for signal acquisition, amplification, filtering, digitalization of the same and DC-DC power supply that will give the necessary autonomy to the equipment and communication for operation in any environment and low cost. In such sense the objectives planted in this investigation were achieved, with the development of an equipment with high operational performance, excellent electronic design in all its phases of design and conditioning, filtering, amplification and digitalization of the bio-potential signals and wireless communication between mobile devices.

Keywords: ECG · Leads · Digitization · Conditioning · Bluetooth communication

1 Introduction

An electrocardiogram is a graphical representation of the bioelectric activity of the cells that make up the heart muscle, measurement of electrocardiographic waves (P, Q, R, S and T) represented by very low level potential differences [1]. It allows us to evaluate the conduction state of the heart and the appearance of pathologies caused by damage to the tissues conduction of electrical signals. Previous research shows that electrocardiography (ECG) is a widely accepted approach for monitoring cardiac activity and clinical diagnosis of heart disease and thus achieve rapid interpretation by implementing a three-padded ECG system (W3ECG) [2]. The market and technology offers a series of high quality and high cost equipment and devices, in addition to being assembled in a compact way that prevents understanding signal processing internally. At the Universidad Militar Nueva Granada, a telemedicine research group worked on the acquisition, processing, visualization and storage of cardiac signals with the project: "Prototype of bipolar electrocardiograph for academic use", with the result of a device

© Springer Nature Switzerland AG 2020
M. Botto-Tobar et al. (Eds.): ICAETT 2019, AISC 1067, pp. 85–96, 2020.
https://doi.org/10.1007/978-3-030-32033-1_9

with low-cost modular technology [3]. Another relevant research was: "A three-lead electrocardiogram acquisition system is constructed with an analog front-end circuit (ADS1293)", which develops an electronic card for data acquisition with visualization through a smart phone via messaging with a computer and visualization in Matlab [4]. It is also important to highlight the project: "An ECG-on-Chip with QRS Detection & Lossless Compression for Low Power Wireless Sensors", which presents the design of a three-lead ECG with low power consumption integrating a chip with QRS detection in real time [5].

The object of study of this article is the design and construction of a three-lead ECG, portable, mobile and with wireless monitoring of a patient's electrocardiogram, for applications in ambulatory health areas [6, 7]. In this case, the first shunt will measure the potential difference between the electrode of the right and left arm; the second, from the right arm to the left leg and the third, from the left arm to the left leg. Since the voltage levels of an ECG signal are very low and within the same range of noisy fluctuations, they need to be eliminated by different filtering techniques [8]. This article proposes the design of the stages of acquisition of bio-potential signals, filtering and digitalization of an electrocardiographic signal, following the main technical and safety standards for the patient, established by ANSI [9]. It is also proposed to develop a graphic interface to visualize the ECG signal for the corresponding medical analysis, in which the necessary parameters can be identified to determine the normality or not of the electrical activity of the heart as well as to separate and obtain, in a precise way, the waves that constitute the electrocardiographic signal. On the other hand, the design and implementation of a communication stage and an Android application will be presented, in such a way that the electrocardiographic signal obtained can be visualized, recorded and printed remotely through mobile devices.

The device to be developed would be a robust, low-cost and portable solution that can be used in a simple manner by medical personnel in ambulances, stretchers or in smaller health centers located in rural or suburban areas that provide sufficient information for the health professional and that allows him/her to make a reliable diagnosis for decision making and subsequent treatment, which are discussed in the following section.

2   ECG Signal Acquisition and Filtering

In order to carry out a correct analysis of cardiac signals, it is necessary to design an acquisition module that allows an electrical signal to be obtained from the heart with as little noise as possible [10]. A data acquisition card was developed whose function is to take the bio-potential signals obtained through three electrodes connected to the patient, amplify them, filter them and transmit them to the digitalization plate using a card based on Arduino Mega for analog-to-digital conversion with accessories for visualization and printing of records [4, 11]. The resulting signal from the implemented analog filters has zero reference, i.e. positive and negative voltage values; it is then necessary to move the reference up to 2.5 V, taking into account that the voltage level recognizable by the Arduino card is TTL: 0 to 5 V. To displace the voltage reference, the same TL084 operational amplifier is used.

The arrangement of the transducers or electrodes for an ECG is established by the Einthoven triangle and its 12 derivations that provide spatial information of the electrical activity of the heart in 3 orthogonal directions: right-left, upper-lower and anterior posterior [12].

Three of the derivations are bipolar and are known as D1, D2 and D3, the other 9 are unipolar and are called VR, VL, and VF, plus V1, V2, V3, V4, V5 and V6 according to the order in which they are taken. For this project, the three standard bipolar derivations will be used:

D1: RA (−) to LA (+) right arm to left arm
D2: RA (−) to LL (+) right arm to left leg
D3: LA (−) LL (+) left arm to left leg

Figure 1 shows the proposed design for the a signal acquisition.

2.1 ECG Signal Filtering Stage

In order to develop the filtering stage, it is necessary to analyze the frequencies of the noise components that affect the electrocardiographic signal. The frequency band of an ECG signal is between 0.1 Hz and 150 Hz with a typical amplitude in the QRS wave (Complex of three vectors in the ECG wave described by Heinthoven) of 1 mV [12]. The design and implementation of the filtering chain was carried out in three stages: the first, by means of a band suppressor filter to eliminate frequencies very close to 60 Hz; then, a high-pass filter with the objective of keeping the frequencies above 0.1 Hz and finally a low-pass filter, which will allow the passage of frequencies below 150 Hz, disregarding or cancelling the 60 Hz network frequency [4]. At the output of the filter chain there is a negative component of continuous signal that must be corrected. This was solved by implementing an offset adjustment circuit with the TL084 operational amplifier, which was also used for the design of the circuits of the instrumentation amplifier, active guard and filter because it has as a characteristic, a high input impedance due to its FET technology.

Within the requirements of an ECG, according to ANSI-AAMI (Association for the Advancement of Medical Instrumentation) EC11-1991 standards, a bandwidth of 0.05 Hz to 100 Hz (+0.5 dB, −3 dB) is recommended and that system noise should be 40 mV, when all inputs are connected [9]. In addition to the IEC (International Electrotechnical Commission) standards, which differ slightly from those of ANSI, it is specified that the frequency response is in the range of 0.05–300 Hz, a high-pass filter is recommended with a frequency range of 0.05 to 0.5 Hz software selection and a low base filter of 40, 100 and 300 Hz software selection, with a sensitivity of 2.44 mV and CMRR > 110 dB at 50/60 Hz. In addition to the IEC (International Electrotechnical Commission) standards, which differ slightly from those of ANSI, it is specified that the frequency response is in the range of 0.05–300 Hz, a high-pass filter is recommended with a frequency range of 0.05 to 0.5 Hz software selection and a low base filter of 40, 100 and 300 Hz software selection, with a sensitivity of 2.44 mV and CMRR > 110 dB at 50/60 Hz [13].

The signals taken from the electrodes are too small for the sensitivity of the ADC (Analog - to Digital - Converter) module of the Arduino Mega board, which was used for the digitization stage [11]. The solution to the problem was to increase the magnitude of the voltage of the bio-signals. Since the signal is too small, ranging from 0.5 mV to 4 mV as a function of time-level, this signal is amplified by giving it a gain of 1000 [14]. For this purpose, an amplification circuit was designed, in which certain important parameters such as the active guard and filters were considered, in order to eliminate the effect of the capacities and increase the CMRR. Several alternatives were studied for the use of low-power, general-purpose operational amplifiers. It was resolved to work with a TL084 operational amplifier, which is characterized by being FET input and meet the needs due to its high input.

2.2   Instrument Amplifier

Instrumentation amplifiers are differential input circuits, which amplify (with precision) very low level signals by eliminating possible common mode noise signals. Among the characteristics that these amplifiers must comply with, the following stand out: stable closed-loop differential gain, externally adjustable without modifying the input characteristics, high common mode rejection both in continuous frequencies and in which noise may appear, high input impedance, low offset voltage and current with few derives, low output impedance and high common mode input voltages [15].

The gain of the instrumentation amplifier is given by the following equation:

$$G = \left(\frac{R18}{R16}\right)\left[\frac{2R7}{(R8+R9)}+1\right] \tag{1}$$

The gain of the previous instrumentation amplifier stage approximates 1000 times the input signal. In this last stage of ECG signal conditioning, with a gain $G = 1000$, and if $R6 = R7 = 680$ KΩ; $R8 + R9 = 22$ KΩ and was obtained $(R13/R16) = 17.64$. If $R7 = 47$ KΩ $\Rightarrow R4 = 2.66$

The signals of alternating current, muscular movement and respiration can interfere in the measurement, to obtain (in essence) the ECG signal, for this the frequency range between 0.1 and 150 Hz is established, in order to avoid external signals to the system. Capacitors were inserted to form a band-pass filter inside the system, according to the calculations made from Eq. 1.

$$f_{cut-off} = \left(\frac{1}{2}\right)\pi RC \tag{2}$$

Considering $f_{lower\,cutoff} = 0.1$ Hz and $f_{upper\,cutoff} = 150$ Hz, it was obtained: $C_{upper} = 22.57$ nF and $C_{lower} = 589.46$ μF.

Figure 1 shows the design adopted for the instrumentation amplifier.

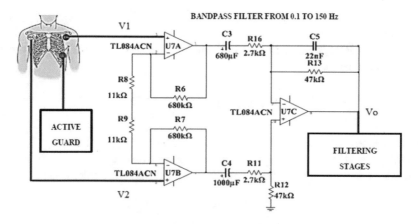

Fig. 1. Designed instrumentation amplifier.

2.3   Filtering Design. Band Suppressor, High-Pass and Active Low-Pass Filter

The design of the band suppressor filter or Notch filter made with the TL084 operational amplifier and the necessary calculations for the components of your external network according to Eq. 3 [15]. For the design of the high-pass filter the passive RC network is chosen the calculations detailed in Eq. 4.

$$f_N = \frac{1}{(4\pi * R * C)} = 60\,\text{Hz}$$
$$\text{with } C = 100\,\text{nF}, R = \frac{1}{(4\pi * 60\,\text{Hz} * 100\,\text{nF})} = 13.26\,\text{K}\Omega \tag{3}$$

$$f_c = \frac{1}{(2\pi * R * C)} \tag{4}$$

With $f_c = 0.1\,\text{Hz}$ and $R = 150\,\text{K}\Omega$, in that case: $C = 10.61\,\mu\text{F}$.

For passing frequencies below 150 Hz, neglecting or overriding the 60 Hz mains frequency, a low-pass filter was designed in the Sallen Key configuration and the calculations are shown in Eq. 5

$$f_c = \frac{1}{(\sqrt{8} * \pi * R * C)} = 150\,\text{Hz} \tag{5}$$

If $C = 50\,\text{nF}$, there is, $R = 15\,\text{K}\Omega$ which is equivalent in the circuit to R22 and R24.

2.4   Signal Conditioning

The resulting signal from the implemented analog filters has zero reference, i.e. positive and negative voltage values; it is then necessary to move the reference up to 2.5 V, taking into account that the voltage level recognizable by the Arduino card is TTL: 0 to 5 V. To displace the voltage reference, the same TL084 operational amplifier is used as an adder. In addition, through RV2, gain will be provided to the applied signal as shown in Fig. 2.

A value of RV1 is assumed for a minimum current: RV1 = 500 KΩ and by means of Eq. 6 we have to:

$$V_{Ref} = VDC2 * \left[\frac{Cursor\,position(\Omega)}{500\,K\Omega} \right] \tag{6}$$

For: $RV1 = 250\,K\Omega\,and\,VDC2 = +5VDC$ we'll have: $V_{Ref} = 2.5VDC$.

In this last stage of ECG signal conditioning, an adjustment is added by means of RV2, which will provide the final gain given by the following calculation:

$$R32 = \frac{Maximum\,position\,RV2}{Gain} \tag{7}$$

A maximum gain of 20 times will suffice and with RV2 = 25 KΩ, R32 = 25 KΩ (Commercial Value = 22 KΩ). The use of a 5.1 V Zener diode as a regulator allows the maximum voltage to be limited to + 5 VDC, to avoid damage to the Arduino Mega 2560 TTL type card.

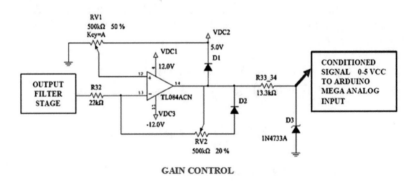

Fig. 2. Signal conditioner design

For the guard circuit, the circuit recommended by Texas Instruments was config-ured with the suggested values for the electronic components and the OPA4131 amplifier used in this stage.

The final result of the amplification and filtering stages was satisfactory according to the requirements and standards proposed from the beginning and was tested by digital simulation with the Multisim 14.1 application as shown in Fig. 3.

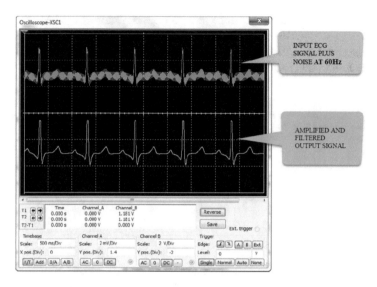

Fig. 3. Amplification response and double-stage suppressor filtering and noise injection.

2.5 Conversor Análogo/Digital ADC

The ATmega328 chip that incorporates the Arduino Mega board, includes an internal ADC, to digitize the ECG signal. This converter has a resolution of 10 bits, delivering integer values between 0 and 1023, expressed in digital as B0000000000 (0 V) and the value of (5 V) analog as B1111111111 (1023). Any intermediate analog value will be represented as a value between 0 and 1023, which is equivalent to adding 1 in binary every 4.88 mV, which is determined as follows $2^N - 1 = 2^{10} - 1 = 1023$, so that $\frac{5V}{1023} = 4,88$ mV and expressed as a percentage would be: $\frac{4,88\,\text{mV}}{5\,\text{V}} * 100\% \approx 0.1\%$ [11, 14]. Using the analogRead function ( ) on the Arduino, the analog input can be read out properly. The default setting for this function is 100 μs, which allows a theoretical sampling rate of 9600 Hz. The internal clock of the ADC is 16 MHz which is divided by a previous scale factor. This prescale is set to 128 by default, so that: $\frac{16\,\text{MHz}}{128} = $ 125 kHz and since each conversion requires 13 clock cycles of the ADC you: $\frac{125\,\text{kHz}}{13} = 9,6$ kHz. With this frequency we would have that 717 clock pulses would be produced at a speed of 1 every 104 μs microseconds, which would give a conversion time for 3.5 V of 74.5 ms.

2.6 Power Supply, Maintenance and Battery Charging Circuit

The conventional power supply of the acquisition and digitization cards with their respective accessories are powered from a conventional 15-volt DC adapter. The power supplied from the VCC adapter feeds the charge maintenance circuit of an 800 mAh battery. In order to maintain the charge in this battery, a circuit was delineated that works as a source of current by means of a voltage regulator of the LM317 type [14]. To perform the slow charge of the 800 mA/h battery, the calculation is reduced to the following expression: Charging current = (800 mAh)/(8 h) = 100 mA.

The load and power supply circuit implemented is shown in Fig. 4.

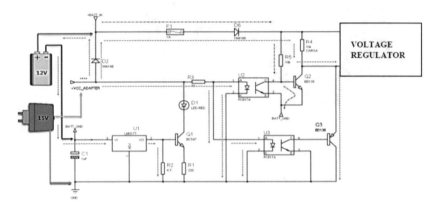

Fig. 4. Power supply and battery charging circuit

2.7   Differential Power Supply

It is based on an isolated DC-DC converter, using the PDQ2-D5-D12-S integrated circuit. The referred chip provides a differential output of +/− 12VDC from a power input of 4.5 to 9VDC. The regulated differential voltage at the converter output supplies the necessary power to the FET input operational amplifiers: TL084 (for instrumentation amplifier and filters) and OPA4131 (for guard circuit) and the power supply voltage for the converter is adopted from the LM7805 regulated output. This integrated circuit, in addition to feeding the DC - DC converter, provides voltage for the reference in the TTL conditioning circuit described above.

The factory recommendations for mounting the chip are adopted, leaving a circuit like the one shown in Fig. 5, obtaining at the output +/− 12VDC with respect to the reference (pin 7).

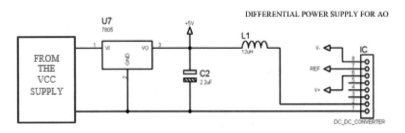

Fig. 5. Polarization of DC-DC converter.

3 Displaying Records and Wireless Communication

A 3.2 in. TFT screen (TFT_320QVT_9341) was used to display ECG signal logs together with the Mega Shield V2.2 TFT LCD. Both are attached to the Arduino Mega 2560 motherboard. All programming was done using Arduino code lines. By means of an application, for Android, called "APP ECG". This application stores a printable image of the electrocardiogram obtained. The communication between the prototype developed and the mobile is carried out with the support of a Bluetooth HC-05 device, coupled to the Mega 2560 [16].

Figure 6 shows TFT- LCD screen, communication devices and the electronic board.

Fig. 6. TFT- LCD screen, communication devices and electronic board

3.1 Results and Tests

After having carried out the digitalization process with the Arduino's ADC, its waveform could be observed through the serial port of the card, by means of the Serial Plotter application, with this it was verified that it adequately fulfills the necessary parameters to be processed in the following stages, in a digital way. Figure 7 shows the signal taken from the serial port of the Arduino using the Serial Plotter.

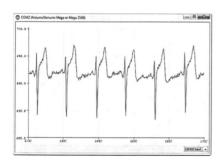

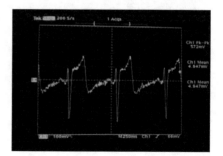

Fig. 7. Signal from Arduino serial port

Fig. 8. Signal from oscilloscope

The results of the obtained signal are ratified through an oscilloscope as shown in Fig. 8, with which it can be validated that the proposed digitalization objectives were met.

The tests were carried out in the medical office of the Israel University with the support of medical personnel and teachers as part of a pilot test for the prototype developed. At the physician's discretion, the ECG signal taken in the two patients complies with the waveform established for a healthy person. Figure 9 shows the tests performed.

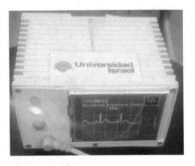

Fig. 9. Functional tests   **Fig. 10.** ECG equipment completed

To conclude with the implementation, the prototype is encapsulated with all its components in a case of dimensions of 13.0×7.0 cm in its front and 10.5 cm deep as shown in Fig. 10, so that it remains a small and portable equipment. In addition, by means of an Android application called "APP ECG" that stores a printable image of the electrocardiogram obtained and by means of Bluetooth communication between the prototype developed and the mobile phone, the process of printing records of the acquired ECG signal is carried out, fulfilling the objectives proposed for a first stage of development. The printed signal is shown in Fig. 11.

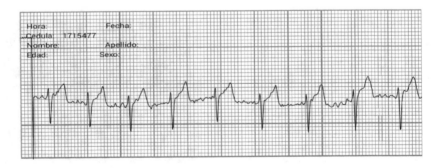

Fig. 11. Printing the electrocardiogram

Based on the results, unresolved problems were identified that should be considered in future developments as a more comprehensive analysis of the ECG signal by developing algorithms to provide other measures such as heart rate and pulse width. To do this you can calculate 6 s of the stroke of the rhythm equivalent to 30 frames, count the number of QRS complexes that fall within the 30 frames and multiply by 10 to obtain the number of beats per minute [17]. On the other hand, it is necessary to work on the graphic part in order to mark the measurement scales of the registers for greater precision of the readings.

4 Conclusions

This work involved the design and development of the stages of acquisition, amplification, filtering and digitalization of an ECG signal, as part of the implementation of a 3-lead electrocardiograph, in accordance with the standards established by ANSI-AAMI EC11-91 standards, which ensure proper management of bioelectric signals and protection of the patient under observation. If this equipment is compared with other previous works mentioned, here we went beyond the visualization of the signal in a computer, in this project we developed the hardware and software to achieve the compact equipment as visualized in Fig. 10. We took advantage of the open features of the Arduino Mega2560 base card that allows the use of multifunction accessories. It was programmed with the appropriate controllers and libraries for the detailed visualization of the graphics generated on a screen of the type TFT_320QVT_9341. The electrocardiographic analog signal was digitized maintaining its shape without appreciable loss of information and complying with the standard parameters of the ADC converters such as resolution and conversion time. It was also necessary to program code lines on the Arduino Mega2560 to open the communication port and make that through the Serial Plotter utility of the IDE transfer and monitor the digitized signal for analysis. The calculations to determine the values of components and gain of the system could be simulated using the software Multisim14.1 and tested with an oscilloscope, which verified a high level of performance and reliability meeting the proposed objectives.

References

1. Pal, A., Gautam, A.K., Singh, Y.N.: Evaluation of bioelectric signals for human recognition. In: International Conference on Intelligent Computing, Communication and Convergence, vol. 48(1), pp. 746–752. Elsevier BV (2015)
2. Cao, H., Li, H., Stocco, L., Leung, V.C.M.: Wireless three-pad ECG system: challenges, design, and evaluations. J. Commun. Netw. 13(2), 113–124 (2011)
3. Ramírez, L., Rodríguez, Y., Yuli, C.: Prototipo de electrocardiógrafo bipolar para uso académico. Ciencia Poder Aéreo 9(1), 115–123 (2014)
4. Fan, M.-H., Guan, M.-H., Chen, Q.-C., Wang, L.-H.: Three-lead ECG detection system based on an analog front-end circuit ADS1293. In: IEEE International Conference on Consumer Electronics - Taiwan (ICCE-TW), pp. 107–108, 27 June 2017

 5. Deepu, C.J., Zhang, X., Heng, C.H., Lian, Y.: A 3-lead ECG-on-Chip with QRS detection and lossless compression for wireless sensors. IEEE Trans. Circ. Syst. II Exp. Briefs **63**(12), 1151–1155 (2016)
 6. Cao, H., Leung, V., Chow, C., Chan, H.: Enabling technologies for wireless body area networks: a survey and outlook. IEEE Commun. Mag. **47**(12), 84–93 (2009)
 7. Jiménez Gómez, J., Palacios Barco, J.C., Potosí Moreno, M.A.: Electrocardiógrafo de tres derivaciones con comunicación inalámbrica. Ingenium **5**(10), 29–36 (2011)
 8. Najeb, J.M., Salleh, S.-H., Yusoff, K.: Two channel data acquisition unit for heart sound analysis. In: 1st International Conference on Computers, Communications, & Signal Processing with Special Track on Biomedical Engineering, pp. 173–175 (2005)
 9. ANSI, ANSI, American National Standars Institute (2018). www.ansi.org
10. Carmel, A.M.S.: Physiological signal processing. In: IEEE 27th Annual International Conference of the Engineering in Medicine and Biology Society, pp. 859–862 (2005)
11. Barrett, S.F.: Arduino Microcontroller Processing for Everyone! 3rd edn. Southern Methodist University, Morgan & Claypool (2013)
12. Yanowitz, F.G.: Introduction to ECG Interpretation, vol. V 10.0, U. o. U. S. o. Medicine, Ed., Utah, Intermountain Healthcare (2018)
13. I. International Electrotechnical Commission, "Webstore, International Electrotechnical Commission," IEC (2019). https://webstore.iec.ch/publication/2638
14. Tocci, R.J., Widmer, N.S., Moss, G.L.: Digital Systems. Principles and Applications, 10th edn. Pearson Education, Columbus (2007)
15. Coughlin, R.F.: Operational Amplifiers and Linear Integrated Circuits, Juares. Prentice Hall Hispanoamericana, S.A. (1999)
16. Sanchez Cortez, W.A.: Implementation of a digital electrocardiograph using bluetooth technology, Bucaramanga (2010)
17. Arámburu, C., Cárdenas Medina, A.: SAPIENS MEDICUS Learn, Think & Apply, Sapiens Medicus, 2 April 2019. https://sapiensmedicus.org/ecg-interpreta-lo-basico-en-7-sencillos-pasos/. Accessed May 2019

Four-Degree Free Spheric Morphology Robotic Arm for Manipulation of Objects by Means of Platform Robotic Operating System (ROS)

Juan David Chimarro Amaguaña$^{(\boxtimes)}$ ⓘ and Fidel David Parra Balza ⓘ

Universidad Tecnológica Israel, Quito, Pichincha, Ecuador
jchimarro@uisrael.edu.ec

Abstract. The present investigation project was intended to evaluate the free software platform Robotic Operating System (ROS) for the control of a robotic arm with spheric morphology, the method that was used is an applied research since it focuses specifically on how they can carry out the general theories for new designs of a robotic manipulator and also the level of accuracy was measured offering an equipment that allowed us to have a clear view and understanding the function and application of robotic theories, such as: inverse and direct kinematics and their different control techniques. The robotic manipulator has a final for pick and place processes. The efficiency of the robot was determined by controls on the platform ROS, with excellent results. The efficiency of the software and the control methods were also shown.

Keywords: Spheric robot · ROS · Robotic manipulator · Direct kinematics · Inverse kinematics

1 Introduction

Many innovative applications have been developed in the robotics area; however, one of the disadvantages is finding appropriate tools suitable for the use of free software that is efficient at the same time. That's why the platform Robotic Operating System (ROS) was implemented so that a module of robotic arm has an accurate integration of all its subsystems.

The spheric robot is a robotic arm that consists of four free degrees: rotational-rotational-lineal-rotational; which is the base, the shoulder, the prismatic elbow, and the wrist respectively. This model of manipulator is developed as a didactic robot with learning purposes, as well as for the food industry, labels, pick and place, assembling, etc.

© Springer Nature Switzerland AG 2020
M. Botto-Tobar et al. (Eds.): ICAETT 2019, AISC 1067, pp. 97–110, 2020.
https://doi.org/10.1007/978-3-030-32033-1_10

2 Design Development

Table 1. Concept definition

Modules	Definition	Selection
Module 1	Mechanic structure	Médium size - Materia: l steel AISI 1060 (TOL) - Articulation base: turning plate - Articulation shoulder: reduction of gears - Articulation prismatic elbow: electric cylinder - Final effector: gripper
Module 2	Actionning power	Newton dynamic model
Module 3	Actionning and control	- Direct current engine (DC)
Module 4	Technical Models for control	- Denavit-Hartenberg. Algorithm Proportional Integral Control Derivative (PID)
Module 5	Sensorial	Internal Sensing - Sensor presence: end of trail - Sensor position and speed: encoder absolute External Sensing: - Sensor presence
Module 6	Software for control	Robotic Operative System (*Robot Operating System*, ROS)

Figure 1 shows the mechanic structure of spheric robot (Table 1).

Fig. 1. Mechanic structure of spheric robot.

Figure 2 shows the mechanism of the articulation base where the structure of the robot spins around an axis by means of a turning plate.

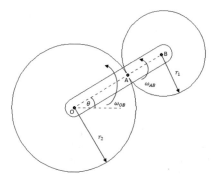

Fig. 2. Mechanism of the articulation base of the spheric robot.

If a pinion rolls on the surface of a gear with an angular speed ω_{AB} as shown in Fig. 2, rolling implies that the velocity of the pinion in its contact spot A in relation to the surface of contact is zero; the velocity in the point O is also zero since it is a fixed spot. Therefore, to find ω_{AB} we use the Eq. 1. Of relative velocity [2].

$$V_B = V_O + \omega_{OB} \times r_{B/O} = V_A + \omega_{AB} \times r_{B/A} \tag{1}$$

Where we obtain:

$$\begin{aligned} \omega_{OB} = \omega_{cintura} \qquad \omega_{AB} = \omega_{motor} \\ \omega_{motor} = \omega_{cintura}(1+N) \\ N = \frac{r_2}{r_1} \quad transmition\ relation \end{aligned} \tag{2}$$

Figure 3 shows the mechanism of articulation shoulder where the movement is carried out direct transmition [3].

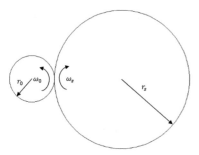

Fig. 3. Mechanism of the articulation base of the spheric robot.

If the pinion is the driving and the mesh is driven they both will have angular velocity ω_0 y ω_s respectively. Where:

$$\omega_s = \omega_{shoulder} \qquad\qquad \omega_0 = \omega_{motor}$$
$$\omega_{motor} = N \cdot \omega_{shoulder} \tag{3}$$
$$N = \tfrac{r_s}{r_0} \quad transmition\ relation$$

The articulation base verifies that an object spins around in relation to a point O (the basis) with a constant angular velocity ω and angular acceleration α equal zero. Once the kind of material and the measures of the mechanic structure have been defined, we proceed to calculate the inertia of the robotic manipulator in relation to the coordinates mentioned before [2].

After that we apply the Eqs. 4 and 5 of movement, using the second law of Newton and the equations of angular movement to calculate the necessary torque to turn the based in relation to point O.

$$\begin{bmatrix} \Sigma Mox \\ \Sigma Moy \\ \Sigma Moz \end{bmatrix} = \begin{bmatrix} Ixx & -Ixy & -Ixz \\ -Iyx & Iyy & -Iyz \\ -Izx & -Izy & Izz \end{bmatrix} \begin{bmatrix} \alpha x \\ \alpha y \\ \alpha z \end{bmatrix} + \begin{bmatrix} 0 & -\Omega z & \Omega y \\ \Omega z & 0 & -\Omega x \\ -\Omega y & \Omega x & Izz \end{bmatrix} \begin{bmatrix} Ixx & -Ixy & -Ixz \\ -Iyx & Iyy & -Iyz \\ -Izx & -Izy & Izz \end{bmatrix} \begin{bmatrix} \omega x \\ \omega y \\ \omega z \end{bmatrix} \tag{4}$$

$$\tau_{waist} = \sqrt{\Sigma Mox^2 + \Sigma Moy^2 + \Sigma Moz^2} \tag{5}$$

$$\tau_{motor} = \frac{\tau_{waist}}{(1+N)} \tag{6}$$

The articulation base verifies that an object spins around in relation to an axis O (the shoulder) with an angular velocity ω, we proceed to calculate the inertia of the robotic manipulator in relation to coordinate system 2 [2].

After that, with the equation seven we proceed to find the inertia moment in relation to an arbitrary axis that goes by the origin and parallel to a unitary vector e which is obtained by:

$$Io = Ixx \cdot ex^2 + Iyy \cdot ey^2 + Izz \cdot ez^2 - 2 \cdot Ixy \cdot ex \cdot ey - 2 \cdot Iyz \cdot ey \cdot ez - 2 \cdot Izx \cdot ez \cdot ex \tag{7}$$

After that we apply Eq. 8 of movement, using the second law of Newton to calculate the necessary torque to turn the shoulder in relation to the fixed turning axis O.

$$\tau_{hombro} = I_o \cdot \alpha_o + M \cdot g \cdot \iota \tag{8}$$

$$\tau_{motor} = \frac{\tau_{hombro}}{N} \tag{9}$$

2.1   Robot Kinematics

Figure 4 shows the coordinate systems and geometric parameters to carry out the solution of direct kinematics. We will try to find the homogeneous matrix by means of Denavit Hartenberg algorithm (D-H) [5].

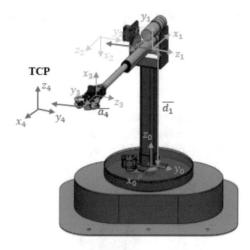

Fig. 4. Geometric parameters to analyze direct kinematics.

where:

$$\overline{d1} = 43\,\text{cm} \qquad \overline{d_3} = 28\,\text{cm} \qquad \overline{a_4} = 10,5\,\text{cm}$$

Therefore the parameters of the algorithm D-H are shown in the following table (Table 2):

Table 2. D-H of spheric robot

Shackle i	θ_i	d_i	a_i	α_i
1	$\theta_1 + 180°$	$\overline{d_1}$	0	90°
2	$\theta_2 - 90°$	0	0	90°
3	180°	$d_3 + \overline{d_3}$	0	90°
4	$\theta_4 + 90°$	0	$\overline{a4}$	90°

Once we have gotten the D-H parameters we calculate the total homogeneous transformation matrix as shown in Eq. 10, in which the rotation matrix is implicit, as well as, translation from the working point known in English as *Tool Center Point* (TCP).

$$T = {}^4 A_0$$

$$= \begin{bmatrix} cos(\theta_1)cos(\theta_2+\theta_4) & -sin(\theta_1) & cos(\theta_1)sin(\theta_2+\theta_4) & cos(\theta_1)\left[(d_3+\overline{d_3})\cos(\theta_2)+\overline{a_4}\cos(\theta_2+\theta_4)\right] \\ sin(\theta_1)cos(\theta_2+\theta_4) & cos(\theta_1) & sin(\theta_1)sin(\theta_2+\theta_4) & sin(\theta_1)\left[(d_3+\overline{d_3})\cos(\theta_2)+\overline{a_4}\cos(\theta_2+\theta_4)\right] \\ -sin(\theta_2+\theta_4) & 0 & cos(\theta_2+\theta_4) & \overline{d_1}-(d_3+\overline{d_3})sin(\theta_2)-\overline{a_4}sin(\theta_2+\theta_4) \\ 0 & 0 & 0 & 1 \end{bmatrix}$$

$$(10)$$

Figure 5 shows the geometrics parameters to solve the inverse kinematics. The problem will be formulated by means of geometrical relations and the angles will be found from the Cartesian Coordinates of TCP [6].

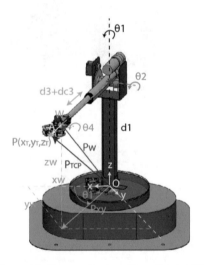

Fig. 5. Geometric Parameters to analyze inverse Kinematics

Steps to find inverse kinematics:

1. To begin the geometric analysis and find θ_1, θ_2, d_3 y θ_4 from x_T, y_T y z_T, to find θ_1 we have to observe the plan xy where we obtain:

$$\theta_1 = \arctan\left(\frac{y_T}{x_T}\right) \tag{11}$$

2. We calculate the vector $P_w = x_w i + y_w j + z_w k$, then we have to observe the plan where the triangle OWP is. It contains the following vectorial addition:

$$P_{TCP} = P_w + A \therefore P_w = P_{TCP} - A \qquad (12)$$

Where:

$$P_{TCP} = x_T i + y_T j + z_T k \qquad (13)$$

Defining that the unitary vector P_{xy} is the same for vector A since it will have the same direction and orientation, but different module, then we see that

$$A = \overline{a_4} \cos(\theta_1) i + \overline{a_4} \sin(\theta_1) j \qquad (14)$$

Then:

$$x_w = x_T - \overline{a_4} \cos(\theta_1) \qquad (15)$$

$$y_w = y_T - \overline{a_4} \sin(\theta_1) \qquad (16)$$

$$z_w = z_T \qquad (17)$$

3. To find angle θ_2 and d_3 we have to observe the plan that contains the triangle OHW, in which we have two different cases. The first case when $z_w \leq d1$ as seen in Fig. 6a, to find θ_4 we need to observe the plan that contains the triangle PWH and knowing that the vector A is parallel to the plan xy as defined in steps 2 as it can be observed in Fig. 6b.

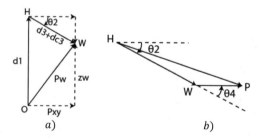

Fig. 6. When $z_w \leq d1$: (a) Triangle OHW (b) Triangle PWH

We obtain that:

$$\theta_2 = \arctan\left(\frac{\overline{d_1} - z_w}{P_{xy}}\right) = \arctan\left(\frac{\overline{d_1} - z_w}{\sqrt{x_w^2 + y_w^2}}\right) \qquad (18)$$

$$d_3 = \sqrt{x_w^2 + y_w^2 + \left(\overline{d_1} - z_w\right)^2} - \overline{d_3} \tag{19}$$

$$\theta_4 = -\theta_2 \tag{20}$$

For the second case when $z_w > d1$, as we can observe in Fig. 7.

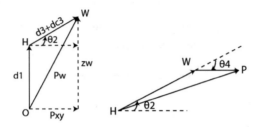

Fig. 7. When $z_w > d1$ (a) Triangle OHW (b) Triangle PWH

We obtain that:

$$\theta_2 = \arctan\left(\frac{z_w - \overline{d_1}}{\sqrt{x_w^2 + y_w^2}}\right) \tag{21}$$

$$d_3 = \sqrt{x_w^2 + y_w^2 + \left(z_w - \overline{d_1}\right)^2} - \overline{d_3} \tag{22}$$

$$\theta_4 = -\theta_2 \tag{23}$$

2.2 Robot Control

To the model of spheric control we took into consideration a proportional, integral and derivative control (PID), considering the position feedbacks and the internal PID that the actuators possess which facilitates the control implementation. However, the lineal actuator does not have its own control; that's why, The PID control is implemented using tuning methods. In Fig. 8 we can appreciate the diagram for control system for direct and inverse kinematics. There we can visualize what was explained before: as we can appreciate in the diagrams, each actuator works in a closed bond Nevertheless, the total control system of the spatial position (TCP) is an open bond. Implementing a control over the whole system is a very complex process; therefore, the Kinematics of the robot is analyzed in order to be able to work in a closed kinematic chain.

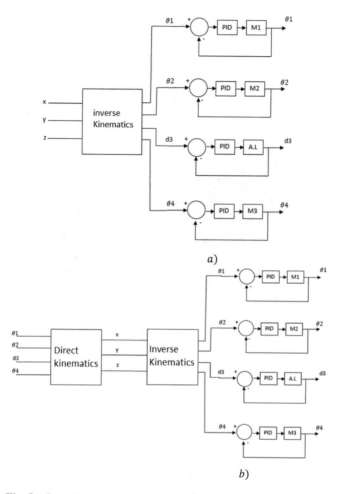

Fig. 8. Control system: (a) inverse Kinematics (b) direct kinematics

2.3   Implementation of the Control System in ROS

To implement the control software an operative Meta system was used. It was an ROS. These systems are completely free or we could say open technology. It can be downloaded from the main page [7]. For the present Project the ultimate version of ROS: JADE will be used. By means of architecture ROS the control can be performed at a graphical level sing nodes. Nodes are feasible processes that carry out the computing and which can be implemented using languages, such as C++ and Python since ROS has direct bookstores with the mentioned programs. To initiate the design, we do the following flow diagram that will represent the control of the robotic manipulator [8].

3 Construction and Implementation of the Spheric Robot

The processes carried out in the construction of the manipulator elements include cutting, bending, drilling, riveting, milling, turning, painting, among others. They were performed in the laboratory of Manufacturing Processes in Universidad de la Fuerzas Armadas – ESPE. In Fig. 9 the assembled mechanic structure is shown.

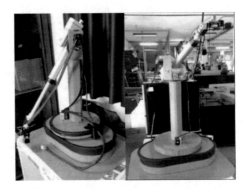

Fig. 9. Mechanic structure of the spheric robot.

3.1 Implementation of the Electrical and Electronic System

The Dynamixel motors used in this project allows us to carry out a bus network that consists of connecting the motors serially, and then by means of the USB2 Dynamixel Robotics adapter each motor is identified by an ID. In addition, wiring is needed for the end effector and an end-of-travel sensor that detects objects in the position to be manipulated. The linear actuator also requires wiring; both for its energization and for its control. Figure 10 specifies the distribution of the electrical and electronic system that connects the UPC with the controllers and these in turn with the sensors, actuators and energy source that are present in the spherical robot.

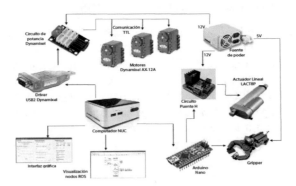

Fig. 10. Electric and Electronics Connections of the spheric robot

4   Analysis of Results

We can observe the implemented nodes in ROS and verify the internal connections, as shown in Fig. 11. By means of the following command ROS:

$ rosrun rqt_graph rqt_graph

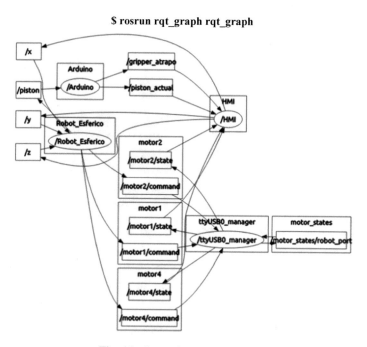

Fig. 11. Internal connections ROS

As we can see in Fig. 11, the diagrams consist of several blocks which represent the nodes that will have a specific task and will be explained in detail as follows:

Node HMI: in this module we will perform a graphic interface of power and control in which we will be able to have Access to the different actions to be performed by the robotic manipulator. Such interface is implemented using the package PyQt4 inside the interpreter Python [9].

Node Spheric Robot: in this node we will perform direct and inverse kinematic actions where the data will be sent to the control nodes of the actuators which are: ttyUSB0 manager and Arduino.

Node ttyUSB0 manager: this node is where the control action of the base, shoulder and wrist articulations will be performed, by means of the device *USB2Dynamixel Robotics*. It is important to notice that these articulations are powered by the engines Dynamixel AX-12-A; each one of which has its own control ID, as it was explained in the engine specifications.

Node Arduino: This node does exactly the same as the previous node with the only difference that it uses the device Arduino to carry out the control action on the prismatic articulation (electric cylinder). This lineal actuator has a PID control.

Figure 12 shows the working space for the spheric control

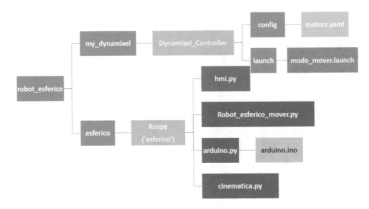

Fig. 12. Work space ROS from spheric robot

The obtained results are presented in the following Table 3 and as it can be seen the percentage errors do not exceed 5%, confirming the effectiveness of the mathematical procedures performed and the functionality of the control system. To be able to watch the video about the operation of the spherical robot it is recommended to enter the following link: https://www.youtube.com/watch?v=iOPwqwilvkw&feature=youtu.be.

Table 3. Kinematics Evaluation

Points	Articulation	Desired position	Achieved position	Error calculation	Error percentage
Point 1	θ_1	180°	179, 47°	0, 52°	0, 29%
	θ_2	47, 8°	48, 58°	0, 78°	1, 63%
	d_3	20 mm	19, 18 mm	0, 82 mm	4, 1%
	θ_4	238, 12°	237, 5°	0, 62°	0, 26%
Point 2	θ_1	70, 03°	69, 68°	0, 35°	0, 5%
	θ_2	17, 4°	17, 5°	0, 35°	0, 57%
	d_3	45 mm	44, 66 mm	0, 44 mm	0, 75%
	θ_4	207, 9	207, 33°	0, 56°	0, 27%

(*continued*)

Table 3. (*continued*)

Points	Articulation	Desired position	Achieved position	Error calculation	Error percentage
Point 3	θ_1	$140,76°$	$140,23°$	$0,53°$	$0,37\%$
	θ_2	$62,17°$	$62,75°$	$0,58°$	$0,94\%$
	d_3	67 mm	$64,44$ mm	$2,56$ mm	$3,82\%$
	θ_4	$252,78°$	$251,9°$	$0,87°$	$0,34\%$
Point 4	x	56 cm	$55,6$ cm	$0,4$ cm	$0,71\%$
	y	0 cm	0 cm	0 cm	0%
	z	43 cm	$42,9$ cm	$0,1$ cm	$0,2\%$
Point 5	x	-45 cm	$-43,9$ cm	$1,1$ cm	$2,4\%$
	y	30 cm	$30,5$ cm	$0,5$ cm	$1,6\%$
	z	56 cm	$56,2$ cm	$0,2$ cm	$0,36\%$

The ROS platform facilitates the development of applications in the field of robotics due to its facility to interact with open technology programs. The control techniques and all the calculations made for the robot allow the positioning of the TCP in the space to be extremely precise. A quantification of the positioning errors was also presented. Practically all the errors were less than 5%. This definitely shows the effectiveness of the direct and inverse kinematics methods, and also the fact that the plant works correctly.

5 Discussions

The development of this robotic manipulator encourages also the development of new research projects whether in the field of robotics with different control strategies, artificial intelligence or dabbling in hybrid robots for instance by combining this manipulator with a mobile robot.

It is possible to work on the project for the development of robotic technology. For instance, it would be advisable to perform a dynamic control where the Levenberg-Marquardt algorithm is applied, so that in case of positioning the TCP in a physically unattainable point, this algorithm recalculates it, positioning the work point in the coordinate closer to the requested one. It is also recommended to complement it with the tools of Rviz and Gazebo, which provide 3D simulations of the robotic model designed.

References

1. University of Brazil. Mechatronic Systems. http://www.mecatronica.eesc.usp.br/mecatronica/index.php/pt/laboratories/laboratorio-de-manipulacao-robotica. Accessed 27 Feb 2015
2. Fowler, B.: Dynamic problems de engineering. Italy (2009)
3. Shigley. Design of Mechanic Engineering. McGraw-Hill, Spain (2008)

4. Trossen robotics (2015). http://www.trossenrobotics.com
5. Barriga, A.: Robotic Fundaments. Mexico (2011)
6. Reyes Cortés, F.: Robotics-Control of Robot Manipulators. McGraw-Hill, Spain (2011)
7. ROS. ROS Installation (2015). http://wiki.ros.org/ROS/Installation
8. ROS. Robotic Operating System (2015). http://www.ros.org/
9. Python. Python & PyQt4 (2015). https://www.python.org/

Prototype of Application and Keyboard for Writing Using the Braille System on a Computer

Pedro Asmal-Arias[1] and Eduardo Pinos-Vélez[2,3(✉)]

[1] GIIATA, Research Group on Artificial Intelligence and Assistive
Technologies, Universidad Politécnica Salesiana, Cuenca, Ecuador
pasmal@est.ups.edu.ec
[2] GIIATA, Research Group on Artificial Intelligence and Assistive Technologies
and GIIB, Research Group on Biomedical Engineering, Universidad Politécnica
Salesiana, Cuenca, Ecuador
epinos@ups.edu.ec
[3] DICOP, Universidad de Piura, Piura, Peru

Abstract. All human beings have the need to communicate, but due to several factors that intervene at the time of communication, this can especially change their code, the set of signs and rules used in the production of a message; The people with visual impairment use the Braille system, as their method of reading and writing, this system tries to form symbols for each letter and number by means of raised points. Currently, in most cases, learning and communication in Braille is limited only to paper and its writing with the help of a punch, so different devices that help this group of people to digital inclusion will be analyzed. participation in the use of a computer for different purposes, already developed by different areas of research, to finally make and present the design of a prototype for the application of this language and functioning as a keyboard to write using the Braille system on a computer, as well as its software and hardware development process, the research and design of all the necessary elements for it and the procedure used to verify and verify its correct operation and tests performed with potential users.

Keywords: Special education · Keyboard · Teaching · Braille system

1 Introduction

In the world there is a great diversity of languages and forms of communication, but in some cases some of these forms of communication are not possible for people who have some kind of disability, such as visual impairment and more at the level of reading and writing, that is why for them there is another type of symbols that represent each of the letters and numbers that we know. This type of communication language is known as Braille, and was invented as a solution for non-vocal communication for people with visual disabilities by Luis Braille in 1829, writing through the use of his hands, focused on reading through the help of the fingertips [1, 2].

© Springer Nature Switzerland AG 2020
M. Botto-Tobar et al. (Eds.): ICAETT 2019, AISC 1067, pp. 111–120, 2020.
https://doi.org/10.1007/978-3-030-32033-1_11

The system consists of printing combinations of raised points on a succession of Braille cells, each with six points arranged in two columns of three points each [3].

With his invention for the first time he allowed himself to write and revise what was expressed by writing quickly, later to invent this writing tool Louis Braille adapted the system to mathematics, music and science [4]. Around this topic several projects that have been developed, in order to offer help to these people, have been made and marketed games such as puzzles or non-electronic dolls that form the sign generator, but with the passage of time from a few years ago several electronic prototypes of more help to people with visual disabilities are being made, due to several functionalities that they have and that are added to them, such as the "Braille teaching electronic prototype", capable of generating an autonomous learning of the Braille reading system [4, 5].

The project will examine the possibility of creating an environment of interaction between the team and the user, to obtain results in learning in a self-taught way, that a person can learn without necessarily having a person to guide them at all times, looking for the device can be used as a gadget, in replacement of the normal keyboard in a computer, used by institutes designated for the care and education of people with visual impairment that in several cases do not have the necessary equipment for the proper teaching of the Braille.

2 The Prototypes and Existing Devices

At present there are several support works for blind people, both in the area of electronic engineering, as well as jobs that do not depend on it, but which are linked to help to obtain better results in learning, teaching and application of the language Braille, mainly developed programs that cover part of the methodologies and teaching methods for people with visual disabilities, such as some that are numbered below.

2.1 Braille Interaction in Back of the Car Steering Wheel

Prototype that allows the entry of text for drivers of cars, in Braille code without using the eyes to write while driving and has the hands on the steering wheel; for data entry there are two sets of three buttons on the back of the steering wheel on both the left and right sides, this approach allows you to enter a character or command with just one input combination without the need to look at the keys, in addition to having visual results and auditory comments to improve the interaction [6].

2.2 Automatic Braille Translation System

Project of construction of a keyboard for the Braille code developed with the objective of allowing teachers of blind people to visualize texts written by their students, an electronic keyboard less noisy and less expensive than traditional mechanics, among its applications is translation of sheets of plain text, mathematics and chemistry written in Braille code, in addition to writing plain text, mathematics or chemistry using the developed keyboard [7].

2.3 Electronic Braille Teaching Prototype

Electronic device capable of generating an autonomous learning of the Braille reading system with the help of the senses of hearing and touch, focused on the teaching of Braille language, mainly in the instruction of the structure of the generating sinus and how each of the characters through a series of actuators that simulate the generator sign and are able to represent all the main characters of the mentioned language [5].

2.4 Braille Keyboard for the Visually Impaired

Keyboard prototype designed and built to provide a simple form of writing for people with visual impairments, by using 6 buttons required to represent characters in Braille, letters, symbols and numbers will be represented by triggering the necessary points to form each of these through the Braille system, which facilitates the use of this keyboard in the writing of texts to people with visual disabilities because they already know all the symbols in Braille and they should only represent them in the matrix of switches [8].

2.5 Single Character Refreshable Braille Display

Development of a braille display that can be updated with a single character, in order to allow a static reading, that is, from character to character, which helps the user to cooperate, this design is implemented with electromagnetic relays and control logic implemented in a programmable field matrix of doors, this braille screen will allow access to the information is being presented continuously in the updatable matrix and also the learning of the characters by the same for beginners [9].

3 Proposed Device

Based on the knowledge on the application and teaching of the Braille language obtained, and with the help for the development of the work and the verification of an Educational Unit in charge of the training of people with different abilities, especially in the area of visual disability, we propose the design in both software and hardware of the prototype of a device for the teaching and application of the Braille language, which can have various functions such as the practice and teaching of characters and words in Braille, also serve as a computer keyboard, clock sound, basic calculator, among other functions that allow the development of people with visual disabilities.

3.1 Important Aspects for the Development of a Prototype of Reading and Writing

For the development of the prototype and to achieve its functionality and support in the areas of teaching and Braille application four axes of interaction, necessary for the formation of motor and hearing skills, have been contemplated.

Technical Characteristics: This aspect seeks to address is that the prototype has the necessary features that allow you to improve the accessibility of children with visual difficulties to learning to read and write, that the device is ergonomic and simplified use, so that the Software and hardware applications can be adapted to the user, facilitating their use [10].

Methodology: In the development of the instruction and application of the Braille language there are several methodologies that are used for the teaching of the literacy process, but there are also complementary tools for learning help such as the case of electronic and computer applications and devices, which could be used both in daily activities in class, as well as for autonomous learning with the need for these devices to guide work to users [10, 11].

Curricular Content: This aspect aims that the prototype has specific contents for each level, avoiding a general education but this one is specialized in each area and level of learning, curricular content will be contained that can be adapted to the level of the student, which will be separated and could be used according to the progress in the learning process, that is to say that for the advance must be fulfilled with the previous activities in a chain, the head of the teaching with the help of the device must arrange which are the most suitable contents to work of all the present in it, but the device will also have a process defined for the purpose of autonomous use.

Software: With the existence of diverse applications focused on the reading and writing of the Braille system, different points and areas of help are available to the beneficiaries, but for these to be mostly accepted by the users, they should be easy to use and resolve the technical aspects already mentioned what will ensure the interaction between users, students and teachers with the application [10].

3.2   Development Process

Once the structure has been studied and known, how communication is developed using the Braille system and how the instruction of this method of communication is carried out in the Educational Units specialized in this type of teaching. We proceed to the general design of a prototype for the teaching and application of this language, the research and design of all the necessary elements for it, as well as the procedure used Fig. 1.

The design of the prototype includes all the necessary factors such as the physical and technical aspect, all the elements are designed, elements such as electronic cards from the electronic circuit, the frame that will contain and protect the electronic elements inside and the programming which will be made from scratch starting from the use of the PIC18F2550 microcontroller and the MP3-TF-16P sound player module.

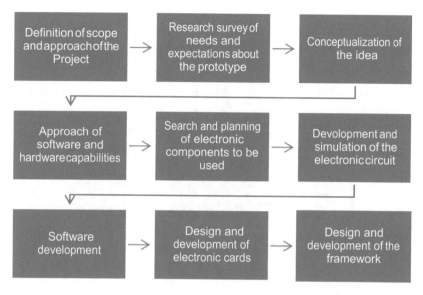

Fig. 1. Development process.

3.3 Technical Description of the Prototype

The prototype revolves around three axes, each of importance for the development of the same, as they are the axis of design, programming and electronic, the latter being the most important since from the elements considered to be used the following two will be routed, as an example when using a specific microcontroller, consideration will be made in programming as its programming language and in turn considerations in its design starting from and considering its dimensions.

Scheme of Electronic Elements: Through the approach of both software and hardware capabilities, the search was made for the elements necessary to comply with these and the design approach based on the interrelation between them, then in Fig. 2, the scheme is shown representative with all its elements, which are considered necessary for the correct functioning according to the expectations raised at the time of design and the initial idea for the device.

From the electronic circuit and the scheme presented in the Figs. 2, 3 electronic cards were designed, the main card will carry the own PIC18F2550 microcontroller for USB connectivity applications, quartz crystals and capacitors needed for operation, RTC circuit (ds1307, battery CR2032) and the USB connector. Card number two will carry the Omron 12 × 12 mm and 6 × 6 mm buttons for user interaction with the device already presented in the scheme, while the last card will contain the MP3-TF-16P module, for the reproduction of sound files, compatible with TF card controller and built-in 3 W amplifier [12].

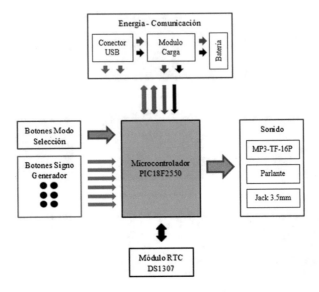

Fig. 2. Scheme of electronic elements of the prototype.

3.4   Software Considerations

The C language will be used for programming the microcontroller, in the case of sending the sonic indications presented by the prototype, these are found on the memory card of the MP3-TF-16P module, the way in which it reproduces is through he sent 10 bytes represented hexadecimal in the programming through the serial port of the microcontroller, these commands can be seen in Table 1 [12].

3.5   Presentation of the Final Design

Fulfilling each of the stages proposed initially we get to obtain the final version of the prototype raised, this can be seen in Fig. 3, which also shows the design that was made in computer can compare it with the final work of printing in third dimension of the frame, it is possible to observe all the external elements and finishes, verifying that they agree with those raised and designed during the process. The final measurements of the frame are 14 × 10 × 2.4 cm, length, width and height respectively, its final design is shown in Fig. 3 and the color considered is black.

Fig. 3. Design and final result of the prototype framework.

Table 1. Description of serial communication commands [12].

Format	Description	Information
$S	Start Byte	0x7E
Ver	Version	0xFF
Len	Data length	Without checksum
CMD	Command	Quick commands
Feedback	Feedback command	1: Feedback 0: No feedback
Para1	Parameter 1	High byte of track
Para2	Parameter 2	Low byte of track
Checksum	Check sum	High byte
		Low byte
$O	End byte	0xEF

After obtaining the final version of the prototype, it will proceed to verify the operation of the same, in different areas, such as the technical, adaptability, performance, proving that the prototype meets the stated capabilities.

4 Experimentation and Validation of the Prototype

The experimentation of the use of the prototype and the validation of its operation was carried out gradually in each phase of electronic design, as well as finally with the finished prototype with the help of the "Claudio Neira Garzón" Educational Unit of the city of Cuenca, its teacher managers and students of the area of visual disability of the first levels of instruction. The procedure that is detailed below was developed with the students of the first levels of basic education mainly of the first year, with the purpose of verifying their comfort with the use of the same, their performance when carrying out each one of the activities, and finally receiving opinions of these and their teachers on the operation and design of the prototype, as well as comments and suggestions that could elevate its functionality. Figure 4.

The way in which the qualifications were obtained in each of the tests was the following, after the process developed 3 qualifications of conformity are obtained in each test one given by the user, in this case the student, another by the teacher who observes and the third by the designer, all on a scale of 0 to 10 that are then averaged so that one table can be shown for each test in the table. Each test is identified by a number that identifies the test user and a letter that is the test number, that is, 2 tests are performed with 3 users. Tables 2 and 3.

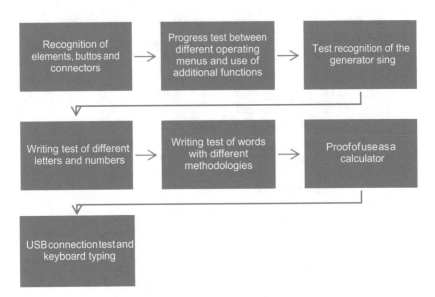

Fig. 4. Experimentation and validation of the prototype.

Table 2. Results of the prototype recognition and adaptability tests.

Description	P1A	P2A	P3A	P1B	P2B	P2B	Prom
Frame recognition	10	10	10	10	10	10	10
Buttons recognition	9	9	8	9	10	9	9
Connectors recognition	8	9	9	9	9	9	9
Operating menus (advance)	8	7	7	8	7	8	8

Table 3. Performance results in writing tests and use of additional functions.

Description	P1A	P2A	P3A	P1B	P2B	P2B	Prom
Recognition generator sign	9	9	9	10	10	10	10
Characters writing	10	9	9	10	10	10	10
Words writing	10	10	9	10	10	10	10
Use as keyboard	8	8	9	9	9	9	9
Calculator	10	9	10	9	9	10	9
Clock	8	8	8	9	9	8	8

5   Analysis of Results

With the design, development and verification of the operation of the prototype, several results are obtained in different aspects, which mostly approve the development of devices for the purpose of communication support in Braille, these aspects are detailed below.

Adaptability: In the tests and reflected in the results which are not less than 9, it is observed that both the design and the location of their control buttons should be as simple, predictable and have the location similar to the generator sign to obtain good results, avoiding the overload of these that can avoid confusion when selecting the correct buttons when performing a task.

Modes of Operation: The opinions on the applications of the prototype are good, but what was not received in a good way is the way of selecting each of these receiving the lowest scores of the tests with 7 on several occasions, which is consider is that the programmer for this same reason already knows how to make the selection between the different menus but for the user is different and more using it for the first time can cause confusion, so it should simplify the access mode to each function.

Writing: The writing tests reflected conformity, and so the process to follow, starting with the knowledge of the sign generator, passing the writing of letters and numbers, to finish writing words and phrases, in addition to the great help provided by having different methodologies designed to carry out the interaction user-device more didactic what achieves an outstanding level of compliance.

Keyboard: The keyboard's use of the prototype was defined as acceptable as the form of use is correctly explained and thus also understood by the user, having a favorable result with opinions that facilitates writing and makes it faster with the practice.

6   Conclusions

Initially the prototype proposed, already had a function for the Braille application, more than those needed for learning and writing practice, which made it an integral device, with which a method could be made step by step of initial learning, to subsequently have various forms and methodologies for practicing the knowledge already obtained by forming letters, initials and words, finally leading to its main activity, which consists of being able to use the prototype as a computer keyboard, which it varies its efficiency in a large percentage since at the end of the teaching process the device can still be used, for writing documents in a computer, sending messages in different social networks, among other activities.

A very important issue is that the applications and functionalities are adapted to people with visual disability and not the other way around, at present, although it is an example of help to blind people, who can use a cell phone with the help of a integrated reader that indicates the selected option, or the number that presses to make calls, and thus be located in the number distribution cell; but nevertheless it is an example that the person with this disability must adapt to the device and have the know the location of each number in the dialing cell, besides being guided by the sound indications, what should be different and the device To adapt to the knowledge of the user, it would be more beneficial for these people to be able to dial a number using the Braille generator sign, that is, using the knowledge they already know, a question that is sought from the beginning with this prototype, and that In the execution we have seen the importance that the applications that we know and use can also be used by people with blind people with the help of technology, taken to their field and not the other way around.

References

1. Mackenzie. La Escritura del Braille en el Mundo. UNESCO, Presidente del consejo mundial del Braille, Paris, pp. 9–19 (1998)
2. BANA, Braille Authority of North America. "Braille Basics". http://www.brailleauthority. org/learn/braillebasic.pdf
3. ONCE, Comisión Braille Española. La didáctica del braille más allá del código, Primera Edición. ONCE, CBE, Madrid, Abril 2015
4. Pierce, B.: The world under my fingers: personal reflections on Braille. Future Reflections, vol. 15, no. 1, Invierno (1996)
5. Aucay, A., Pinos, E., Serpa, L.: Braille teaching electronic prototype. In: 2016 IEEE International Autumn Meeting on Power, Electronics and Computing (ROPEC), Ixtapa, Mexico
6. Osswald, S., et al.: Back of the steering wheel interaction: the car Braille keyer. In: Paternò, F., de Ruyter, B., Markopoulos, P., Santoro, C., van Loenen, E., Luyten, K. (eds.) Ambient Intelligence, AmI 2012. Lecture Notes in Computer Science, vol. 7683. Springer, Berlin, Heidelberg (2012)
7. Shahbazkia, H.R., Silva, T.T., Guerreiro, R.M.: Automatic Braille code translation system. In: Sanfeliu, A., Cortés, M.L. (eds.) Progress in Pattern Recognition, Image Analysis and Applications, CIARP 2005. Lecture Notes in Computer Science, vol. 3773. Springer, Berlin, Heidelberg (2005)
8. Manohar, P., Parthasarathy, A.: An innovative braille system keyboard for the visually impaired. In: 2009 11th International Conference on Computer Modelling and Simulation, Cambridge, UK (2009)
9. Ignat, M.C., Faragó, P., Hintea, S., Roman, M.N., Vlad, S.: A single-character refreshable Braille display with FPGA control. In: Vlad, S., Roman, N. (eds.) International Conference on Advancements of Medicine and Health Care through Technology; 12th–15th October 2016, Cluj-Napoca, Romania. IFMBE Proceedings, vol. 59. Springer, Cham (2017)
10. Sabino, B.: TIC's como herramientas de apoyo en el proceso de lectoescritura Braille. In: Revista Iberoamericana para la Investigación y el Desarrollo Educativo, publicación #11, Mexico (2013)
11. Rodríguez, M., Arroyo, M.: Las TIC al servicio de la inclusión educative. In: Digital Education Review, 25 June 2014. http://greav.ub.edu/der/
12. DFRobot Members. dfrobot.com, 15 August 2018. https://www.dfro-bot.com/wiki/index. php/DFPlayer_Mini_SKU:DFR0299. Accessed 06 Mar 2019

Sliding Mode Controller Applied to a Synchronous DC/DC Power Converter

Oscar Gonzales[1(✉)], Cristian Amaguaña[1], Marcelo Pozo[1], Oscar Camacho[1], and Jorge Rosero-Beltrán[2]

[1] Escuela Politécnica Nacional, Ladrón de Guevara, E11.253 Quito, Ecuador
{oscar.gonzalesz,cristian.amaguana,marcelo.pozo,oscar.camacho}@epn.edu.ec
[2] Universidad de las Américas, Calle Queri S/N y Av. De los Granados, Quito, Ecuador
jorge.rosero@udla.edu.ec

Abstract. In the present work, a Sliding Mode Controller (SMC) was implemented in a Synchronous Buck Converter (SBC) to regulate output power. Several applications in renewable energies such as, power stability in photovoltaic panels, motivated this paper realization. The SMC was compared with a benchmark regulator such as, Proportional–Integral (PI) controller. Tracking and robustness results were analyzed through the Integral Absolute Error (IAE) criterion.

Keywords: Synchronous Buck Converter · SMC · Power control applications · IAE

1 Introduction

In the last years, renewable energy applications have been struggling against the intense climate change effects, produced by carbon dioxide emissions. In this field, research proposed different alternatives to use renewable energies efficiently [1]. Many countries have established policies to use sustainable energy, especially in far locations from the big cities to benefit the people from isolated communities [2]. These policies suggest the utilization of hybrid energy systems to decrease the consumption of fossil fuels [3]. In electrical energy systems, engineers have been developing technological projects at high, as well as, at low voltage levels with renewable energy sources penetration. These systems use power electronic devices in the static energy conversion. These devices operate at high frequency and perform a large amount of energy conversion [4].

Research in power electronics has generated several topologies for high and low power conversion applications. For example, circuits implemented in photovoltaic panels employ DC/DC and DC/AC power converters. The first converter, steps up or down the energy from the photovoltaic panels and the second one, drives the DC energy to the grid. In [5], a DC/DC Single–Ended Primary–Inductor Converter (SEPIC) performed a Maximum Power Point Tracking

© Springer Nature Switzerland AG 2020
M. Botto-Tobar et al. (Eds.): ICAETT 2019, AISC 1067, pp. 121–130, 2020.
https://doi.org/10.1007/978-3-030-32033-1_12

(MPPT). This technique used the maximum energy available from photovoltaic panels although the changing environment conditions around.

In [6], an inverter was implemented in order to perform the MPPT on the photovoltaic panels. MPPT strategies are wide, significant and widely used in electric power systems. In [7], a sensorless technique in photovoltaic cells to MPPT was applied. On the other hand, control algorithms also improve the performance of photovoltaic systems, [8] implemented an optimum regulation technique such as, Model-based Predictive Control to achieve MPPT. Other applications like resonant microinverters, developed the control system and optimization in a resonant converter [9]. The converter optimization was accomplished by an on-line change of impedances in the circuit according to the switching frequency of the power semiconductors in order to care its service life.

Despite all the existing research and projects on power electronics converters, more proposals are focused in areas such as closed loop control. Many regulation techniques seek objectives such as: zero error in stable state, short settling time and robustness against disturbances. Thus, Sliding Mode Control (SMC) has gained an important position in industrial applications and research. Due to its advantages against other controllers, some of which are: steady state stability, non-linear algorithm design and robustness, make the SMC implementation as an attractive election for control engineers [10]. Even though the SMC technique is widely applied in different slow-dynamic processes, new studies are seeking to use the SMC in power electronics applications. For instance, in renewable energy systems, such as photovoltaic plants, research projects work on MPPT.

In this context, the main purpose of the present work is to implement an advance control technique, such as SMC, in a Synchronous Buck Converter which is used in renewable energy applications [11,12]. This paper is organized as follows: the Sect. 2, a description of a Synchronous Buck Converter is presented, the Sect. 3, the system identification is described, the Sect. 4, the control algorithm is designed, the Sect. 5 the experimental results are showed, and the Sect. 6 the conclusions are presented.

2   Synchronous Buck Converter

The DC/DC buck converters generate an output voltage lower or equal than the supply voltage. Several applications utilize these power converters topology, especially for a high value of the output current at a specific value of voltage. Due to the simplicity in the construction and control of DC/DC buck converters, and cost, its use is extensive in areas such as control, automation and energy applications [13]. However, a variant for the conventional buck converter known as a DC/DC synchronous buck converter (SBC) is a better option for applications which require high efficiency (Fig. 1(a)). The SBC replaces the diode of the conventional buck converter by a power transistor which reacts quickly in the on/off switching. On power-up state, switch Q1 is in saturation while Q2 is in a cut-off. Energy flow from the source to the load through the inductance L and the capacitor C (Fig. 1(b)). In the off state, the transistor Q2 closes the

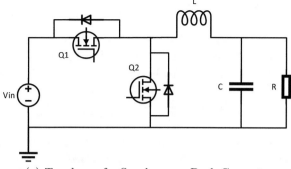

(a) Topology of a Synchronous Buck Converter

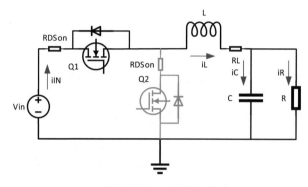

(b) Q1 on and Q2 off

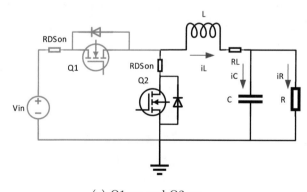

(c) Q1 on and Q2 on

Fig. 1. Synchronous buck converter switching stages

circuit, so the inductance current is discharged to the capacitor and the load (Fig. 1(c)). The simultaneous operation of the switches Q1 and Q2 require a dead time between the commutations in order to prevent the short-circuiting in real applications.

3   Converter Model

To model the plant, the identification algorithm takes measurements of the output signals (voltage at the load) and the input signals (step Pulse Width Modulation - PWM). From [14], the second-order function modelling is used.

$$G(s) = \frac{K}{(\tau s + 1)(t_0 s + 1)} \tag{1}$$

Where K is the transfer function gain, τ is the time constant of the process, and t_0 is the delay time from the following approximation:

$$e^{-t_0 s} \approx \frac{1}{t_0 s + 1} \tag{2}$$

Then, the transfer function for the SBC is:

$$G(s) = \frac{16966.51}{(3720.8s + 1)(372.1s + 1)} \tag{3}$$

4   Sliding Mode Control (SMC)

The SMC links mathematically two functions: a discontinuous function which brings closer the output signal to the reference as quickly as possible, and a continuous function, which preserves the output signal at the set-point despite of possible undesired effects of disturbances. Figure 2 shows how the discontinuous (1) and continuous (2) functions work in a sliding surface that contains the error and the time change of the error as variables.

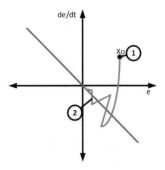

Fig. 2. Sliding surface in SMC

The error and its derivative part must tend to zero in the time. However, Fig. 2 shows a basic idea of a sliding surface, because the surface can consider other structures as in [15] where the sliding surface is PID-type. This surface was chosen for this work due to its robustness characteristics. The control scheme

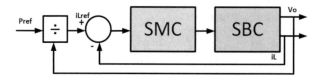

Fig. 3. Control scheme for SMC implementation

for the SBC is shown in Fig. 3, where a cascade scheme is used to regulate the output power of the load.

The control law is defined as:

$$U_{SMC} = U_C + U_D \tag{4}$$

The discontinuous function U_D acts aggressively in order to reduce the error in the transient state, and the continuous function U_C keeps the controlled variable at the desired reference.

The continuous function has a two-pole transfer function:

$$G(s) = \frac{K}{(\tau s + 1)(t_0 s + 1)} \tag{5}$$

where K is the gain, τ and t_0 are the poles of the transfer function.

The transfer function converted into a differential equation can be represented as:

$$\tau \, t_0 \frac{d^2 x(t)}{dt} + (\tau + t_0) \frac{dx(t)}{dt} + x(t) = K u(t) \tag{6}$$

The chosen sliding surface is a PID-type.

$$s(t) = \frac{d}{dt} e(t) + \lambda_1 e(t) + \lambda_0 \int_0^t e(t) \tag{7}$$

The derivative of the sliding surface must be equal to zero to achieve a control action which seeks zero error and zero derivative of the error in stable state.

$$\frac{ds(t)}{dt} = 0 \tag{8}$$

$$e(t) = R(t) - X(t) \tag{9}$$

$$\frac{ds(t)}{dt} = \frac{d^2 e(t)}{dt^2} + \lambda_1 \frac{de(t)}{dt} + \lambda_0 e(t) = 0 \tag{10}$$

By replacing the previous expressions in (4):

$$\frac{d^2(R(t) - x(t))}{dt^2} + \lambda_1 \frac{d(R(t) - x(t))}{dt} + \lambda_0 (R(t) - x(t)) = 0 \tag{11}$$

Second derivatives of the reference are not considered because they are constant terms, the continuous control action is:

$$U_C(t) = \left(\frac{t_0 \tau}{K}\right) * \left\{ \left[\frac{t_0 + \tau}{t_0 \tau} - \lambda_1\right] \frac{dX(t)}{dt} + \frac{1}{t_0 \tau} X(t) + \lambda_0 e(t) \right\} \tag{12}$$

Considering as tuning constants:

$$\lambda_1 = \frac{t_0 + \tau}{t_0 \tau} \tag{13}$$

$$\lambda_0 \leq \frac{{\lambda_1}^2}{4} \tag{14}$$

For the discontinuous function analysis, Lyapunov's candidate function is defined as:

$$V = \frac{\sigma^2}{2} \tag{15}$$

For $\sigma \neq 0$, where V is a positive definite function and $\dot{V} < 0$ should be a negative definite function, so the reachability condition is:

$$\dot{V} = \sigma \dot{\sigma} < 0 \tag{16}$$

To satisfy the condition of (14), K_D must comply that:

$$K_D > 0 \tag{17}$$

The discontinuous function with the best performance is the sign function. However, due to its aggressive response, it generates high frequency switching known as chattering [16], and real electrical and mechanical elements cannot follow this fast action. To avoid so fast chattering, the discontinuous function is smoothed by means of a sigmoid function.

$$U_D = K_D sign(S) \tag{18}$$

$$sign(S) = \frac{S}{|S| + d} \tag{19}$$

The control low with continuous and discontinuous functions is:

$$U_{SMC} = \left(\frac{t_0 \tau}{K}\right) * \left\{ \frac{1}{t_0 \tau} X(t) + \lambda_0 e(t) \right\} + \frac{K_D S}{|S| + d} \tag{20}$$

5 Experimental Results

As mentioned in Sect. 2, the SMC implementation was carried out in a real SBC, Fig. 4 shows the converter. The elements used in the experiment are: 1. SBC controlled by an STM32F407VGT6 Discovery microcontroller, 2. Tektronix oscilloscope, current clamp and fluke multimeter, 3. rheostat from 0–86 Ω, 4. Computer with MATLAB^{TM} 2018a.

The laboratory SBC module contains the following characteristics Table 1:

The constants of the SMC are: $K_D = 0.1118$, $\lambda_0 = 72.73$, $\lambda_1 = 0.0005375$ and $d = 0.68$.

The test performed was the tracking of power references of 4 W and 9W, as shown in Fig. 5. To evaluate the SMC, a PI controller is firstly implemented as

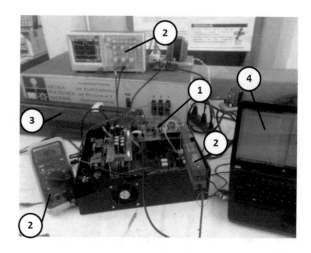

Fig. 4. SBC test module

in [17] is showed in order to contrast the performance of each controller. The constants of the PI are: $K_P = 0.00021488$ and $K_I = 5.39$. These values where obtained in a trial and error test.

The SMC presents better response when reaching the two reference levels. In the 9 W value, the SMC response is better than the PI because the last one oscillates in steady state. In Fig. 6, the power values are shown by means of a laboratory oscilloscope.

The second test verified the robustness of each controller at a value of 5.8 W. The disturbance generation consisted in a sudden change of the lever at the load. Two perturbations were added with values above and below the stable state value of the system which is 22 Ω at the load.

Figure 7 shows that the SMC is more robust by sudden load changes, while the PI controller delayed its response in reaching the desired reference.

Table 1. Synchronous buck converter parameters

Parameter	Characteristics
Switching frequency	100 MHz
MOSFET	IRFP460N
Inductance	1.8 mH
Capacitance	200 uF
Maximum output power	35 W

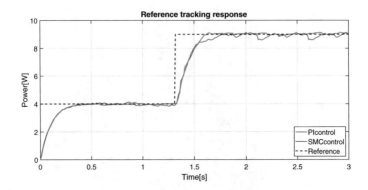

Fig. 5. Reference tracking test

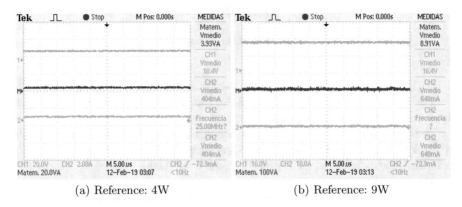

(a) Reference: 4W   (b) Reference: 9W

Fig. 6. Power measurement through an oscilloscope

To evaluate numerically the performance of each controller, the criterion of the Integral of the Absolute value of the Error (IAE) is used [18]:

$$IAE = \int_0^t |e(t)| \, dt \tag{21}$$

The performance results from the IAE are shown in Table 2.

Table 2. IAE performance

Controller	Reference	Disturbance
PI	2.021	3.011
SMC	0.938	0.655

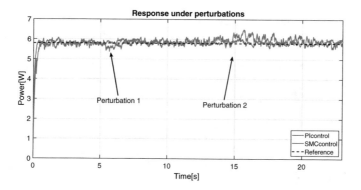

Fig. 7. Disturbance response

Table 2 shows the robustness of the SMC tests. For MPPT at a photovoltaic system, which suffer a kind of disturbances due to the unstable environment conditions, such as the radiation and the temperature changes.

6 Conclusions

An SMC controller were implemented in an SBC in order to test the robustness and advantages over other algorithms.

Although the power devices are very fast, it is important to avoid the SMC chattering by choosing a discontinuous sigmoid-type function.

The range for reference values were 4 W to 9 W. The SMC showed a better performance especially in the reference tracking. The PI failed in this purpose because this controller may have left the area of operation in which it was designed.

The second-order system approximation for the SBC was a practical option for modelling these type of converters which have non-linearities.

The SMC showed robustness under disturbance effects. This is a prominent advantage over PI regulator because the converter keeps constant the output power under a range of operation.

References

1. Shyam, B., Kanakasabapathy, P.: Renewable energy utilization in India—policies, opportunities and challenges. In: 2017 International Conference on Technological Advancements in Power and Energy (TAP Energy), pp. 1–6, December 2017
2. Lamnadi, M., Trihi, M., Boulezhar, A.: Study of a hybrid renewable energy system for a rural school in Tagzirt, Morocco. In: 2016 International Renewable and Sustainable Energy Conference (IRSEC), pp. 381–386, November 2016
3. Anoune, K., Bouya, M., Ghazouani, M., Astito, A., Abdellah, A.B.: Hybrid renewable energy system to maximize the electrical power production. In: 2016 International Renewable and Sustainable Energy Conference (IRSEC), pp. 533–539, November 2016

4. Wang, H.: Microgrid generation planning considering renewable energy target. In: 2016 IEEE International Conference on Power and Energy (PECon), pp. 356–360, November 2016
5. Latif, T., Hussain, S.R.: Design of a charge controller based on SEPIC and buck topology using modified incremental conductance MPPT. In: 8th International Conference on Electrical and Computer Engineering, pp. 824–827, December 2014
6. Dong, Y., Ding, J., Huang, J., Xu, L., Dong, W.: Investigation of PV inverter MPPT efficiency test platform (2015)
7. Sher, H.A., Murtaza, A.F., Noman, A., Addoweesh, K.E., Al-Haddad, K., Chiaberge, M.: A new sensorless hybrid MPPT algorithm based on fractional short-circuit current measurement and P&O MPPT. IEEE Trans. Sustain. Energy **6**(4), 1426–1434 (2015)
8. Morales-Caporal, M., Rangel-Magdaleno, J., Morales-Caporal, R.: Digital simulation of a predictive current control for photovoltaic system based on the MPPT strategy. In: 2016 13th International Conference on Power Electronics (CIEP), pp. 295–299, June 2016
9. Alamir, N., Abdel-Rahim, O., Ismeil, M., Orabi, M., Kennel, R.: Fixed frequency predictive MPPT for phase-shift modulated llc resonant micro-inverter, In: 2018 20th European Conference on Power Electronics and Applications (EPE 2018 ECCE Europe), pp. P.1–P.9, September 2018
10. Vaidyanathan, S., Lien, C.-H.: Applications of Sliding Mode Control in Science and Engineering, June 2017
11. Rekioua, D., Achour, A., Rekioua, T.: Tracking power photovoltaic system with sliding mode control strategy. Energy Procedia **36**, 219–230 (2013)
12. Khemiri, N., Khedher, A., Mimouni, M.F.: A sliding mode control approach applied to a photovoltaic system operated in MPPT. In: 10th International Multi-Conferences on Systems, Signals Devices 2013 (SSD 2013), pp. 1–6, March 2013
13. Sreedhar, J., Basavaraju, B.: Design and analysis of synchronous buck converter for UPS application, In: 2016 2nd International Conference on Advances in Electrical, Electronics, Information, Communication and Bio-Informatics (AEEICB), pp. 573–579, February 2016
14. Chen, L., Li, J., Ding, R.: Identification for the second-order systems based on the step response. Math. Comput. Modell. **53**(5–6), 1074–1083 (2011)
15. Bordoloi, N., Buragohain, M.: Bacteria foraging optimized and modified PSO optimized PD-SMC and PID-SMC controller for inverted pendulum system. In: 2017 International Conference on Smart grids, Power and Advanced Control Engineering (ICSPACE), pp. 171–176, August 2017
16. Herrera, M., Gonzales, O., Leica, P., Camacho, O.: Robust controller based on an optimal - integral surface for quadruple-tank process. In: 2018 IEEE Third Ecuador Technical Chapters Meeting (ETCM), pp. 1–6, October 2018
17. Doubabi, H., Chennani, M., Essounbouli, N.: Modeling and design of synchronous buck converter for solar-powered refrigerator. In: 2017 International Renewable and Sustainable Energy Conference (IRSEC), pp. 1–5, December 2017
18. Kumar, K.R.: Implementation of sliding mode controller plus proportional double integral controller for negative output elementary boost converter. Alexandria Eng. J. **55**(2), 1429–1445 (2016)

Automation of the Feeding System for Washing Vehicles Using Low Cost Devices

Fernando Saá[(⊠)], José Varela-Aldás, Fernando Latorre,
and Belén Ruales

Facultad de Ingeniería y Tecnologías de la Información y la Comunicación,
Grupo de Investigación en Sistemas Industriales, Software y Automatización,
Universidad Tecnológica Indoamérica, 180103 Ambato, Ecuador
{fernandosaa, josevarela, belenruales}@uti.edu.ec,
flatorre@teojama.com

Abstract. Vehicle washing is a profitable business today, but it produces unwanted environmental effects that should be considered; autonomous systems are a viable solution for the optimization of resources in production processes and services. This paper presents the implementation of an automatic resource feeding system for vehicle washing using low cost devices. The system consists of an Arduino Mega controller, ultrasonic sensors to measure the level of liquids, several actuators, Bluetooth communication devices and a mobile application. The results present the consumption of resources in a month of operation, demonstrating the benefits of this proposal.

Keywords: Automation · Resource optimization · Vehicle washing · Mobile application · Arduino

1 Introduction

Automatic systems are present in the industrial and service sectors, since they provide efficiency and flexibility. Current processes are automated taking into account more parameters and considering all the factors to be controlled [1]. In addition, it is a multidisciplinary task and requires skills that are developed with practice [2]. Even with all the benefits of automation there is still a sector with a culture of distrust for its development, playing a substantial role in daily life because the modern world is saturated with autonomous systems [3], this demands safer and more reliable autonomous processes. On the other hand, technology has invaded our homes, improving our living conditions and facilitating interaction with household components [4].

The main reason to implement an autonomous system is the optimization of production elements [5], because it provides automatic machines that perform the work continuously and without loss of resources, reducing time and movements [6]. In addition, it allows to know the states of the process through the implementation of SCADA systems, delivering information in real time for a timely decision making [7]. Normally these improvements represent high installation costs, although the current market provides a wide variety of low cost equipment that makes process automation more feasible [8].

© Springer Nature Switzerland AG 2020
M. Botto-Tobar et al. (Eds.): ICAETT 2019, AISC 1067, pp. 131–141, 2020.
https://doi.org/10.1007/978-3-030-32033-1_13

Regarding autonomous home systems, the price is more convenient and the equipment more diverse. It is common to find a low-cost home automation system that uses a basic microcontroller (Arduino) and the most important sensors to condition the home [9]. The tendency of these systems is to use mobile applications that facilitate the remote operation of a smart home, generally connected to the system through Bluetooth, Wifi or the worldwide network (internet) [10]. Smartphones are not only found in home automation, with the advent of the internet of things, every automated process is compatible with a mobile application, generating uncertainty in the future of technology and the misuse of it. Mobile applications are available to anyone, therefore, the design of engineering software in mobile applications should contribute to social awareness [11].

Car washing is a very profitable business today, and more with the introduction of mobile technology in the offer of this service [12], directly affecting the environment, with the use of chemicals that harm flora and fauna, as well as excessive use of water. In response to this problem, wastewater treatment and, in some cases, reutilization in the washing process are carried out [13]. On the other hand, there are proposals to automate the vehicle washing process, using 3D scanner, water cannon angle control and liquid optimization [14], these systems contribute in the reduction of environmental impact in the development of this service.

This work proposes the optimization of the resource feeding system of the vehicle washing device in the Talleres y Servicios S.A company, through the implementation of an automatic system using a Mega arduino and a mobile application as a control interface. The document is made up of the following sections: *2. Current situation,* the vehicle washing process is reviewed before optimization. *3. Proposed system,* the design of the electromechanical system and the functions of the main program are detailed. *4. Results,* results obtained after the implementation of the autonomous system are reported, and *5. Conclusions,* the conclusions of the project are presented.

2   Current Situation

The process of washing vehicles within a company is important as it is part of preventive and corrective maintenance, since dust, mud or dirt deteriorate the paint layer, and therefore it is important to carry out this task occasionally.

Talleres y Servicios S.A. offers three types of vehicle washing: courtesy washing, express and full. Courtesy washing is done to the customer's vehicle after a maintenance or in promotions, where the exterior is cleaned with water pressure for an approximate time of 10 min, no supplies are used and no cleaning is done inside. Express washing has an additional cost and uses pressure water and soap for vehicles on the exterior, it lasts approximately 20 min and the inside is cleaned with a damp cloth. The complete washing, is a deeper work in terms of vehicle cleaning, it uses all the inputs indicated by the manufacturer of the vehicle, using water pressure, soap and degreaser, with an approximate time of 40 to 45 min. Table 1 shows the types of washes over time and the amounts used.

Table 1. Types of vehicle washing.

Type of washing	Washing time (minutes)	Supplies
Courtesy	10	Pressurized water
Express	20	Pressurized water and soap
Full	40–45	Pressurized water, soap, degreaser, absorbing material and panel polish

Figure 1 shows how an employee performs vehicle washing inside the company, using water as a cleaning resource, which through a water washer expels the liquid to remove the dirt embedded in the vehicle.

Fig. 1. Cleaning of vehicles with a water washer.

Applying the times and movements study technique for the improvement of the work methodology, it is observed that the worker in charge of the process works in unequal work, that is, it differs in the times and activities independently of the vehicle that enters, without having a specific routine. It is evident that there are repetitive, short and long-term work cycles, without adequately controlling the necessary resources in the operation, resulting in the following observations:

- The personnel in charge of the operation has no awareness of water consumption and the importance of saving.
- It is evident that the staff performs the work thoroughly and investing as much water as possible, when they have a lot of time; conversely, when the operator does not have much time, he uses water to a lesser extent and uses little or no supplies. That is, this process is not done in a standardized way.
- Taking the previous paragraph as a base, the operator does not know the different types of washing (courtesy, express and complete) and the duration time of each of them.

Table 2 shows the expenses by type of washing, defining a valuation by vehicle model and type of washing, taking into account the averages of washings in 2017 at the Ambato branch of Talleres y Servicios SA, with the expenses in proportion for water,

Table 2. Expenses by type of washing.

Type of washing	Model	Vehicles served / month	Water expenditure / cm^3	I spend liquid soap / cm^3	Degreasing expense / cm^3	Time spent / min
Courtesy	H300	57	$ 9,22	$ 10,14	$ 17,52	3876
Express	H300	28	$ 5,17	$ 7,75	$ 9,82	1944
Full	H300	6	$ 2,02	$ 3,23	$ 3,83	412
Courtesy	H500	28	$ 5,17	$ 7,75	$ 9,82	1944
Express	H500	12	$ 2,52	$ 3,52	$ 4,78	818
Full	H500	6	$ 2,07	$ 2,69	$ 3,94	423
Courtesy	H700	18	$ 3,85	$ 6,17	$ 7,32	1252
Express	H700	2	$ 0,35	$ 0,55	$ 0,66	103
Full	H700	1	$ 0,46	$ 0,74	$ 0,88	80
Mean			**$ 3,43**	**4,73**	**$ 6,51**	**1205**
Standard deviation			**$ 2,8**	**$ 3,37**	**$ 5,32**	**1229**

liquid soap, and degreaser according to the result of the study. The time taken to wash 158 units is provided, regardless of the type of washing.

3 Proposed System

3.1 Electromechanical Design

The construction of an automatic machine for the resource feeding system for washing vehicles is proposed, controlled by a mobile application, that controls the amount of water, soap and degreaser used in the washing process of the vehicles of the 300, 500 series and 700; Likewise, fixed operating times are established, which avoids downtime and low productivity in the process.

For the structure, metallic recycled material that exists in the company is used. The sizing of the structure is calculated according to the weight of the water washer, the volume and weight of the supply tanks of the washing supplies, in this case, water, soap and degreaser. Correct management of order and verticality that regulate good storage practices is considered.

For the electrical and signals system, low costs for the company and the ease of use of the elements are considered. In the search for the best criteria an Arduino MEGA 2560 card is used, which has multiple outputs and inputs, and is versatile to establish communications. Then the use of ultrasound sensors HC-SR04 is defined to detect the levels in tanks and the use of Bluetooth modules HC-05 to establish communication. A phone or tablet that uses an Android operating system is necessary to define the HMI interface.

Since the water washer works with 220 V and the solenoid valves work with 110 V, relay modules from 5 V dc to 110–220 V 10 A are used, having into account that the pumps at the outlet of the tanks consume a of current greater than 1 A.

The electrical diagram was developed using the FRITZING software, (see Fig. 2). In this design the digital outputs of the "Arduino Mega 2560" card are taken from pin 22 to 31 to control the relays, while pins 2-4-6 are established as inputs and 3-5-7 as outputs for the sensors. The remaining pins (0-1-14-15) are used for communication with Bluetooth cards (sending and receiving).

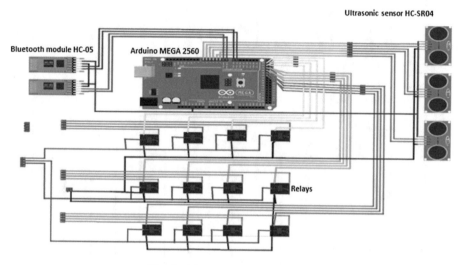

Fig. 2. Electrical signals diagram.

The electrical design is divided into two parts: the power circuit and the signals circuit, the power circuit is related to the ignition of the water washer that has an electric motor of 5.5 KW at 220 V. Figure 3, consists of three phases that are directly connected to the breaker as a first protection, after the breaker the three phases are connected to the contactor, connecting the thermal relay to the output of the contactor, and it is also connected directly to the motor. The contactor coil is supplied with 110 V to control the ignition directly or with the relay in the Arduino.

The relay R1 is activated with the output # 22 that acts with M1 which is the motor of the water washer. Relay R2 is activated with output # 23 which acts with E1, which is the water filling electrovalve. Relay R3 is activated with output # 24 which acts with E2, which is the liquid soap filling solenoid valve. Relay R4 is activated with output # 25 that acts with E3 that belongs to the degreaser filling solenoid valve. Relay R5 is activated with output # 26 which acts with E4 which is the water discharge solenoid valve. Relay R6 is activated with output # 27 that acts with E5 which is the liquid soap discharge solenoid valve. Relay R7 is activated with output # 28 that acts with E6 which is the degreaser discharge solenoid valve. Relay R8 is activated with output # 29 which acts with M2 which is the water discharge pump motor. Relay R9 is activated with output # 30 that acts with M3 which is the liquid soap pump discharge motor. Relay R10 is activated with output # 31 that operates with M4 which is the degreaser

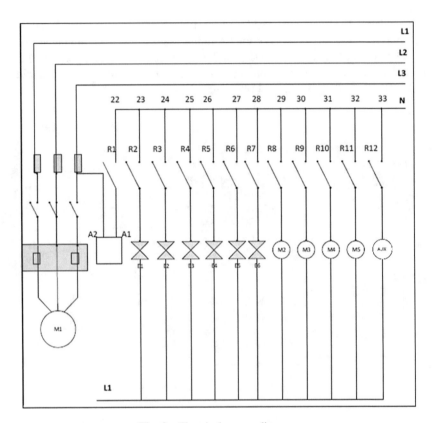

Fig. 3. Electrical power diagram.

discharge pump motor. Relay R11 is activated with output # 32 that acts with M5 which is the siren motor. Relay R12 is activated with output # 33 that acts with AUX which is the emergency stop.

3.2   Mobile Application

Using the mobile application, the vehicle type is selected according to the model, then the program on the Mega Arduino card executes a sequence according to the operation diagram of Fig. 4. The courtesy wash is the simplest, the express wash uses detergent, and the full washing includes washing with water under pressure, double application of detergent and degreaser.

On the other hand, the mobile application is developed in APP Inventor. Figure 5 shows the start and end screens of the process, in the first one (start) the type of vehicle is chosen and sent to wash, in the second screen (end) the washing report is observed, this data can be exporter or deleted.

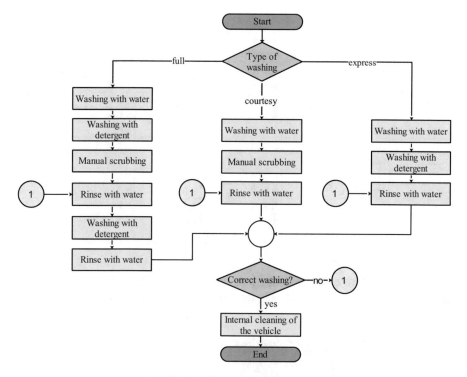

Fig. 4. Operation diagram.

Fig. 5. Control interface in the mobile application.

4 Results

Once the automatic system is implemented, a new table of expenses is obtained, Table 3 indicates the expenses after installing the automation system, using the data of a full month, with the detail of costs for water, liquid soap and degreaser dosed

Table 3. Vehicle wash expenses with the system implemented.

Type of washing	Model	Vehicles served / month	Water expenditure / cm^3	I spend liquid soap / cm^3	Degreasing expense / cm^3	Time spent / min
Courtesy	H300	57	$ 1,22	$ 0,00	$ 0,00	1348
Express	H300	28	$ 1,16	$ 1,85	$ 0,00	818
Full	H300	6	$ 0,35	$ 0,55	$ 0,66	188
Courtesy	H500	28	$ 1,25	$ 0,00	$ 0,00	961
Express	H500	12	$ 0,49	$ 0,78	$ 0,00	358
Full	H500	6	$ 0,74	$ 1,18	$ 1,40	287
Courtesy	H700	18	$ 0,96	$ 0,00	$ 0,00	722
Express	H700	2	$ 0,06	$ 0,10	$ 0,00	49
Full	H700	1	$ 0,13	$ 0,22	$ 0,26	64
Mean			**$ 0,71**	**0,52**	**0,26**	**533**
Standard deviation			**$ 0,47**	**$ 0,65**	**$ 0,48**	**451**

according to the time of assortment by the automatic washing system, which had an implementation cost of USD 562. Figure 6 shows the system in operation.

Fig. 6. System in operation

Figure 7, indicates a comparison between manual washing vs automatic vehicle washing, this shows a reduction in water consumption, liquid soap, degreasing, time and mainly cost reduction, this occurs with the implementation of the automatic car washing system.

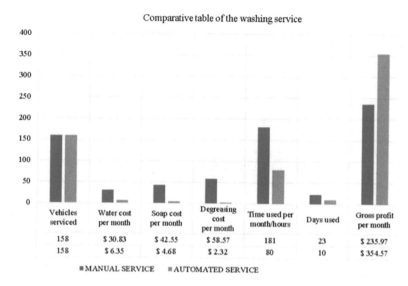

Fig. 7. Comparison of vehicle washing service.

Figure 8 shows a comparison between the manual washing service of vehicles with the automatic washing service, where the study is carried out for 158 vehicles resulting in a high cost in the payment of water, in manual service.

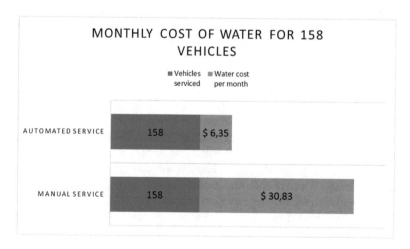

Fig. 8. Monthly cost of water.

On the other hand, Fig. 9 shows an analysis of the gross income for the 158 vehicles, resulting in a greater profit for the automatic service.

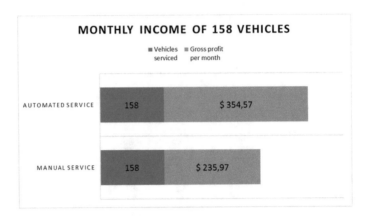

Fig. 9. Monthly gross income

5 Conclusions

The automation of the resource feeding system for vehicle washing is carried out to optimize the consumption of water, detergent and degreaser, in addition, the working time is standardized. To reach this aim, low cost devices are used, which includes an Arduino microcontroller, several sensors and actuators, and a mobile application. The control and power electrical systems are designed considering the guidelines of the proposal, and a mobile application is developed for devices with Android operating system. Although this work does not perform an advanced control system as in the related works presented in the introduction, it has the purpose of optimizing resources, focusing on economizing money in the installation of the automatic system, creating the possibility of replicating the service in small businesses and a quick solution to the excessive consumption of resources in vehicle washing.

For an average value of 158 units of vehicles entered per month, the results show a reduction in the monthly water cost by 79%; in liquid soap, this is reflected in a decrease of 89%; and in degreaser, a decrease of 96%. The time used in the operation is reduced by 56%; and in the same amount of vehicles is spent 10 days of water consumption. In addition, a gross revenue increase by 50% is evidenced, demonstrating the benefits of the implemented system. Finally, the statistical analysis of mean value and standard deviation corroborate the positive effects of the proposal, denoting that prior to the installation of the proposal there were scattered values in the consumption of resources and washing times, with the new process are observed few scattered data, due to the standardization.

Future work is intended to carry out a more comprehensive analysis of the benefits of these systems, including a statistical analysis to demonstrate a hypothesis proposed, that responds to the needs of optimization in the process.

References

1. Zou, M., Ocker, F., Huang, E., Vogel-heuser, B., Chen, C.: Design parameter optimization of automated production systems. In: 2018 IEEE 14th International Conference on Automation Science Engineering, pp. 359–364 (2018)
2. Cordeiro, A., Abraços, M., Monteiro, L., Andrade, E., Pires, V.F., Foito, D.: Industrial automation self-learning through the development of didactic industrial processes. In: Advanced Intelligent System Computing, vol. 715, pp. 872–883 (2018)
3. Huang, H., Bashir, M.: Users' trust in automation: a cultural perspective. In: Advanced Intelligent System Computing, vol. 595, pp. 282–289 (2018)
4. Naseera, S., Sachan, A., Rajini, G.K.: Design of smart home using internet of things. In: Advanced Intelligent System Computing, vol. 668, pp. 349–357 (2018)
5. Nagiev, A.G., Aliyeva, F.A., Nagiyev, H.A.: Problem of optimal management of resources of industrial production with given statistical data of disturbance parameters. In: Advanced Intelligent System Computing, vol. 502, pp. 997–1007 (2017)
6. Roberto, M., Araújo, A., Varela, M.L., Machado, J., Mendonça, J.P.: Methods time measurement on the optimization of a productive process: a case study. In: 4th International Conference on Control, Decision and Information Technologies (CoDIT 2017), Barcelona, pp. 980–985 (2017)
7. Ruiz Carmona, A.R., Muñoz Benítez, J.C., García-Gervacio, J.L.: SCADA system design: a proposal for optimizing a production line. In: International Conference on Electronics, Communications and Computers, CONIELECOMP 2016, pp. 192–197. Institute of Electrical and Electronics Engineers Inc., Cholula (2016)
8. Brusco, G., Barone, G., Burgio, A., Menniti, D., Pinnarelli, A., Scarcello, L., Sorrentino, N.: A smartbox as a low-cost home automation solution for prosumers with a battery storage system in a demand response program. In: EEEIC 2016 - International Conference on Environment and Electrical Engineering. Institute of Electrical and Electronics Engineers Inc., Florence (2016)
9. Bhatt, A., Saxena, A., Chauhan, S., Jaiswal, U., Verma, Y.: Android-based home automation using bluetooth and ESP8266. In: Advanced Intelligent System Computing, vol. 624, pp. 1767–1776 (2018)
10. Asadullah, M., Ullah, K.: Smart home automation system using bluetooth technology. In: International Conference on Innovations in Electrical Engineering and Computational Technologies 2017. Institute of Electrical and Electronics Engineers Inc., Karachi (2017)
11. Barn, R., Barn, B.S.: Integrating values into mobile software engineering. In: ACM International Conference Proceeding Series, p. 196. Association for Computing Machinery, PilaniGoa (2016)
12. Zhong, S., Zhang, L., Chen, H., Zhao, H., Guo, L.: Study of the patterns of automatic car washing in the era of internet of things. In: 31st IEEE International Conference on Advanced Information Networking and Applications Workshops, pp. 82–86 (2017)
13. Ucar, D.: Membrane processes for the reuse of car washing wastewater Deniz Uçar. J. Water Reuse Desalin. **8**, 169–175 (2018)
14. Yu, Y., Kurnianggoro, L., Jo, K.: Design of intelligent car washing system. In: 2015 54th Annual Conference of the Society of Instrument and Control Engineers of Japan, pp. 1447–1450. Institute of Electrical and Electronics Engineers Inc., Hangzhou (2015)

Application of Telematic Instrumentation for Monitoring and Control in the Treatment of Neuromuscular Diseases

Nelson Sinchiguano$^{(\boxtimes)}$ ⓘ, Fidel Parra ⓘ, Javier Guaña ⓘ,
and Pablo Recalde ⓘ

Universidad Tecnológica Israel, Quito, Ecuador
n.sinchiguano@gmail.com, {fparra,jguana,
precalde}@uisrael.edu.ec

Abstract. Thanks to the development of information and communication technologies, the application of telematic instrumentation allows the implementation of a virtual instrumentation system that contains an electronic interface with remote co-communication that gives medical personnel the facility to communicate at a distance between the nursing assistant and the patients, in such a way that the medical professional can perform monitoring and control in the treatment of neuromuscular diseases, regardless of the place of origin where the patient is located.

This article describes the development of a remote virtual instrumentation system oriented to support the area of physiatry and rehabilitation of a health establishment. Due to modern technological advances in biomedical instruction, there are technological devices that can be optimized as instruments for the acquisition, monitoring, and processing of bioelectric signals, to such an extent that it has facilitated the implementation of medical equipment consisting of an electromyograph and an electro-stimulator; this allows data to be sent and received over long distances, which causes radical changes with certain advantages that contribute to the quality of medical care.

In this sense, it can be asserted that the execution of the research allows the development of a functional technological strategy for the diagnosis and treatment of neuromuscular diseases with the sole objective of covering vulnerable rural areas where the medical accessibility of their inhabitants is limited.

Keywords: Health informatic · Telehealth · Instrumentation · Telematics

1 Introduction

The scientific advances in the field of telematic instrumentation are unstoppable, which leads to exploit the Communication and Information Technology (ICT) as a fundamental part of medicine. Today underdeveloped countries in the area of technology apply their knowledge in the innovation of the centers. doctors or hospitals taking advantage of the benefits of telehealth, telemedicine [1].

To understand the gaps and benefits of telehealth it is necessary to begin by understanding that computer-assisted medicine, from the analysis of the patient's health

© Springer Nature Switzerland AG 2020
M. Botto-Tobar et al. (Eds.): ICAETT 2019, AISC 1067, pp. 142–153, 2020.
https://doi.org/10.1007/978-3-030-32033-1_14

care, is a method that allows the medical professional to diagnose, record, and treat a person's illness no matter how distant they are [2]. Es relevante mencionar que el presente documento contiene análisis de varios trabajos investigativos realizado por determinados autores que brindan nuevas brechas en la esencia para el desarrollo de la investigación.

It is important to mention that the present document contains analyses of several research projects carried out by certain authors that provide new gaps in the essence for the development of research.

Knowing that the basis of medicine is the diagnosis [1]; since through it is possible to establish the appropriate treatment, and it is precisely because of its easy and direct way of processing health information that it is more efficient to apply ICT. It is for this reason that interest has arisen in analyzing the current situation regarding the contribution of technology in monitoring and control in the treatment of neuromuscular diseases oriented to the area of physiatry, in response to the question that 12779 cases suffering from neuromuscular diseases have been identified in Ecuador [3].

The advantages offered by modern technology have been an important gap in the implementation of a remote virtual instrumentation system for rural areas, physiatrics and rehabilitation, providing safety, integrity and quality of service [3] to patients who require timely medical support, regardless of geographic location.

2   Research Strategy and Methods

The research has been developed through four fundamental stages that show its dynamics: exploratory and diagnostic stage, stage of establishment of needs, stage of implementation and finally the stage of evaluation of the scientific result.

During the research, a quali-quantitative approach is applied, driving the implementation process for the development of a remote virtual instrumentation system based on the application of telematic instrumentation, the purpose of which is to deliver a technological strategy that allows monitoring and control in the treatment of neuromuscular diseases.

For the successful advancement of research, certain areas of scientific practice are applied, such as field research, since it is necessary to carry out a systematic study of the facts where the problem is generated, in order to obtain direct information that is useful in the development of the proposal. In the same way, bibliographic and documentary research is applied through books, scientific journals and electronic publications in order to deepen different approaches with respect to the subject of research.

2.1   Exploratory Stage

Advances in health information technology (health IT) hold promise for helping small and rural communities overcome health care delivery challenges such as distance and staff shortages. Despite progress in recent years, adoption rates remain low for clinical health information technologies [4].

In the article "Neurophysiology in the study of neuromuscular diseases, development and current limitations" is mentioned [5], that through the result of the scientific-technical advance [5], there has been a process of specialization and improvement in Clinical Neurophysiology. These methods provide a functional diagnosis of diseases and injuries of the nervous system, although they are currently in a state of profound revision and search for new diagnostic strategies, especially in the leading countries of Neurophysiology, the United States and the most developed European countries [5].

Given that neuromuscular diseases (NMD) are characterized by a progressive decrease in muscular strength accompanied by secondary symptoms such as fatigue, fatigue, cramps [6], it can be mentioned that in the article Im-plementing electromyographic equipment through USB interfaces carried out by the authors [7], the design and implementation of a hardware capable of acquiring and processing surface EMG signals has been carried out, basically consisting of instrumentation amplifiers, analogue filters, a microcontroller among others, leaving as a first step that through the results of the research future projects can be proposed that conclude in a user interface for quadriplegic people.

A decade ago it was only thought as science fiction in the technology we currently have, since it was impossible to imagine the technological wonders that exist in 2019, in this sense on the advance of technology it is necessary to mention that the use of the telehealth system in Ecuador began with a pilot plan designed to work in three phases. The first phase was scheduled to run from 2009 to 2011 all applied in the provinces of Morona Santiago, Pastaza and Napo. The second phase is focused on working on telehealth as projects aimed at the Amazon provinces. Finally, the third phase aims to cover the entire national area, [8]. However, as described by [9], Ecuador is experiencing problems due to the lack of medical services, especially when it is necessary to cover rural areas located far away from health care centers; All of this reflects the lack of specialist doctors and the scarcity of resources, which is why there is a need to design a medical instrument that is accessible to patients and that allows diagnosis and rehabilitation of neuromuscular diseases through remote communication with the sole purpose of improving the quality of medical care by responding in a timely manner to various needs arising in the area of physiatrics and rehabilitation.

2.2 Needs Establishment Stage

For the development of the present work the mixed approach was used, since its stage contains a qualitative analysis and a quantitative analysis that, through a field work, a survey was applied to the area of physiatrics and rehabilitation in the medical attention centers, being able to identify in an appropriate way the necessities of innovating the medical attention process. The research has begun in the medical care centers of the province of Cotopaxi - Ecuador, for this reason the applied survey has as reference to 10 medical professionals.

Once the data has been tabulated, different parameters are set as shown below.

Of those surveyed, 40% have an electronic interface for Neuromuscular Disease Monitoring and Control, while 60% do not. According to the data obtained, certain centres require an electronic interface for the treatment of neuromuscular diseases (Table 1).

Table 1. Availability of an electronic interface for the physiatrics.

Response	Quantity/Frequency	Rate %
Yes	4	40
Not	6	60
TOTAL	10	100

In the following Table 2, 90% of those surveyed state that it is necessary to implement a virtual instrumentation system for the treatment of neuromuscular diseases in order to broaden the scope of medical care, while due to their lack of knowledge, 10% of those surveyed state that it is not necessary to implement the system.

Table 2. Need for virtual instrumentation system

Response	Quantity/Frequency	Rate %
Yes	9	90
Not	1	10
TOTAL	10	100

According to the data obtained, doctors do not have a virtual instrumentation system. Most of the doctors state that it would be advantageous to have a virtual system for the treatment of neuromuscular diseases (Table 3).

Table 3. Need to implement an Electromyograph

Response	Quantity/Frequency	Rate %
Yes	8	80
Not	2	20
TOTAL	10	100

Of those surveyed, 80% consider that it is necessary to implement an electrosmograph with remote communication for patients, while 20% state that it is not necessary. The data acquired in Table 4 corroborates the need to implement a medical instrument that allows the diagnosis of neuromuscular diseases of the patient regardless of where they are found.

Table 4. Implementation of a medical instrument.

Response	Quantity/Frequency	Rate %
Yes	10	100
Not	0	0
TOTAL	10	100

In its entirety 100% of the respondents state that it would be appropriate to implement a precise and exact medical instrument given that the area of medicine is demanding in health care, especially when it comes to acquiring EMG signals.

Of those surveyed, 80% say that an electronic interface will help raise the quality of medical care in monitoring neuromuscular diseases, while 10% say it is not necessary due to their lack of knowledge (Table 5).

Table 5. Electronic interface

Response	Quantity/Frequency	Rate %
Yes	8	80
Not	2	20
TOTAL	10	100

Of those surveyed, 100% state that the implementation of a virtual instrumentation system will help the personal and professional development of health personnel in urban areas as well as in remote and distant places, thus improving the quality of service (Table 6).

Table 6. Professional development

Response	Quantity/Frequency	Rate %
SI	10	100
Not	0	0
TOTAL	10	100

The majority of those surveyed represent 100% who agree to innovate the area of rehabilitation through information and co-communication technologies by adopting technological systems to provide immediate medical care (Table 7).

Table 7. Innovar área de rehabilitación

Response	Quantity/Frequency	Rate %
Yes	10	100
Not	0	0
TOTAL	10	100

2.3   Implementation Phase

Presentation of the proposal. Since research development has different technologies and architectures, which allow the application of telematic instruction for monitoring and control in the treatment of neuromuscular diseases, it is relevant to present the stages that make up the practical development of the virtual instrumentation system.

Figure 1 shows the elements that interact together to facilitate the monitoring and control of bioelectric signals produced by muscles.

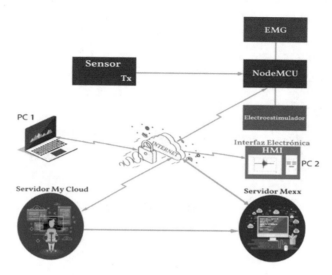

Fig. 1. General outline of the proposal.

The block of the electronic instrumentation system is composed of a sensing stage, this stage contains a primary element known as a sensor, which through surface electrodes captures the small electrical pulses resulting from muscular activity and sends it to the EMG stage itself which consists of a con-versor ADC responsible for converting the signal from analog to digital.

The EMG stage makes it possible to determine the real patterns of muscular activation by measuring the electrical signals, in this way treatment plans can be established and the progress of the treatment can be monitored.

The electrostimulator stage is an electronic current generating plate that produces electrical impulses with energy to generate an action potential (PA) in excitable cells such as muscle or nerve.

Presentation of the Proposal. The selected hardware architectures have input and output points of analogical signals that allow the acquisition, processing, and transmission of the bioelectric signals coming from the muscular contraction. It also allows the communication and control of the electro-stimulator with the electronic interface that through analogical signals allows to activate the electrical pulses for the treatment of neuromuscular diseases.

System Software Architecture. In order for the selection of the architecture to be adequate in the creation of the electronic interface, an architecture must be chosen that contains a set of pertinent patterns and abstractions that provide a logic of instructions, necessary to guide the construction of the software, allowing communication between the data acquisition modules and the PC.

Flowchart of the functional structure of the electronic interface. The flowchart allows understanding the functional process of the electronic interface as part of the remote virtual instrumentation system. Figure 2, it shows the previous process as an analyzed solution in relation to the identified problem.

Fig. 2. Flowchart of the electronic interface.

3   Results

Once the strategy and research methods have been applied, the result is a remote virtual instrumentation system that forms part of an innovative technological gap for the process of care in the area of physiatrics and rehabilitation of a health facility that indirectly allows covering vulnerables rural areas where the medical accessibility of its inhabitants is limited. However, in order for the system to be functional, it must be evaluated.

3.1   Validation of the Proposal

Once the remote virtual instrumentation system is concluded and implemented, it is necessary to verify the functionality of the system through evaluations, which must allow the acquisition, monitoring and processing of bioelectric signals. To do this, the system is validated through a standard instrument or it is also validated according to the sector and beneficiary, since the tool will be successful as long as it is easy to use and apply.

3.2   Evaluation of the Implemented System

Electromyograph (EMG) Test. Three types of tests are performed to determine that the EMG is functioning properly. As the signals are acquired, the EMG medical instrument can be validated in two ways: The first option is through the use and judgement of a physiatric medical specialist or through the use of a standard instrument.

Acquisition of the Electromyograph signal. Prior to the acquisition of the bioelectric signals, the patient must be prepared considering certain criteria:

- The patient must not have metal objects that can cause electro-magnetic noises.
- Clean with antiseptic alcohol the possible impurities of the area of the skin where the electrodes will be placed.
- The electrodes should be placed parallel to each other at a distance of 5 cm, then connect the cable to the electrode, and press it against the patient's skin (Fig. 3).

Fig. 3. Placement of electrodes.

Electromyogram Normal. In reference to the electromyogram, there are certain tests that are performed according to muscle contractions.

Muscle Signal at Rest. The first test is performed with the muscle relaxed in this case tends to stabilize the signal to a straight line corresponding to an isoelectric line that represents a state of muscle balance [12]. Figure 4; it represents a suitable isometric line in the electrophysiological register.

Fig. 4. EMG of resting muscle.

When comparing the EMG signals obtained by means of the implemented electromyograph, and that of the instrument pattern, the signals show similarity when delivering a voltage signal, so the first continuous test is acceptable. Figure 5 the recorded signal of the muscle at rest can be observed by means of the standard instrument.

Light Muscle Contraction Test. The patient performs a slight isometric contraction so that the amplitude of the EMG signal tends to increase causing the phenomenon of a

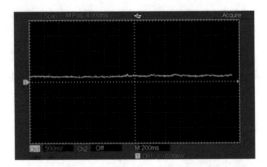

Fig. 5. Recording of muscle at rest.

partial wave pattern. Figure 6, it shows the EMG signal when a slight muscle contraction is performed.

Fig. 6. Mild muscle contraction record.

In the Fig. 7, the recording of the signal can be visualized by performing a slight contraction which, when compared with the signal visualized by the electronic interface, maintains the static characteristics that the implemented electromyograph should have.

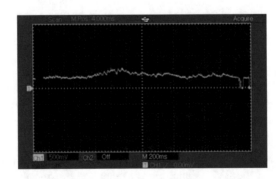

Fig. 7. Recording of mild muscle contraction with the standard instrument.

Maximum Muscle Contraction Test. The patient must perform an isotonic muscle contraction by applying an energetic force that allows maximum tension to be recorded during the entire movement. Figure 8, it shows the recording of the EMG signal captured by the implemented electromyograph.

Fig. 8. Contracción muscular máxima.

By making a comparison between the signal acquired by the implemented EMG and the standard instrument it is possible to validate that the instruments deliver similar signals that demonstrate integrity, safety and quality. Figure 9, it shows the EMG signal of the maximum muscle contraction obtained with the standard instrument.

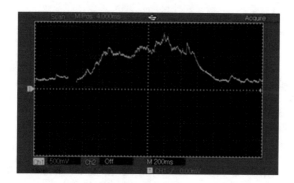

Fig. 9. Recording the maximum muscle contraction with the standard instrument.

4   Discussion

- The methods, instruments and techniques used in the investigation facilitated the identification of the needs required by the area of physiatrics and rehabilitation of medical care centers, allowing in turn to deliver a technological and innovative solution to the problem identified.
- Once the evaluation of the implemented system has been carried out, it can be assured that a functional prototype has been implemented that complies with the

static characteristics of a medical instrument, being at the same time accessible for both the medical professional and the patient. In the possible case that the government does not wish to invest in the acquisition of such an instrument, the patient who suffers from neuromuscular diseases can acquire the equipment without difficulty since the cost of acquiring the medical equipment is around 80 dollars, considering it accessible and economical even more knowing that the equipment has two functions at the same time, which is to allow the diagnosis and treatment of neuromuscular diseases.

- With the implementation of the virtual instrumentation system, it is possible to expand access to medical services, saving time and money for poor people who belong to rural areas, regardless of location or distance, and to interact in real time with doctors, nursing assistants and patients, especially those suffering from severe neuromuscular diseases that require regular control.
- The electronic interface that is part of the implemented system is functional and user-friendly, allowing the medical professional to interact between the device and the system with the function of monitoring and sensory control of patients suffering from neuromuscular diseases.
- The successful implementation of hardware and software has allowed to contribute to health with a medical instrument accessible to users that allows the diagnosis and treatment of neuromuscular diseases through wireless communication through an electronic interface which guarantees the integration, simplicity and quality, in the monitoring and control of the functioning of the peripheral nervous system muscular.

References

1. Sainz de Abajo, B., Rodrigues, J., García Salcines, E., Burón Fernández, J., López Coronado, M., Castro Lozano, C.: M-Health y T-Health. La Evolución. La Evolución Natural del E-Health, vol. 7, no. 25, p. 10 (2011). RevistaeSalud.com
2. Ibañes, C., Zuluaga, Á., Trujillo, A.: TELEMEDICINA: Introducción, aplicación y principios de desarrollo. Redalyc.org, p. 15 (2007)
3. Castiglioni, C., Bevilacqua, J., Hervias, C.: Clinical approach to the diagnostic evaluation of hereditary and acquired neuromuscular diseases in adolescence. Revista Médica Clínica Las Condes (2015). https://www.elsevier.es/es-revista-revista-medica-clinica-las-condes-202-articulo-enfermedades-neuromusculares-el-adolescente-sintomas-S0716864015000115
4. Álvarez Fiallo, R., Medin, E.: La neurofisiología en el estudio de las enfermedades neuromusculares, desarrollo y limitaciones actuales. Revista Cubana de Medicina Militar, vol. 33, no. 3 (2004)
5. INEC. Registro Estadístico de Camas y Egreso Hospitalarios (2017). http://www.ecuadorencifras.gob.ec/camas-y-egresos-hospitalarios/. Último acceso 02 Mayo 2019
6. Singh, G.: Departamento de Salud y Servicios Humanos de los Estados Unido (2017). https://healthit.ahrq.gov/key-topics/health-it-small-and-rural-communities
7. Barros, G., Moreira, I., Ríos, R.: Treatment – rehabilitation and management of neuromuscular diseases. Revista Médica Clínica Las Condes **29**(5), 498–584 (2018)
8. Urbina Rojas, W.F., Martínez Santa, F.: Implementing electromyographic equipment through USB interfaces. Tecnura **16**(33), 15 (2012)

9. Asociación Iberoamericana de Telesalud y Telemedicina (AITT). Informe Telesalud, 3 Junio 2015.   http://teleiberoamerica.com/publicaciones/InformeTelesaludTelemedicina-version-03Junio-2015a.pdf
10. Mintel y MSP, Expansión del Programa de Telemedicina a Nivel Nacional, Marzo 2013. https://www.telecomunicaciones.gob.ec/wp-content/uploads/2016/01/Expansion-del-Programa-de-Telemedicina-Nivel-Nacional.pdf. Último acceso 09 agosto 2017
11. Santos, A., Fernández, A.: Desarrollo de la telesalud en America Latina Aspectos conceptuales y estado actual, Octubre 2013. https://repositorio.cepal.org/bitstream/handle/11362/35453/S2013129_es.pdf;jsessionid=7EBF5B2B53CDB4F85714870C1F1EA022? sequence=1. Último acceso 06 06 2018
12. Sinchiguano, N.: Repositorio Institucional de la Universidad de las Fuerzas Armadas ESPE (2015). https://repositorio.espe.edu.ec/bitstream/21000/10116/1/T-ESPEL-ENI-0355.pdf
13. MedlinePlus. Biblioteca Nacional de Medicina de los Estados Unidos, 04 Junio 2018. https://medlineplus.gov/spanish/ency/patientinstructions/000919.htm.   Último   acceso   20 Junio 2018

Test System for Control Algorithms on a DC Motor

Fernando Saá[1]([✉]), Andrés Gordón[2], Darío Mendoza Chipantasi[2],
Pamela Monserrath Espejo Velasco[2], and Nancy Velasco E.[2]

[1] Universidad Tecnológica Indoamérica, Ambato, Ecuador
fernandosaa@uti.edu.ec
[2] Universidad de las Fuerzas Armadas ESPE, Sangolquí, Ecuador
{amgordon,djmendoza,pmespejo,ndvelasco}@espe.edu.ec

Abstract. This work implements a system for testing control algorithms. The objective is to find an optimal regulator for the closed-loop speed control of a DC Motor. The system consists of a DC motor, an encoder, the Arduino card and a computer. The speed of the motor is the control variable. First, the transfer function is obtained based on the electrical and mechanical characteristics of the system. The function is reduced based on data obtained experimentally from plant input and output using Matlab's IDENT tool. The equation that governs the system found allows to simulate the different controllers: proportional, proportional derivative, proportional integral, proportional integral derivative and fuzzy control. The results are displayed in a graphical interface in Labview. In the experiments, closed-loop and open-loop control subsystems are simulated. The experiments indicated that the proportional integral derivative controller has a better response.

Keywords: Control · Fuzzy · PID

1 Introduction

Technology has made great strides in the area of control. Currently there are intelligent control techniques, real time monitoring systems and better data processing capabilities. On the other hand, the didactic test benches of motors are scarce.

Bhushan, B. and Singh, M., in 2011 at the Department of Electrical Engineering, Delhi University of Technology, India proposes adaptive control of the DC motor using the bacteria food search algorithm. It consists of a bacterial feed search algorithm for a high-performance speed control system (BFA) for a DC motor. The continuous current motor rotor speed follows an arbitrary selected path. The unknown non-linear dynamics of the motor and load are captured by BFA. The trained BFA identifier is used with a desired reference model to achieve control of the DC motor path. The algorithm is applied for identification and control of DC motor. The simulation study in the proposed system has been conducted at MATLAB. The operation of the DC motor has been carried out through the genetic algorithm (GA) as well. A comparison of the performance analyses used by the BFA and GA regulators for trajectory tracking shows

© Springer Nature Switzerland AG 2020
M. Botto-Tobar et al. (Eds.): ICAETT 2019, AISC 1067, pp. 154–163, 2020.
https://doi.org/10.1007/978-3-030-32033-1_15

that the adaptive BFA-based controller works effectively to track the desired trajectory in the DC motor with less calculation time [1].

Fang, L. conducts research on DC Motor with a binding-integral control algorithm in 2013 at Nanjing Institute of Technology, Nanjing, China. He proposes the analysis of the mathematical model of the brushless motor and deficiencies of the traditional PID DC control, combining with the intelligent algorithm and the conventional PID, offers an improved algorithm and simulations in the MATLAB environment. A brushless DC motor speed regulation system PWM is designed based on high-performance MCU. Improved algorithm and control effects of conventional algorithm were analyzed and compared in the experiment [2].

Hudy, W. and Jaracz, K., in 2011 at the Institute of Technology of the Pedagogical University of Krakow, Poland propose a study on the selection of control parameters in a control system with a DC electric motor series using the evolutionary algorithm. A method of selection of the regulator parameters in a control system using the evolutionary algorithm was presented. The control system has a PI controller and a hysteresis controller. The value of the proportional band and the value of the time integral are defined by the evolutionary algorithms. The control object was a Brown Boveri GS10A motor. The working functions were the step change of the rotation speed and step change of the motor torque. The control system with the parameters selected by the evolutionary method was verified using MATLAB/Simulink environment [3].

Lv, H., Wei, G., Ding, Z. and Ding, X. in 2014 at the Department of Science and Control Engineering, Shanghai University for Science and Technology, Shanghai, China propose a no-feedback control for the brushless DC motor: a Kalman filter algorithm. The proposed UKF algorithm is used to estimate the speed position and rotor of the BLDC motor using only terminal voltage measurements and three-phase currents. In order to observe unit performance, two simulation examples are given and the feasibility and effectiveness of the UKF algorithm are verified through the simulation results, and the precise estimated performance is shown in the simulation figures [4].

Rashidi, B., Esmaeilpour, M. and Homaeinezhad, M. in 2015 at the Department of Electrical and Computer Engineering, University of Alberta, Edmonton, Canada; Department of Mechanical Engineering, K. N. Toosi University of Technology, Tehran, Iran Mechatronic Laboratory Mechanisms (MML), K. N., M., M., M., M. in 2015 at the Department of Electrical and Computer Engineering, University of Alberta, Edmonton, Canada; Department of Mechanical Engineering, K. N. Toosi University of Technology, Tehran, Iran Mechatronic Laboratory Mechanisms (MML), K. N. Toosi University of Technology, Tehran, Iran propose an Accurate control of the angular velocity of permanent magnet DC motors in the presence of high modeling uncertainties through sliding adaptive algorithm mode based on the reference of the observer model. The study is the development of a control scheme with parameterization and insensitive load functions capable of regulating the precise angular velocity of a permanent magnet motor (PM) DC in the presence of modelling uncertainties. An appropriate non-linear dynamic friction model, the modified LuGre model, was chosen and incorporated into the mathematical model of a DC permanent magnet motor.

The sliding model (SMO) is designed to calculate the state variable of the friction model, in an adaptive model-reference control system in which the friction state values are estimated and the parameters are fed is designed to track the desired velocity trajectory while alleviating the adverse effects of model uncertainties and friction. Stability of the proposed SMO-based MRAC system is discussed through Lyapunov's stability theorem, and its asymptotic stability is verified. The algorithm is implemented in a new variable structure test bench that gives us the ability to simulate variations of desired parameters and changes in external disturbances in the experiment [5].

Zhao, P. and Shi, Y. propose a design of DC Motor Control System based on clonal algorithm selection in 2011 at the Ministry of Education, Northwest Polytechnic University, China. According to multivariate, strong coupling and non-linearity of DC motor speed control system, and traditional PID control is difficult to achieve the highest speed performance. Combining the clonal selection algorithm (CSA) with traditional PID control, a type of intelligent PID control system was designed. The controller accounts for real-time and in-line adjustment of the PID parameters, thus improving the dynamic and static performance and robustness of the DC motor control system. The simulation experiment was processed based on MATLAB environment. Simulation results show that this intelligent PID controller achieves better performance than the traditional PID controller [6].

We propose the speed control of a DC motor using the classic control (Proportional, Proportional Differential, Proportional Integral, Proportional Integral Derivative) and the modern control (Fuzzy Control). On the one hand, we perform a system simulation based on the intrinsic physical characteristics of the motor to obtain a representation of the mathematical model. Then control algorithms are applied in a virtual environment and to the physical system to contrast the results. The system can be used to solve closed loop control problems. Project development time and costs will decrease.

2   Development

The physical design of the system has to be applied to the engine DC presented in Fig. 1. Where um(t) Motor input voltage, i(t) is the armature current, eb(t) is the Counter-electromotrical Force, Rm is the Terminal Resistance, lm is the Rotor Inductance, τ m (t) is the Motor Torque, θm(t) is the Motor Angular Speed, $\dot{\theta}$m(t) is the Motor Angular Acceleration, Jm the Inertia of the motor, τ l(t) is the Load torque seen from the motor shaft, τ f (t) is the Friction torque, Bm is the biscose friction constant, kb is the electromotive force constant, km is the torque constant.

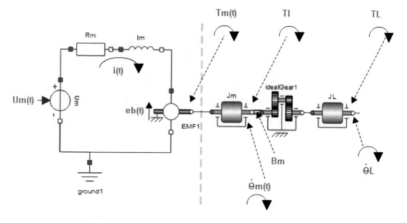

Fig. 1. DC motor with gearbox and a JL load

2.1 Mathematical Design of the PID System

An analysis is made in place of the roots to observe if the system is stabilized in a target pole, calculated based on ts and Mp, where the objective poles of Eq. 1 are located.

$$P_{1,2} = -\sigma \pm \omega_d j \tag{1}$$

$$ts(2\%) = 4/\sigma \tag{2}$$

$$\sigma = 4/1 = 4$$

$$M_p = e^{\frac{-\pi}{\tan\phi}} \tag{3}$$

$$\phi = \frac{-\pi}{\ln(0,05)} = 60°$$

$$\tan(60°) = \frac{4}{\sigma}$$

$$\sigma = 2,31$$

In Eqs. 2 and 3 the data are replaced and the objective poles located in P1,2 = (4, ±2, 31j) are obtained, which is represented in the S plane as in Fig. 2

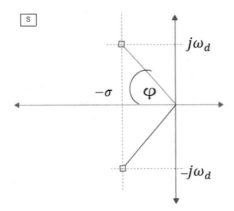

Fig. 2. S plane

The place of the roots of the system must pass through the objective poles, which in this case when analyzing Fig. 3 it is observed that it does not happen since the system has a pole located on the real axis of the S plane at −0.554, so we look for a Transfer function GC (s), which causes the system to cross through the objective poles.

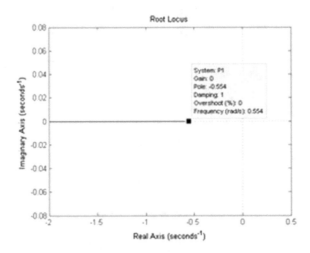

Fig. 3. Place of the roots of the system to be controlled

An analysis is made instead of the roots to see if the system is stabilized at an objective pole. If not, look for a GC Transfer Function(s), which makes the system cross the target poles (See Fig. 4). This function is described in Eq. 4.

$$G^1(S) = \frac{(1{,}04s + 1)(3559)}{(0{,}3934s + 1)(1{,}805s + 1)} \tag{4}$$

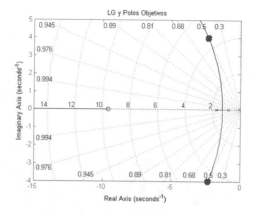

Fig. 4. Place of roots G1

The test circuit consists of mechanical, electrical and software to acquire data, control and monitor the system. The electromechanical part is represented by the motor of 6 V DC and 210 RPM for the tests. The electrical part is given by the controller card, with digital input pins for the connection of an incremental encoder for speed acquisition, as well as output pins of the type PWM (Pulse Width Modulation). The software part consists of control and simulation algorithms.

2.2 Mechanical and Electrical Part

The physical part of the system is shown in Fig. 5.

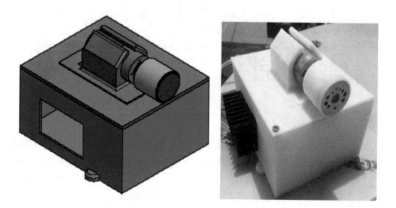

Fig. 5. Designed system

The DC motor is coupled to an incremental encoder. The encoder allows to measure the speed of the motor. The data is acquired through the Arduino card. The Arduino card communicates via serial to a monitoring and storage software. The required circuit connections are shown in Fig. 6.

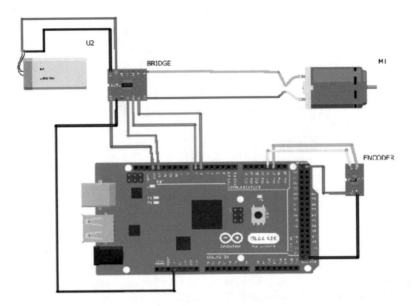

Fig. 6. Electrical connections

Table 1 lists the complete hardware requirements.

Table 1. Hardware requirements.

	Description
DC motor	Pololu 210 RPM motor at 6 Vdc with reducer
Speed sensor	Incremental Encoder 48ppr
Control target	Arduino Mega
Personal computer	Intel (R) Core (TM) i5-4210U CPU @ 2.4 GHz
Motor controller	IBT-2
Power supply	DC source, 1800 mA, 3 V, 6 V, 9 V, 12 V

2.3   Software Part

The control algorithms are in the software. The speed remains stable at a value indicated by the user. Table 2 lists the software requirements of the implemented system.

Table 2. Software requirements.

	Description
Monitoring	LabVIEW 2012
Arduino programming	Arduino IDE
Simulation	Wolframe system modeller
Mathematical calculation	Matlab

The user test interface performed in Labview is shown in Fig. 7.

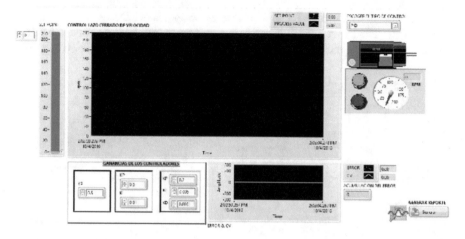

Fig. 7. User test interface

The simulation is done using the Wolfram System Modeller (See Fig. 8).

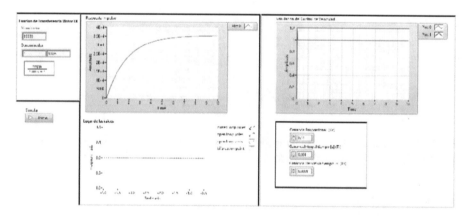

Fig. 8. User simulation interface

3 Tests and Results

The tests and results are carried out by testing the different regulators one by one. The final calibration of the most suitable regulator is done using the error test method, starting from the simulation values after the mathematical analysis. The error of each controller is measured for 10 s.

The proportional control is insufficient to correct the error and reach the given set point. Derivative proportional control has a high oscillatory characteristic and does not decrease the error in steady state. The integral proportional control presents better result, but still has an error of 7%. The Integral Proportional Derivative PID control presents better results by reducing oscillation and error in the stable state. The fuzzy control gives an acceptable result. The best answer is the PID control (See Fig. 9).

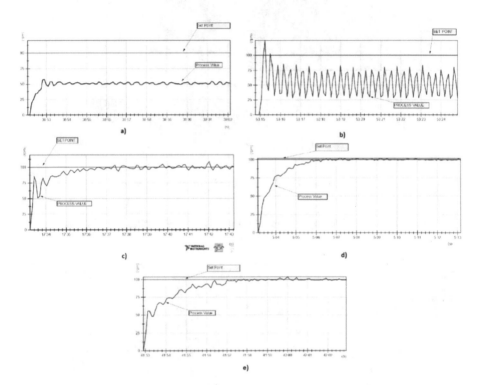

Fig. 9. Controller response: (a) Proportional, (b) Proportional Derivative, (c) Proportional Integral, (d) Proportional Integral Derivative and (e) Diffuse.

4   Conclusions

The Integral Proportional Derivative PID control presents a better response, because it reduces oscillations, shorter response time and reduces the error in a stable state.

The data was obtained during 10 s of operation and the taking of 114 samples.

The man-machine interface HMI created allows to control, simulate, monitor and document the system in real time.

The Fuzzy control has a good response in this application system, but with a longer setup time than a PI or PID controller.

References

1. Bhushan, B.: Adaptive control of DC motor using bacterial foraging algorithm. Appl. Soft Comput. **11**, 4913–4920 (2011)
2. Fang, L.: Research on DC motor intelligent control algorithm. In: AMM (2013)
3. Hudy, W.: Selection of control parameters in a control system with a DC electric series motor using evolutionary algorithm. Archives of Electrical Engineering (2011)
4. Ding, Z., Lv, H., Wei, G., Ding, X.: Sensorless control for the brushless DC motor: an unscented Kalman filter algorithm. Systems Science & Control Engineering, pp. 8–13 (2014)
5. Rashidi, B., Esmaeilpour, M., Homaeinezhad, M.R.: Precise angular speed control of permanent magnet DC motors in presence of high modeling uncertainties via sliding mode observer-based model reference adaptive algorithm. Mechatronics **28**, 79–95 (2015)
6. Zhao, P., Shi, Y.: Design of DC motor control system based on clonal selection algorithm. In: AMR, pp. 2343–2348 (2011)
7. MatÃa, A.R.F., JimÃ©nez, A.: TeorÃa de sistemas. E.T.S., Madrid (2003)

Forecasting Building Electric Consumption Patterns Through Statistical Methods

Xavier Serrano-Guerrero[1]([⊠]), Luis-Fernando Siavichay[1],
Jean-Michel Clairand[2], and Guillermo Escrivá-Escrivá[3]

[1] Grupo de Investigación en Energías, Universidad Politécnica Salesiana,
010105 Cuenca, Ecuador
jserranog@ups.edu.ec, lsiavichay@est.ups.edu.ec
[2] Facultad de Ingenería y Ciencias Agropecuarias,
Universidad de las Américas, 170122 Quito, Ecuador
jean.clairand@udla.edu.ec
[3] Institute for Energy Engineering, Universitat Politècnica de València,
46022 Valencia, Spain
guieses@die.upv.es

Abstract. The electricity sector presents new challenges in the operation and planning of power systems, such as the forecast of power demand. This paper proposes a comprehensive approach for evaluating statistical methods and techniques of electric demand forecast. The proposed approach is based on smoothing methods, simple and multiple regressions, and ARIMA models, applied to two real university buildings from Ecuador and Spain. The results are analyzed by statistical metrics to assess their predictive capacity, and they indicate that the Holt-Winter and ARIMA methods have the best performance to forecast the electricity demand (ED).

Keywords: ARIMA · Electric demand · Forecast · Load
Uncertainties · Statistical methods · Winter

1 Introduction

The electricity is an essential element for the development of the current world, it promotes the progress of societies and raises the standard of living of people. However, the increasing population and new generation solutions, such as renewable energy, create new challenges for the operation, planning and energy management of power systems [10, 19]. Thus, it is essential to have proper tools to forecast electricity demand (ED), considering its significant uncertainties, which have widely been studied for several years [8].

There are three kinds of load forecasting depending on the time scale: short-term, medium-term, and long-term. The short-term carries out a forecast of the ED many hours or days ahead. The medium-term carries out a forecast of various

© Springer Nature Switzerland AG 2020
M. Botto-Tobar et al. (Eds.): ICAETT 2019, AISC 1067, pp. 164–175, 2020.
https://doi.org/10.1007/978-3-030-32033-1_16

weeks and months ahead, and it is of particular interest for negotiating energy contracts. The long-term forecasting corresponds for a study of several years ahead [7]. Short-term forecasting allows managing generation, distribution, and transmission operation efficiently [18, 20].

Researchers in this area have developed several techniques. For example, fuzzy procedures have been performed for short-term [9], and medium-term forecasting [6]. In [3], Echo State Networks and Principal Component Analysis were used for a day horizon. Another technique significantly used are neural networks. For example, the authors of [5] studied a neural-network-based model for the short-term load forecast of the distribution grid. In [17], a method based on backpropagation neural networks and election of important variables as inputs were studied. The authors of [15] proposed an ensemble forecast framework, combining three neural network predictors.

In [14], advanced metering infrastructure was used to improve the load forecasting, based on clustering techniques of a group of customers.

Most of these techniques suffer from a lack of an efficient feature selection technique. To address this, the authors of [1] studied the load and price forecast based on new feature selection techniques.

Fewer works have investigated probabilistic load forecasting techniques. For example, in [11] the load forecasting problem includes quantile regression averaging on a set of sister point forecasts.

Although these works and others have studied the forecast of demand in distribution systems, fewer works have explored the demand forecast in buildings, which could allow managing efficiently the energy. For example, in [16] a statistical methodology to assess changes in the electric demand of buildings was proposed. The authors of [2] compared various techniques to forecast electricity consumption of buildings. The aim of this paper is to assess probabilistic forecasting methods in university buildings.

The rest of the paper is organized as follows: Sect. 2 describes an overview of various probabilistic forecasting techniques. Section 3 presents the Case Study and the Overall procedure. The results are discussed in Sect. 4. Finally, Sect. 5 highlights the main conclusions.

2   Background

This section describes probabilistic forecasting techniques.

2.1   Simple Linear Regression (SLR)

The SLR models the ED from an independent variable. The information of this independent variable is used for performing a forecast of the ED, based on the following mathematical expression:

$$\widehat{y} = \beta_0 + \beta_1.x + e \tag{1}$$

where $\widehat{y}$ is the forecast value of the ED, β_0 the intercept, β_1 the slope, x the independent variable, and e the error.

2.2   Multiple Linear Regression (MLR)

The MLR is an extension regression of the SLR. The forecast is obtained based on the information of two or several k independent variables:

$$\widehat{y} = \beta_0 + \beta_1.x_1 + \ldots + \beta_k.x_k + e \qquad (2)$$

To determine the best relation between the independent variables, the least-squares approach has to be implemented, which minimizes the sum of squared residuals.

2.3   Time Series

Time series models consist of data recorded in an orderly manner over time. They can be stationary or non-stationary. The time series is stationary if the mean and variance are constant. The trend, seasonal variation, cyclical and irregular variation, are factors that have time series. The trend shows a uniform behavior that grows or decreases within time. Seasonal variation exists when the data series shows a pattern that varies in a similar way year after year. The cyclic variation is described as the fluctuation of the time series data in defined periods. Finally, the random variation presents fortuitous changes in the time series.

2.4   Exponential Smoothing (ES)

Exponential smoothing (ES) methods involve standard procedures for continuously revising a forecast in light of more actual information corresponding to the estimated data. In brief terms, the methods relegate exponentially decreasing weights as the observation gets older [12]. Various approaches of exponential smoothing exists, of which the Holt and Winter methods are detailed:

Holt Method (HM). The HM appropriates for series including a linear time trend and additive seasonal variation. The series are expressed as follows:

$$A_t = \alpha.Y_t + (1 - \alpha).(A_{t-1} + T_{t-1}) \qquad (3)$$

The trend estimation is defined by:

$$T_t = \beta.(A_t - A_{t-1}) + (1 + \beta).T_{t-1} \qquad (4)$$

The forecast is computed by:

$$\widehat{Y}_{t+p} = A_t + p.T_t \qquad (5)$$

where $\widehat{Y}$ is the new series value.

Winter Method (WM). The WM includes a linear trend and multiplicative seasonal variation. The smoothed series are defined:

$$\widehat{Y} = (A_t + p.T_t).S_{t-L+p} + e_t \tag{6}$$

where S is the seasonal factor, the coefficients are defied by the recursions:

$$A_t = \alpha.\frac{Y_t}{S_{t-L}} + (1-\alpha).(A_{t-1} + T_{t-1}) \tag{7}$$

$$T_t = \beta(A_t - A_{t-1}) + (1-\beta).T_{t-1} \tag{8}$$

$$S_t = \gamma.\frac{Y_t}{A_t} + (1-\gamma).S_{t-L} \tag{9}$$

2.5   ARIMA Models

The ARIMA Models (p, d, q) are a class of stochastic processes used to analyze time series [4]. They include autoregressive processes, which are univariable models whose value in time depends on its data in a previous time series and on a random term. The time series data are used for searching and prediction [13]. The obtained forecasts are aggregated for each bootstrapped time series to generate the final output Y_t [12]. It is defined as follows:

$$Y_t = \mu + \beta_0.u_t + \beta_1.u_{t-1} + \beta_2.u_{t-2} + \ldots + \beta_q.u_{t-q} \tag{10}$$

where μ is a constant term, β_q are the means, and u_t the static errors.

3   Case Study and Overall Procedure

3.1   Case Study

The ED data corresponds to two university buildings: the Edificio Cornelio Merchán (ECM) of the Universidad Politécnica Salesiana (UPS), Cuenca, Ecuador, and the 8E building (8EB) of the Universitat Politècnica de València (UPV), Spain. The ED data was adjusted in order to be hourly. Besides, the analysis of variance (ANOVA) allowed classifying the ED according to the similarity of the consumption, as well as relating them with working and non-working days. The data of the ED from the ECM corresponds from March 14th 2017 to December 8th 2017. The data of the ED from the 8EB corresponds from July 1st 2014 to November 27th 2016.

3.2   Evaluation Measures

To evaluate the effectiveness of each method, various evaluation measures are used:

Mean Absolute Error (MAE).

$$MAE = \frac{1}{n} \sum_{i=1}^{n} |y_i - \widehat{y}_i| \tag{11}$$

Mean Absolute Percentage Error (MAPE).

$$MAPE = 100.\frac{1}{n} \sum_{i=1}^{n} |\frac{y_i - \widehat{y}_i}{y_i}| \tag{12}$$

Root Mean Squared Error (RMSE).

$$RMSE = \sqrt{\frac{1}{n} \sum_{i=1}^{n} (y_i - \widehat{y}_i)^2} \tag{13}$$

3.3   Overall Procedure

The studied ARIMA model is estimated based on the methodology of Box-Jenkins. Firstly, the Autocorrelation Function (ACF) is obtained to identify the possible parameters: p, d, and q. The procedure used includes four steps: identification, estimation, validation, and evaluation.

Identification. Due to the behavior of the time series, an integration has been applied. Figure 1 shows the ACF and the partial autocorrelation function (PACF). The results of the functions shown in the figures already have a differentiation to guarantee seasonality in the data. The first ACF presents a decrease toward zero but is not erased, approaching the behavior of first-order autoregressive (AR) model. While, when reviewing the PACF in Fig. 1(b), it can be observed that the behavior is similar to a moving average (MA) process.

Estimation. The model is approximated to the theoretical patterns of the ACFs, which are shown in Table 1. First, order parameters are estimated for the ECM model of the UPS, that is, an AR (1), I (1) and MA (1). Due to the initial integration, the final model is an ARIMA.

Validation. Various considerations are taken to select the best alternative. The ARIMA model has to reflect the most effective results in RMSE, Bayesian information criteria (BIC), and having an $R^2 > 0.8$, with a $p > 0.05$. Tables 1 and 2 show the ARIMA model parameters for the ECM, in working and in a non-working day respectively. Based on Table 1, the ARIMA (100) (011) is the most suitable and is used for the forecast of the ED in working days. Based on Table 2, the ARIMA is the ideal for non-working days.

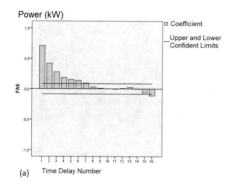

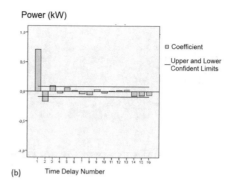

Fig. 1. (a) Autocorrelation function; (b) Partial autocorrelation function.

Table 1. Arima model parameters for the ECM in a working day

Model	Param.	Param.	RMSE	R^2	p	CIB
(100)(111)	AR	SARMA	1,227	0,974	0	0,437
(100)(011)	AR	SMA	1,226	0,974	0	0,43
(001)(110)	MA	SAR	1,344	0,968	0	0,613
(101)(101)	ARMA	SARMA	1,33	0,969	0	0,606
(102)(111)	ARMA	SARMA	1,235	0,970	0.001	0,433

Table 2. Arima model parameters for the ECM in a non-working day

Model	Param.	Param.	RMSE	R^2	p	CIB
(100)(011)	AR	SMA	0,771	0,796	0,0414	−0,456
(101)(111)	ARMA	SARMA	0,764	0,808	0,056	−0,432
(101)(011)	ARMA	SMA	0,767	0,799	0,102	−0,446
(001)(111)	MA	SAR	0,954	0,688	0	−0,31
(001)(011)	AR	SARMA	0,769	0,798	0,02	−0,441

Evaluation. To evaluate the ARIMA model, several predictions were performed in the SPSS software, for working and non-working days. The ARIMA model (100)(011) with $R^2 = 0.974$ and RMSE $= 1.266$ kW is used, and its result is depicted in Fig. 2(a). For the prediction of non-working days, the ARIMA (101)(111) is performed, with a RMSE $= 0.764$ kW and $R^2 = 0.808$, and the forecast is illustrated in Fig. 2(b).

Concerning the 8EB building, the ACF shows a similar behavior than the ECM. The ACF defines an AR. As a result, the ARIMA model (101) (111) is selected for predicting both working and non-working days. In Fig. 3 the prediction of a working day with the chosen model is shown, the coefficient of determination is $R^2 = 0.985$, and RMSE $= 6.686$ kW.

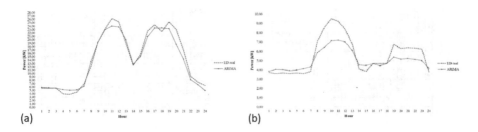

Fig. 2. ARIMA forecast of ECM: (a) in a working day; (b) in a non-working day.

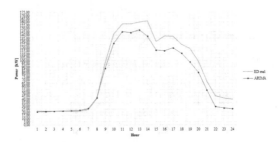

Fig. 3. ARIMA forecast of 8EB Building in a non-working day.

4   Results and Discussion

4.1   Results Comparison

To assess the different forecasting methods, their results are compared. The methods used are SLR, MLR, MHW, and ARIMA. The ES and MH were not considered since its forecasting capacity is weak and the errors are high enough. Figure 6 illustrates the forecasting patterns of the different methods of a working day of the two university buildings studied. The forecast errors of the ECM and the 8EB are shown in Table 3.

The results indicate that the ARIMA method and the HW have better forecast capacity. Although SLR and MLR have a low prediction error, these models have difficulty following the shape of the electrical source profile.

Then, to assess the forecast capacity of the ARIMA and HW methods, a non-working and a working day were also simulated. Figure 5 depicts the ED forecast of a non-working day in the two studied university buildings for the corresponding methods. Table 4 shows the forecast errors of ARIMA and HW of a non-working day in the ECM. It is observed that the HW does not present proper results and the ARIMA is significantly more accurate comparing to the real demand (Fig. 5).

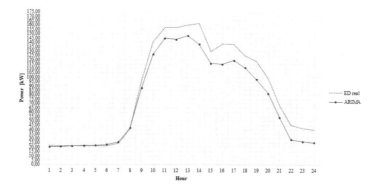

Fig. 4. ED forecast of a working day: (a) In the ECM; (b) In the 8EB.

Table 3. Forecast errors of a working day in the ECM

	Holt-Winter	SLR	MLR	ARIMA
MAE (ECM) [kW]	11.77	3.6493	6.47	0.9001
MAPE (ECM) [%]	78.79	24.1601	65.19	7.2902
RMSE (ECM) [kW]	14.10	5.3338	7.71	1.1763
MAE (8EB) [kW]	35.21	30.58	46.8818	8.3135
MAPE (8EB) [%]	40.27	58.49	75.0664	12.8791
RMSE (8EB) [kW]	44.38	38.18	51.7378	10.2593

Table 4. Forecast errors of ARIMA and HW of a non-working day in the ECM

	Holt-Winter	ARIMA
MAE (ECM) [kW]	13.1100	0.8469
MAPE (ECM) [%]	239.2056	13.6818
RMSE (ECM) [kW]	14.6200	1.0692
MAE (8EB) [kW]	13.7052	1.1036
MAPE (8EB) [%]	63.6774	5.1225
RMSE (8EB) [kW]	15.4199	1.2538

The ED forecast of ARIMA and HW of a working day is depicted in Fig. 6, in the two university buildings. The forecast errors are shown in Table 5. Note also that the ARIMA model presents the most accurate results, and the HW is not proper to forecast the ED of the university buildings.

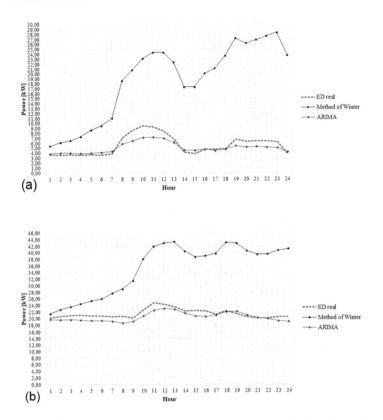

Fig. 5. ED forecast of ARIMA and HW a non-working day: (a) In the ECM; (b) In the 8EB.

Table 5. Forecast errors of ARIMA and HW of a working day in the ECM

	Holt-Winter	ARIMA
MAE (ECM) [kW]	17.8018	1.1365
MAPE (ECM) [%]	107.9935	9.9044
RMSE (ECM) [kW]	22.6148	1.4875
MAE (8EB) [kW]	20.8968	8.6825
MAPE (8EB) [%]	61.1155	12.0744
RMSE (8EB) [kW]	24.2468	13.2142

4.2   Discussion

The ARIMA model obtains the best short-term prediction results. The average value of the MAPE for five working days (one week) is 9.90% and 12.07% for the ECM and 8EB respectively while the HW method obtains a MAPE of 107.99% and 61.12% for the same case.

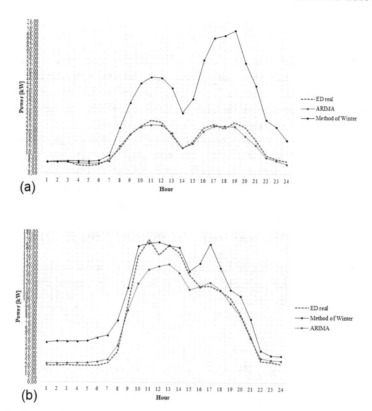

Fig. 6. ED forecast of ARIMA and HW of a working day: (a) In the ECM; (b) In the 8EB.

Regarding the predictions for non-working days, the HW method evidences a MAPE of 239% and 65% for the ECM and the 8EB respectively. The ARIMA model for this same case presents MAPE values of 13.68% and 5.12% respectively.

The results obtained indicate that the simple exponential smoothing methods and the Holt method cannot predict electrical source profiles. The models based on SLR and MLR did not achieve either good results because the dependent variables have a coefficient of determination (R^2) less than 0.6 concerning the output of the model. On the other hand, the HW method obtains better results, but the prediction errors increase with time.

The application of the Box-Jenkins methodology allows estimating suitable ARIMA models for the prediction of electrical sources for any ED. An adequate theoretical approximation of the ACF allows estimating the order of the parameters of the model. In comparison with other statistical models, the ARIMA models are the ones that best adjust to the shape of the real electrical consumption, obtaining low prediction errors.

5   Conclusions

In this paper, various statistical forecasting methods have been studied for two university buildings, to assess the effectiveness of these methods. This study was performed in working and non-working days in the Universidad Politécnica Salesiana and Universitat Politècnica de Valencia, which are universities from Ecuador and Spain, and present different load patterns.

Forecasting methods have presented strengths and weakness. The forecasts from Winter methods have strong similarities with the real electricity load pattern, but the power values differ significantly. Linear Regression methods have presented severe disadvantages because the external variables do not have a significant relationship with electricity consumption.

Finally, the ARIMA models are the most suitable, comparing to the other studied models. However, the forecast requires an organized experimental process to select the best model.

The prediction of the ARIMA models could be improved by the incorporation of advanced data clustering techniques, seasonality analysis of time series data and artificial intelligence techniques.

References

1. Abedinia, O., Amjady, N., Zareipour, H.: A new feature selection technique for load and price forecast of electrical power systems. IEEE Trans. Power Syst. **32**(1), 62–74 (2017). https://doi.org/10.1109/TPWRS.2016.2556620
2. Amber, K.P., Ahmad, R., Aslam, M.W., Kousar, A., Usman, M., Khan, M.S.: Intelligent techniques for forecasting electricity consumption of buildings. Energy **157**, 886–893 (2018). https://doi.org/10.1016/j.energy.2018.05.155
3. Bianchi, F.M., Santis, E.D.E., Rizzi, A., Sadeghian, A.: Short-term electric load forecasting using echo state networks and PCA decomposition. IEEE Access **3**, 1931–1943 (2015). https://doi.org/10.1109/ACCESS.2015.2485943
4. Contreras, J., Espinola, R., Nogales, F.J., Conejo, A.J.: ARIMA models to predict next-day electricity prices. IEEE Trans. Power Syst. **22**(9), 57–57 (2002). https://doi.org/10.1016/j.energy.2018.05.155. http://ieeexplore.ieee.org/document/4312577/
5. Ding, N., Benoit, C., Foggia, G., Bésanger, Y., Wurtz, F.: Neural network-based model design for short-term load forecast in distribution systems. IEEE Trans. Power Syst. **31**(1), 72–81 (2016). https://doi.org/10.1109/TPWRS.2015.2390132
6. Elias, C.N., Hatziargyriou, N.D.: An annual midterm energy forecasting model using fuzzy logic. IEEE Trans. Power Syst. **24**(1), 469–478 (2009)
7. González-Romera, E., Jaramillo-Morán, M.Á., Carmona-Fernández, D.: Monthly electric energy demand forecasting based. IEEE Trans. Power Syst. **21**(4), 1946–1953 (2006)
8. Hippert, H.S., Pedreira, C.E., Souza, R.C.: Neural networks for short-term load forecasting: a review and evaluation Full Text as PDF Full Text in HTML. IEEE Trans. Power Syst. **16**(1), 4333 (2014)
9. Khosravi, A., Nahavandi, S.: Load forecasting using interval type-2 fuzzy logic systems: optimal type reduction. IEEE Trans. Ind. Inform. **10**(2), 1055–1063 (2014). https://doi.org/10.1109/TII.2013.2285650

10. Kroposki, B., et al.: Achieving a 100% renewable grid: operating electric power systems with extremely high levels of variable renewable energy. IEEE Power Energy Mag. **15**(2), 61–73 (2017). https://doi.org/10.1109/MPE.2016.2637122

11. Liu, B., Nowotarski, J., Hong, T., Weron, R.: Probabilistic load forecasting via quantile regression averaging on sister forecasts. IEEE Trans. Smart Grid **8**(2), 730–737 (2017). https://doi.org/10.1109/TSG.2015.2437877

12. Meira, E., Oliveira, D., Luiz, F., Oliveira, C.: Forecasting mid-long term electric energy consumption through bagging ARIMA and exponential smoothing methods. Energy **144**, 776–788 (2018). https://doi.org/10.1016/j.energy.2017.12.049

13. Park, S., Han, S.: Demand power forecasting with data mining method in smart grid. In: 2017 IEEE Innovative Smart Grid Technologies - Asia (2017). https://doi.org/10.1109/ISGT-Asia.2017.8378423

14. Quilumba, F.L., Lee, W.J., Huang, H., Wang, D.Y., Szabados, R.L.: Using smart meter data to improve the accuracy of intraday load forecasting considering customer behavior similarities. IEEE Trans. Smart Grid **6**(2), 911–918 (2015). https://doi.org/10.1109/TSG.2014.2364233. http://ieeexplore.ieee.org/document/6945384/

15. Raza, M.Q., Mithulananthan, N., Li, J., Lee, K.Y.: Multivariate ensemble forecast framework for demand prediction of anomalous days. IEEE Trans. Sustain. Energy **3029**(c), 1–9 (2018). https://doi.org/10.1109/TSTE.2018.2883393

16. Serrano-Guerrero, X., Escrivá-Escrivá, G., Roldán-Blay, C.: Statistical methodology to assess changes in the electrical consumption profile of buildings. Energy Build. **164**, 99–108 (2018). https://doi.org/10.1016/j.enbuild.2017.12.059

17. Serrano-Guerrero, X., Prieto-Galarza, R., Huilcatanda, E., Cabrera-Zeas, J., Escriva-Escriva, G.: Election of variables and short-term forecasting of electricity demand based on backpropagation artificial neural networks. In: 2017 IEEE International Autumn Meeting on Power, Electronics and Computing (ROPEC), ROPEC 2017, January 2018, pp. 1–5 (2018). https://doi.org/10.1109/ROPEC.2017.8261630

18. Tao, S., Li, Y., Xiao, X., Yao, L.: Load forecasting based on short-term correlation clustering. In: 2017 IEEE Innovative Smart Grid Technologies - Asia, pp. 1–7. IEEE (2017). https://doi.org/10.1109/ISGT-Asia.2017.8378416

19. Vartanian, C., Bauer, R., Casey, L., Loutan, C., Narang, D., Patel, V.: Ensuring system reliability: distributed energy resources and bulk power system considerations. IEEE Power Energy Mag. **16**(6), 52–63 (2018). https://doi.org/10.1109/MPE.2018.2863059

20. Ye, X.Z., Ji, T.Y., Li, M.S., Wu, Q.H.: A morphological filter-based local prediction method with multi-variable inputs for short-term load forecast. In: 2017 IEEE Innovative Smart Grid Technologies - Asia (2017). https://doi.org/10.1109/ISGT-Asia.2017.8378323

Dynamic State Estimation of a Synchronous Machine Applying the Extended Kalman Filter Technique

Jorge Ninazunta[✉] and Silvana Gamboa

Escuela Politécnica Nacional, Quito, Ecuador
{jorge.ninazunta, silvana.gamboa}@epn.edu.ec

Abstract. The purpose of this work is to implement a dynamic state estimator (DSE) of a synchronous machine by applying the extended Kalman filter (EKF). This work is focused on the synchronous generator because it constitutes a fundamental component of the Electric Power Systems. The dynamic behavior of the generator in an electric grid is simulated in Matlab using the One Machine-Infinite Bus System (OMIB). On the other hand, two standard dynamic models of the synchronous generator are used to develop the dynamic state estimator: the Two-Axis fourth order model and the sixth order GENROU model. The EKF estimation algorithm is applied to this two dynamic models in order to design the estimator block, and the filter's parameters are tuned by simulating the ensemble "OMIB System-Estimator". Finally, the performance of the developed state estimators is evaluated and compared.

Keywords: Synchronous generator · Dynamic state estimation · Kalman filter

1 Introduction

Currently, the evolution and development of modern electric grids continue at an accelerated pace [1]. The massive integration of renewable sources and the introduction of active loads result in the increasingly frequent appearance of dynamic phenomena in the power system operation [1, 2]. Therefore, it is necessary to establish a new paradigm for the modern electric grid, which allows to improve the monitoring and control of the power system under dynamic operating conditions. It has been proposed to adopt a dynamic operation paradigm where the dynamic state estimation (DSE) tool constitutes the fundamental component, since it can provide a complete dynamic view of the system [1, 2]. Wide Area Monitoring Systems (WAMS) are being implemented worldwide, based on a massive deployment of phasor measurement units (PMU) [3]. PMUs are capable of directly measuring the voltage and current phasors of the electric grid with a high update rate, high precision, and synchronized, which allows them to capture the power system dynamic behavior [1]. Therefore, this synchrophasor measurement technology provides a solid foundation to enable DSE in power systems [1]. It is important to note that the system dynamic behavior is fundamentally determined by the response of the generators operating within the grid [4, 5]. Therefore, the power system dynamic state is defined as the set of the generator electromechanical variables.

© Springer Nature Switzerland AG 2020
M. Botto-Tobar et al. (Eds.): ICAETT 2019, AISC 1067, pp. 176–188, 2020.
https://doi.org/10.1007/978-3-030-32033-1_17

In the literature, two dynamic models of different order have been widely used for DSE purposes. The first one is the classical model which only includes the dynamics of the mechanical variables [5, 6]. Other studies have incorporated the Two-Axis fourth order model which includes the dynamics of the d-q components of the transient voltage, in addition to the mechanical variables [1, 3, 7–10]. In spite of the satisfactory results obtained using these two dynamic models, several recent studies have pointed out the need to incorporate more complex generator dynamic models in order to improve the results of the estimation [1, 8]. In this paper, it is proposed to use a sixth order dynamic model to develop the state estimator, thus it is intended to contribute in the analysis and assessment of the impact of model order of the synchronous generator on the DSE problem. A diagram of the DSE problem is shown in Fig. 1. The work is divided in the study of a simulation model that is considered to describe the real machine behavior, and the study of some standard generator dynamic models used for state estimation purposes. The OMIB system is used to simulate the generator dynamic response to external disturbances in the electric grid, and it is used to generate simulation data in Matlab. The state estimator receives as input signals the measurements/data obtained from a PMU that is assumed to be installed at the machine terminal bus. Under this approach, the estimator only processes local data from the generation unit. Since PMU field measurements were not available, the simulation dataset generated in Matlab was used.

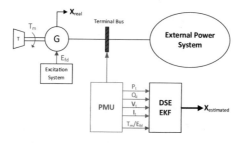

Fig. 1. Synchronous generator dynamic state estimation problem

Several DSE methods can be found in the literature, among which the techniques based on the Kalman filter are the most widely used [3, 7, 8, 10]. Under the assumptions of a linear system and Gaussian noise, the linear Kalman filter defines a recursive method to produce optimal estimates of the system state [11]. The concepts of the Kalman filter can be extended to non-linear systems through several non-linear state estimation methods, such as [1, 10]: extended Kalman filter (EKF), unscented Kalman filter (UKF), particle filter (PF), among others. It is worth mentioning that all available non-linear estimation methods only produce sub-optimal state estimates [11]. The main difference between these estimation algorithms is the approach that each one uses to propagate de probability density function (PDF) of the system state [10]. The EKF propagates the mean and covariance of the state using the Jacobian matrices of the linearized system state-space model [10, 11]. The EKF has been established as the standard technique for implementing non-linear state estimators because of its high estimation accuracy and high computational efficiency [10, 11]. However, the states estimates produced by this filter tends to have poor accuracy and can even diverge for highly non-linear systems [10]. The UKF propagates the PDF of the state using a deterministic sampling method to pass a set of points (known as sigma points) through the non-linear system function [3, 10]. The UKF tends to outperform the EKF in terms of accuracy and robustness when the system presents severe nonlinearities [3, 10]. The EKF and the UKF are based on the assumption of additive Gaussian noise present in the system, thus the application of these algorithms is limited to this type of systems [10]. In order to overcome this limitation, the PF is a general state estimation algorithm that is not based on the Gaussian noise assumption, and therefore it can be applied directly to non-linear systems with Gaussian or non-Gaussian noise. The PF uses a set of weighted random samples (known as particles) to represent the PDF of the system state, and it propagates its mean and variance through Monte Carlo simulation [9, 10]. The PF requires a large number of particles in order to obtain a sufficiently good estimation accuracy and to achieve convergence of the algorithm [9]. However, the computation time of the PF increases as the number of samples is increased, thus this algorithm is computationally expensive and is not suitable for real-time DSE [10]. It is important to note that these non-linear estimation methods are recursive and its implementations have a similar structure. Recent studies have shown the feasibility of applying the EKF to estimate the dynamic state of a power system [7, 8]. The generator dynamic models are expressed as a set of non-linear differential equations. For this reason, the EKF is applied to non-linear models in order to design the generator DSE algorithm.

2   Simulation Model

The synchronous generator is a fundamental component of the electrical grids and its electromechanical response is a determining factor in power system dynamics [5]. A model structure with two rotor circuits for each of the d-q axes is considered, which in most cases, it is considered adequate to represent the effects of the damper circuits in the machine [4, 5]. The state-space mathematical model of the synchronous generator is presented in Eqs. (1) to (4). This model is expressed in terms of the machine funda-mental parameters and the rotor circuits magnetic fluxes as state variables. The variable $\Delta\omega$ is the rotor angular velocity deviation with respect to the synchronous speed (in pu) and δ is the rotor angle (electrical radians). The flux variables ψ_{fd}, ψ_{1d}, ψ_{1q}, and ψ_{2q} are the rotor circuits fluxes (in pu). The signals T_m and E_{fd} are the mechanical torque and the exciter voltage, respectively. The vector x is the machine state vector and u is the input vector. The d-q components of the subtransient voltage (e_d'' and e_q'') are expressed in terms of the machine state variables, according to Eq. (4). The nomenclature of the other parameters that appear in Eq. (2) is detailed in Table 1. The generator terminal variables are chosen as the model outputs: active and reactive power (P_t and Q_t); terminal voltage and current (V_t and I_t). Note that the output variables are expressed in

Table 1. Fundamental and Standard parameters of the synchronous machine

Fundamental parameters		Value	Standard parameters		Value
Stator-rotor mutual inductance	L_{ad}	1.97	Synchronous	X_d	2.1
	L_{aq}	1.87	reactance	X_q	2.0
Field circuit inductance and resistance	L_{fd}	0.1861	Transient reactance	X'_d	0.30
	R_{fd}	0.0008		X'_q	0.50
Damper circuit $1d$ inductance and resistance	L_{1d}	0.0708	Subtransient	X''_d	0.18
	R_{1d}	0.0087	reactance	X''_q	0.22
Damper circuit $1q$ inductance and resistance	L_{1q}	0.4613	Transient OC time	T'_{d0}	7.0 s
	R_{1q}	0.0082	constant	T'_{q0}	0.75 s
Damper circuit $2q$ inductance and resistance	L_{2q}	0.1189	Subtransient OC	T''_{d0}	0.073 s
	R_{2q}	0.0185	time constant	T''_{q0}	0.07 s
Stator leakage inductance	L_l	0.13	Inertia constant	H	3.0
Stator resistance	R_a	0.0	Damping coefficient	K_D	0.0

terms of the subtransient d-q reactances which are related to the fundamental parameters by the expressions $X''_d = L''_{ad} + L_l$ and $X''_q = L''_{aq} + L_l$.

$$x = \left[\Delta\omega, \delta, \psi_{fd}, \psi_{1d}, \psi_{1q}, \psi_{2q}\right]^T; u = \left[T_m, E_{fd}\right]^T; y = [P_t, Q_t, I_R, I_I, V_R, V_I]^T \quad (1)$$

$$\begin{cases} \dot{x}_1 = \frac{1}{2H}\left[u_1 - e''_d i_d - e''_q i_q - \left(X''_q - X''_d\right)i_d i_q - K_D x_1\right] \\ \dot{x}_2 = \omega_0 x_1 \\ \dot{x}_3 = \frac{\omega_0 R_{fd}}{L_{fd}}\left[\frac{L_{fd}}{L_{ad}}u_2 - x_3 + e''_q - L''_{ad}i_d\right] \\ \dot{x}_4 = \frac{\omega_0 R_{1d}}{L_{1d}}\left[-x_4 + e''_q - L''_{ad}i_d\right] \\ \dot{x}_5 = \frac{\omega_0 R_{1q}}{L_{1q}}\left[-x_5 - e''_d - L''_{aq}i_q\right] \\ \dot{x}_6 = \frac{\omega_0 R_{2q}}{L_{2q}}\left[-x_6 - e''_d - L''_{aq}i_q\right] \end{cases} \quad (2)$$

$$\begin{cases} y_1 = e''_d i_d + e''_q i_q + \left(X''_q - X''_d\right)i_d i_q \\ y_2 = e''_q i_d - e''_d i_q - X''_d i_d^2 - X''_q i_q^2 \\ y_3 = i_d s_{x_2} + i_q c_{x_2} \\ y_4 = i_q s_{x_2} - i_d c_{x_2} \\ y_5 = \left(e''_d + X''_q i_q\right)s_{x_2} + \left(e''_q - X''_d i_d\right)c_{x_2} \\ y_6 = \left(e''_q - X''_d i_d\right)s_{x_2} - \left(e''_d + X''_q i_q\right)c_{x_2} \end{cases} \quad (3)$$

$$e''_q = L''_{ad}\left(\psi_{fd}/L_{fd} + \psi_{1d}/L_{1d}\right); e''_d = -L''_{aq}\left(\psi_{1q}/L_{1q} + \psi_{2q}/L_{2q}\right) \quad (4)$$

A representation of the OMIB system is shown in Fig. 2, which consists of a generator unit connected to an infinite bus via a transmission network [4]. The external electrical system can be considered as an infinite bus due to its large size relative to the machine [4]. Based on the equivalent circuit of the OMIB system, it is possible to express the currents i_d and i_q as a function of the machine state as expressed in Eq. (5). Equations (1) to (4), together with Eq. (5), completes the OMIB simulation model that is considered to describe the real machine behavior and it is used to generate dynamic simulation data in Matlab. The simulation model can be expressed as a set of equations in state space form that can be solved by applying numerical integration methods using the computational tools available in Matlab/Simulink. A Simulink model was implemented to simulate the dynamic response of the OMIB system applying the second order Runge-Kutta integration method (RK-2) to the model described by Eqs. (1) to (5), with a time step $\Delta t = 1$ms. The OMIB system simulation provides the machine state vector x, which it is considered to be the real generator state, and the machine output vector y. These signals are used as inputs to the dynamic state estimators described in the following sections.

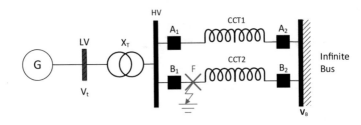

Fig. 2. OMIB System

$$i_d = \left(e_q'' - V_B cos(\delta)\right)/(X_d'' + X_E); i_q = \left(V_B sen(\delta) - e_d''\right)/\left(X_q'' + X_E\right) \qquad (5)$$

3   The Extended Kalman Filter

The Kalman filter is a mathematical tool that generates an optimal state estimation of a dynamic system [11]. In general, the input/output signals are the only observable variables of the system by using measurement devices [11]. These measurements can be affected by noise due to the distortions introduced by any sensor [11]. On the other hand, every system is affected by perturbations that cannot be modeled in a deterministic manner [11]. For these reasons, the estimation model must incorporate certain notion of uncertainty/noise in the mathematical model and in the measurements [12]. The Kalman filter is optimal in the sense that it minimizes the error between the real and estimated state, under the assumptions of a linear system and that the noise/uncertainty has Gaussian white noise statistical properties [11]. Thus, the Kalman filter can only be directly applied to linear nature physical systems [11]. Nevertheless, the concepts of linear estimation can be extended to non-linear systems by using the extended Kalman filter (EKF), which involves linearizing the system model around the estimated state, in each iteration of the algorithm [11, 12]. Given a system expressed in the form of a stochastic nonlinear dynamic model in Eq. (6).

$$x_k = f_k(x_{k-1}, u_{k-1}) + w_{k-1}; y_k = h_k(x_k, u_k) + v_k \qquad (6)$$

Where, x_k is the system state, u_k contains the inputs, and y_k is the measurement vector. The function f_k is the nonlinear state transition function, and the function h_k is the measurement model. The variables w_k and v_k represent the process and measurement noise, respectively, and they are assumed to be Gaussian white noise described by Eq. (7). The Q_k and R_k matrices are the process noise covariance and the measurement noise covariance, respectively. The EKF algorithm is described in Fig. 3. The filter equations define a recursive state estimation algorithm, which significantly facilitates its practical implementation in a digital computer [12].

$$w_k \sim \mathcal{N}(0, Q_k); v_k \sim \mathcal{N}(0, R_k) \qquad (7)$$

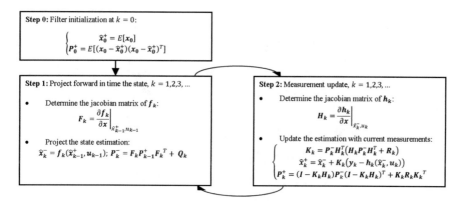

Fig. 3. EKF state estimation algorithm

4   Estimation Models

The estimator receives the measurements/data from a PMU installed at the machine terminal bus. Under this configuration, the PMU is able to provide the following set of measurements: V_t, I_t, P_t, Q_t. Also, it is assumed that T_m and E_{fd} can be locally obtained, since these signals are usually available in power plants for local control [4, 7, 9]. Therefore, the input vector u and the measurement vector y of the estimation models are defined according to Eq. (8). These vectors are defined in this matter to use all the information available from the measurement device. In a typical state estimation problem, the measured variables provide complementary information of the system.

By combining more measurements using the Kalman filter, the estimation accuracy is improved because we are incorporating more information.

The standard dynamic models of the generator are expressed in terms of the machine standard parameters, which results in an approximate representation of the actual behavior of the machine [13]. Two standard dynamic models are used to develop the state estimator: the first one is the GENROU model [6, 14]; and the second one is the Two-Axis model [6]. The subtransient sixth order GENROU model has been widely used to represent cylindrical rotor generators in stability studies [6, 15]. This model is described by Eqs. (9), (10) and (11). The Two-Axis fourth-order transient model neglects the dynamics of the fluxes ψ_{1d} and ψ_{2q} [6]. This model is described by Eqs. (12), (13) and (14).

$$\boldsymbol{u} = \left[T_m, E_{fd}, I_R, I_I\right]^T ; \boldsymbol{y} = \left[P_t, Q_t, V_R, V_I\right]^T \tag{8}$$

$$\boldsymbol{x} = \left[\Delta\omega, \delta, e_q', e_d', \psi_{1d}, \psi_{2q}\right]^T \tag{9}$$

$$\begin{cases} \dot{x}_1 = \frac{1}{2H}\left[u_1 - s_1 T_1 - c_1 T_2 - \left(X_q'' - X_d''\right)\left(\frac{U_1}{2}s_2 - U_2 c_2\right) - K_D x_1\right] \\ \dot{x}_2 = \omega_0 x_1 \\ \dot{x}_3 = \frac{1}{T_{d0}'}\left[u_2 - (1 + k_{d0}k_{d2})x_3 + k_{d0}k_{d2}x_5 - k_{d0}k_{d1}(u_3 s_1 - u_4 c_1)\right] \\ \dot{x}_4 = \frac{1}{T_{q0}'}\left[-(1 + k_{q0}k_{q2})x_4 - k_{q0}k_{q2}x_6 + k_{q0}k_{q1}(u_3 c_1 + u_4 s_1)\right] \\ \dot{x}_5 = \frac{1}{T_{d0}''}\left[-x_5 + x_3 - k_{d3}(u_3 s_1 - u_4 c_1)\right] \\ \dot{x}_6 = \frac{1}{T_{q0}''}\left[-x_6 - x_4 - k_{q3}(u_3 c_1 + u_4 s_1)\right] \end{cases} \tag{10}$$

$$\begin{cases} y_1 = s_1 T_1 + c_1 T_2 + \left(X_q'' - X_d''\right)\left(\frac{U_1}{2}s_2 - U_2 c_2\right) \\ y_2 = s_1 T_2 - c_1 T_1 - \left(X_q'' - X_d''\right)\left(\frac{U_1}{2}c_2 + U_2 s_2\right) - \frac{X_q'' + X_d''}{2}U \\ y_3 = E_R'' + \frac{X_q'' - X_d''}{2}(u_3 s_2 - u_4 c_2) + \frac{X_q'' + X_d''}{2}u_4 \\ y_4 = E_I'' - \frac{X_q'' - X_d''}{2}(u_4 s_2 + u_3 c_2) - \frac{X_q'' + X_d''}{2}u_3 \end{cases} \tag{11}$$

$$\boldsymbol{x} = \left[\Delta\omega, \delta, e_q', e_d'\right]^T \tag{12}$$

$$\begin{cases} \dot{x}_1 = \frac{1}{2H}\left[u_1 - s_1 T_1 - c_1 T_2 - \left(X_q' - X_d'\right)\left(\frac{U_1}{2}s_2 - U_2 c_2\right) - K_D x_1\right] \\ \dot{x}_2 = \omega_0 x_1 \\ \dot{x}_3 = \frac{1}{T_{d0}'}\left[u_2 - x_3 - \left(X_d - X_d'\right)(u_3 s_1 - u_4 c_1)\right] \\ \dot{x}_4 = \frac{1}{T_{d0}'}\left[-x_4 + \left(X_q - X_q'\right)(u_3 c_1 + u_4 s_1)\right] \end{cases} \tag{13}$$

$$
\begin{cases}
y_1 = s_1 T_1 + c_1 T_2 + \left(X'_q - X'_d\right)\left(\frac{U_1}{2} s_2 - U_2 c_2\right) \\
y_2 = s_1 T_2 - c_1 T_1 - \left(X'_q - X'_d\right)\left(\frac{U_1}{2} c_2 + U_2 s_2\right) - \frac{X'_q + X'_d}{2} U \\
y_3 = E'_R + \frac{X'_q - X'_d}{2}\left(u_3 s_2 - u_4 c_2\right) + \frac{X'_q + X'_d}{2} u_4 \\
y_4 = E'_I - \frac{X'_q - X'_d}{2}\left(u_4 s_2 + u_3 c_2\right) - \frac{X'_q + X'_d}{2} u_3
\end{cases}
\tag{14}
$$

5 State Estimator Design

The EKF algorithm is applied to the two estimation models in order to design the dynamic state estimator. The two estimation models are discretized using the Euler method with a sampling rate of $T_s = 1$ ms. At $k = 0$, an initial estimate of the state $(\hat{x}_0^+)$ and the state covariance matrix (P_0^+) must be specified. The initial state estimate is set to be equal to the initial value of the generator state (x_0). Thus, $\hat{x}_0^+$ is $[0; 0.9089; 1.1298; 0.5330; 0.9614; -0.6645]$ for the GENROU model and $[0; 0.9089; 1.1298; 0.5330]$ for the Two-Axis model. Moreover, it is assumed that P_0^+ is a diagonal matrix by setting the covariance between two sate variables equal to zero. Thus, this matrix is set to be $P_0^+ = (10^2) * I_{nxn}$, where n is the number of state variables. The values of the R_k and Q_k matrices have a great impact on the performance of the Kalman filter [16, 17]. The matrix R_k characterizes the precision of the sensors [12]. The matrix Q_k reflects the level of uncertainty in the system model used in the state estimation [7, 11]. In practice, the values of R_k and Q_k are adjusted in a filter tuning process in order to obtain an appropriate performance of the state estimator [12]. The methodology used for the tuning process consists in selecting approximate base values of R_k and Q_k, based on prior knowledge of the system under study. Then, the values of R_k and Q_k are tuned by simulating the ensemble "OMIB system-Estimator", until achieving an adequate performance of the state estimator. It is assumed that R_k is a diagonal matrix under the consideration that the measured variables are independent [16]. The diagonal elements of matrix R_k are related to the precession of the sensors used [12]. For both estimation models, R_k is a 4×4 matrix. According to the IEEE C.37.118.1-2011 standard [18], the PMU measurement error must be less than 1% of the total vector error (TVE), therefore the standard deviation of the measurements is set to be equal to 0.01 in per unit. Thus, the base value of the matrix R_k is given by Eq. (15).

$$
R_k = \sigma_y^2 I_{4\times4} = (0.01)^2 I_{4\times4}
\tag{15}
$$

The matrix Q_k is considered to be diagonal, under the assumption that the noise in the state variables is independent from each other [16]. The value of the Q_k matrix diagonal is established based on the maximum value of the state vector expected variation, between two consecutive sampling instants. This is an approximate method to model the matrix Q_k, which allows to weight more heavily the fastest-changing machine states [9, 10, 16]. The dynamic response of the synchronous generator to various fault/disturbances scenarios was analyzed in order to determine the maximum value of the state vector variation. The results of this analysis showed that the scenario of a three-phase fault at the sending end of the transmission network causes the steepest response of the machine. Thus, the following base values of Q_k are used for the estimation models: $Q_{0_6} = \text{diag}\left([0.0002, 0.0095, 0.0004, 0.0016, 0.0062, 0.0114]^2\right)$ for the GENROU model and $Q_{0_4} = \text{diag}\left([0.0002, 0.0095, 0.0004, 0.0016]^2\right)$ for the Two-Axis model. Two estimator blocks were implemented in Matlab/Simulink: "EKF_4Ord" and "EKF_6Ord". These two blocks implement the EKF applied to the two estimation models: The Two-Axis transient model (fourth order) and the subtransient GENROU model (sixth order). The input signals to the estimators are the input vector u and the measurement vector y, defined in Eq. (8). Note that the measurement vector y is affected by the measurement noise. The sensor noise added to y is characterized by $\tilde{v}_k \sim \mathcal{N}\left(0, (0.01)^2 I_{4\times4}\right)$, corresponding to 1% of the TVE. Several simulations of the ensemble "OMIB System-Estimator" are performed using the base values of the EKF parameters previously determined, and then, these values are varied in each simulation. The simulation of the complete system is performed for a time interval from 0 to 100 s for a steady-state operating condition. To evaluate and compare the performance of the two estimators, the mean square error (MSE) metric is used [19]. The value of R_k is varied according to the expected noise level in the measurements. In this case, the variation goes from 0.1% to 50% of the TVE. The value of Q_k is varied by multiplying the base value of this matrix (Q_{0_4} and Q_{0_6}) by a given scaling factor that goes from 0.01 to 100, in multiples of 10. The calculated MSE values of the rotor angle estimation for the two estimators are summarized in Table 2.

Table 2. MSE –EKF_4Ord and EKF_6Ord

R\Q	EKF4 - Delta						EKF6 - Delta					
	0.1 %	0.5 %	1%	5%	10 %	50 %	0.1 %	0.5 %	1%	5%	10 %	50 %
0.01x	6,5E-04	2,0E-04	1,5E-04	2,5E-05	1,2E-05	8,9E-07	3,9E+01	7,4E-05	7,0E-05	2,5E-04	1,7E-04	6,4E-06
0.1x	4,4E-03	4,9E-04	2,4E-04	2,7E-05	1,2E-05	9,3E-07	2,9E-02	8,7E-05	5,0E-05	3,3E-04	1,9E-04	6,0E-06
1x	1,2E-01	1,0E-03	5,7E-04	3,3E-05	1,4E-05	1,2E-06	7,1E-02	1,0E-03	7,7E-04	6,0E-04	2,7E-04	6,2E-06
10x	2,0E-01	5,9E-02	4,3E-03	6,7E-05	2,1E-05	2,3E-06	7,8E-02	5,9E-02	2,8E-02	8,4E-04	4,0E-04	7,6E-06
100x	2,2E-01	1,8E-01	1,2E-01	4,5E-04	5,5E-05	5,5E-06	7,7E-02	7,5E-02	6,8E-02	8,2E-03	4,7E-04	9,0E-06

For both estimators, it is shown that by increasing the value of R_k, the MSE is reduced. On the other hand, when the value of Q_k is reduced by a scaling factor, the MSE is also reduced. It should be noted that the selected values of these matrices should reflect the actual conditions of uncertainty and/or noise present in the system and should not deviate much from the base values previously determined [16]. The tuned values of these matrices are chosen following the selection criteria reported in the literature [12, 16]: By increasing the value of R_k, the estimation algorithm weighs more heavily the state prediction with respect to the system measurements, which is equivalent to assuming large noise in the measurements. Selecting a larger value of R_k makes the estimator more robust to measurement noise. However, if chosen too large, the filter underestimates the information provided by the measurements, thus reducing the filter's feedback from the system [17]. On the other hand, by reducing the value of Q_k, the algorithm assumes that there is little uncertainty in the machine state space model, thus it relies heavily in the state predicted by the machine model and ignores the measurements [16]. If chosen too small, the estimator can become overconfident of its state prediction, and end up ignoring the system measurements. As a result of the tuning process, it was obtained that the two state estimators have an adequate per-formance by setting the value of $R_k = (0.1)^2 I_{4\times4}$ and a scaling factor of 0.1 that multiplies the base values of matrix Q_k.

6 Results

The generator dynamic response is simulated under a fault event in the external electric grid. The fault event considered is shown in Fig. 2, where the system response to a three-phase-to-ground fault at the sending end of the transmission line is studied. Initially, it is assumed that the generator is in steady-state. At $t = t_f$, a short-circuit-to-ground occurs at point "F". The fault is cleared by simultaneously opening the circuit breakers of the faulted line. The stability of the generator response depends largely on the fault-clearing time (t_c) [4]. The initial operating condition, expressed in pu on a 100 MVA/13.8 kV base, is [6, 14]: $P_t = 1.0$, $Q_t = 0.5724$, $V_t = 1.095\angle11.59°$, and $V_B = 1.0\angle0°$. The initial state value (x_0) can be determined from the given initial operating condition. The values of the fundamental and standard parameters of the generator are specified in Table 1. The state estimation results using the data and measurements generated by the OMIB system simulation model are presented for the analyzed fault scenario and using the tuned values of R_k and Q_k. The generator response to the fault scenario and the estimation results obtained using the two state estimators are shown in Fig. 4. The fault occurs at $t_f = 10$ s and the clearing time is

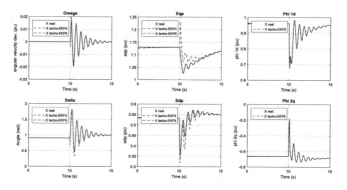

Fig. 4. State estimation

$t_c = 0.11$ s. Prior to the fault event, both state estimators converge to the real value of the machine steady-state. After the fault occurs, the estimation produced by the EKF_4Ord block is not able to adequately follow the evolution of the real state during the first cycles of the machine response transient period. This is due to the rapid change of the machine state variables that is triggered by the short-circuit fault in the trans-mission network. In contrast, the estimated state of the EKF_6Ord block adequately follows the evolution of the real state, during this same time period. After few seconds posterior to the fault-clearing, both state estimators converge to the new value of the machine steady-state.

The estimation error is shown in Fig. 5. The EKF theoretical performance is described by the estimation error covariance matrix, P_k [16]. Under steady-state, the estimation error of both estimators converge to values close to zero and are within the limits of EKF theoretical performance. However, when the fault occurs, the EKF_4Ord estimation error suddenly changes and it doesn't remain within the filter's theoretical limits during the first cycles after the fault. Conversely, the EKF_6Ord estimation error remains within the filter's theoretical limits, even during the fault event.

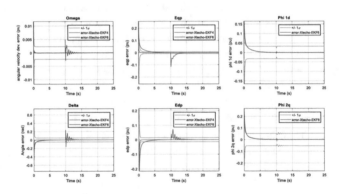

Fig. 5.   State estimation error

7   Conclusions

This work was focused on the dynamic state estimation of a synchronous generator using the measurements available from a PMU that is assumed to be installed at the machine terminal bus. Under this approach, the state estimation problem formulation can be decoupled from the external electrical network, i.e. without the need for measurements/parameters of the power system. The EKF was applied in order to develop two state estimators that are based on two different dynamic models of the synchronous machine. The first one, denoted as "EKF_4Ord", is based on the Two-Axis transient model (fourth order), and the second one, denoted as "EKF_6Ord", is based on the subtransient GENROU model (sixth order). The performance of the EKF greatly depends on the values of the matrices R_k and Q_k. The methodology for tuning the filter consist of selecting base values of these parameters based on the prior knowledge of the system under study. Then, these values are systematically varied by performing several simulations of the ensemble "OMIB System-Estimator" until achieving an adequate performance of the state estimator. Based on the simulation results, it is concluded that both state estimators are able to adequately estimate the electromechanical state variables of the synchronous generator. In steady-state, the estimation produced by both estimators converges to the real value of the machine state. In both cases, the estimation error converges to a value near zero within the EKF theoretical limits. Better dynamic state estimation results were obtained when using the EKF_6Ord estimator block, especially during the first cycles of the generator dynamic response to a severe fault in the transmission network. This is because the GENROU model includes the machine subtransient dynamics (used in the EKF_6Ord), while the Two-Axis model (used in the EKF_4Ord) is not able to capture the subtransient phenomena of the generator dynamic response. Therefore, it is concluded that the sixth order subtransient GENROU model is adequate to be incorporated on the dynamic state estimation problem in power systems.

References

1. Zhou, N., Elbert, S., Huang, Z., Wang, S., Meng, D., Diao, R.: Capturing Dynamics in the Power Grid: Formulation of Dynamic State Estimation through Data Assimilation. Richland (2014)
2. Huang, Z., et al.: Dynamic paradigm for future power grid operation. IFAC Proc. **45**(21), 218–223 (2012)
3. Ghahremani, E., Kamwa, I.: Online state estimation of a synchronous generator using unscented Kalman filter from phasor measurements units. IEEE Trans. Energy Convers. **26**(4), 1099–1108 (2011)
4. Kundur, P.: Power System Stability and Control. McGraw Hill, Nueva York (1994)
5. IEEE Power Engineering Society. IEEE guide for synchronous generator modeling practices and applications in power system stability analyses (2003)
6. Overbye, T.: Lecture 10: Reduced Order and Commercial Machine Models. University of Illinois, Illinois (2014)
7. Ghahremani, E., Kamwa, I.: Dynamic state estimation in power system by applying the extended Kalman filter with unknown inputs to phasor measurements. IEEE Trans. Power Syst. **26**(4), 2556–2566 (2011)
8. Ghahremani, E., Kamwa, I.: Local and wide-area PMU-based decentralized dynamic state estimation in multi-machine power systems. IEEE Trans. Power Syst. **31**(1), 547–562 (2016)
9. Zhou, N., Meng, D., Lu, S.: Estimation of the dynamic states of synchronous machines using an extended particle filter. IEEE Trans. Power Syst. **28**(4), 4152–4161 (2013)
10. Zhou, N., Meng, D., Huang, Z., Welch, G.: Dynamic state estimation of a synchronous machine using PMU data: a comparative study. IEEE Trans. Smart Grid **6**(1), 450–460 (2015)
11. Simon, D.: Optimal State Estimation. Wiley, Hoboken (2006)
12. Welch, G., Bishop, G.: An Introduction to the Kalman Filter. Carolina del Norte (2001)
13. Kundur, P., Dandeno, P.: Implementation of advanced generator models into power system stability programs. IEEE Trans. Power Appar. Syst. **PAS-102**(7), 2047–2054 (1983)
14. Overbye, T.: Lecture 11: Commercial Machine Models and Exciters. University of Illinois, Illinois (2014)

15. Weber, J.: Description of Machine Models GENROU, GENSAL, GENTPF and GENTPJ (2015)
16. Labbe, R.: Kalman and Bayesian Filters in Python (2018)
17. Schneider, R., Georgakis, C.: How To NOT make the extended Kalman filter fail. Ind. Eng. Chem. Res. **52**(9), 3354–3362 (2013)
18. IEEE Power and Energy Society. IEEE Standard for Synchrophasor Measurements for Power Systems (2011)
19. Akhlaghi, S., Zhou, N.: Adaptive multi-step prediction based EKF to power system dynamic state estimation. In: 2017 IEEE Power and Energy Conference at Illinois (PECI), pp. 1–8 (2017)

Automation of a Universal Testing Machine for Measuring Mechanical Properties in Textile Fibers

Liliana Topón-Visarrea$^{(\boxtimes)}$, Mireya Zapata,
and Bernardo Vallejo Mancero

Research Center in Mechatronics and Interactive Systems, Universidad
Tecnológica Indoamérica, Machala y Sabanilla, Quito, Ecuador
blancatopon@uti.edu.ec

Abstract. The purpose of this paper is to design a control system to perform tensile and compression tests on textile fibers and sponges in a universal testing machine. In order to design the control system, a supervisory and front panel was implemented to enter the parameters used to develop the tensile and compression tests. Control and force circuits were designed to control the motor and acquire data from the position sensors and load cell. The communication with the computer was established by an Atmega 16 microcontroller through a data acquisition device. In addition, using the LabVIEW software, a human machine interface was designed to monitor and control the parameters that intervene in the process. Finally, in order to validate the results, comparison tests were developed in a different machine with similar characteristics in order to test the reliability of the presented prototype.

Keywords: Universal testing machine · Tensile test · Compression test · Textile mechanical properties

1 Introduction

The textile industry is primarily concerned with the production of textile fibers, clothes and footwear. The textile sector is important due to the fact that these are first products of necessity [2]. In this context, the breaking of the thread is the biggest cause of machine stoppage. Therefore, properly manufacturing textile fibers and stringent quality control process could vastly improve the productivity of this industry. Increasingly, the textile industry has more and more customer demands because of the quality of the products. Consequently, the fibers that make up a textile garment are subjected to tests and analysis to determine clotting resistance and other physical and chemicals properties of the material [3]. A Universal Testing Machine (UTM) is a tool used to perform tensile and compression tests to measure mechanical properties of textile fibers [6]. In order to preserve the required properties of the tested material, scaled test tubes are used. Some of the evaluated properties are strength, elongation, tenacity and deformation. The UTM aims to test the resistance of various materials. This is due to a system that applies controlled loads on a sample in order to obtain parameters like deformation and force applied at the moment of rupture [1, 7].

© Springer Nature Switzerland AG 2020
M. Botto-Tobar et al. (Eds.): ICAETT 2019, AISC 1067, pp. 189–199, 2020.
https://doi.org/10.1007/978-3-030-32033-1_18

The presented work presents the automation of a UTM T5000 analog model, which was obsolete and out of service. The desire to repower and fix this machine, improving the monitoring and supervision system through a computer controller is based on its importance within the textile industry. A sustainable design strategy was used, which consists in the re-engineering of a machine that was in the process of being scrapped. With this upgrate, waste was avoided, and the machine's life was extended to 10 years. However, with an adequate maintenance plan the UTM T5000 lifetime could extend. The upgrade performed involved the following tasks: signal conditioning of a strain gauge, redesign of speed control, printed circuits, design of supervision front panel, and human machine interface to monitor variables and export data. To validate the results, a comparison was made with a LLoyd Istrument LF plus equipment with similar characteristics where an error of 2.99% in maximum force and 2.87% in maximum deformation were obtained.

In the following section, a description of the Hardware and Software System Design is presented, followed by the presentation of the design implemented along with tensile and compression tests in sample of nylon and polyester thread. Section 4 include the discussion. Finally, Sect. 5 provides conclusions and future work.

2   System Design

The UTM performs tests to determine the properties of various textile materials, and it is designed to evaluate the deformation of these materials when they are subjected to traction and compression efforts. It has many mechanical elements, and from all of them, in this work we hardware redesign and create the required software to control the UTM through an HMI.

The presented prototype has two operating modes: local and interface. The local mode allows for the input of the configuration data from the front panel, and the numerical results are displayed on a liquid crystal display (LCD). The interface mode employ a HMI that permit the user to enter the test configuration parameters. The results are displaying in numerical and graphic form for tensile and cyclic compression testing.

2.1   Hardware Design

The Fig. 1 shows the hardware architecture of the system and the design and implementation considerations are detailed below.

UTM T5000: As is shown in the picture, it has endless screws, a fixed crosshead and a mobile crosshead, crosshead position filament, position sensor output, upper and lower limit switches, load cell, DC motor, and optical sensor. To carry out a tests, a sample of textile fiber must be attached between a fixed plate and a mobile crosshead in order to measure the tensile stress through a load cell that can also operate in compression. Additionally, the system makes measurements through a position sensor that shows the place where the crosshead is located. The movement of the crosshead is done by two parallel endless screws that rotate by means of a motor and fastening bands.

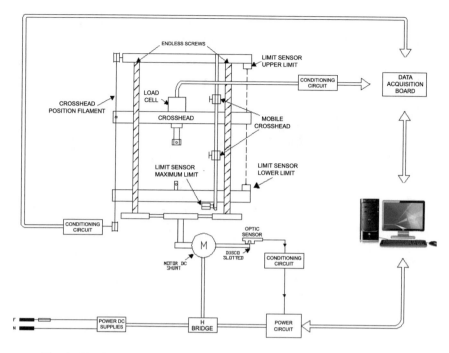

Fig. 1. Hardware architecture - block diagram of the implemented prototype

Power Circuit: It has two parts constituted by the AC-DC power supply (field DC bus and armature) and the motor control circuit created by a H bride.

– Power supply: As shown in Fig. 2, the resistors that make up the voltage divider of the comparator circuit were dimensioned to disconnect the capacitor load resistance. In the implementation, an LM324 operational amplifier, a 2N3904 transistor and a 12 V relay are used. This last element is the one that handles the change of the load and discharge resistance. The power supply output provides the voltages of 5 and 12 V required for this prototype.

– Motor control circuit: The UTM motor speed control is performed by an H bridge, which has four MOSFETS IRF740. In order to have the correct dimensions, it is important to consider current, voltage, and frequency of switching, as well as four opto-couplers 6N136 that acts as insulation elements to avoid damage produced by the power circuit. An ATMega16 microcontroller generates PWM signals to control the DC motor through the H bridge and establishing communication with the HMI using RS232 protocol. The prototype has a keyboard and a liquid crystal display (LCD) that allows the user to enter configuration options as is illustrated in Fig. 3. Keyboard operation is detected by the external interruption INT1 of the microcontroller.

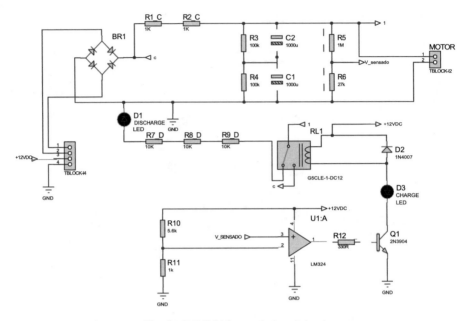

Fig. 2. DC field bus and armor circuit

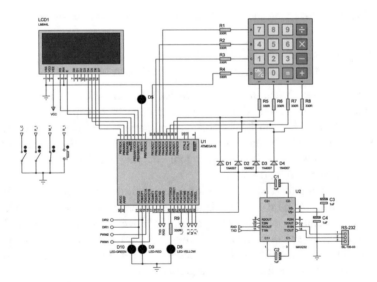

Fig. 3. Motor control circuit

Load Cell: This element must transform a force signal into a voltage signal that is acquired by a data acquisition device and processed in the interface. The capacity of the load cell is 100 N, with a variation range of 0–60 mΩ, and a nominal resistance of 120 Ω with a sensitivity of +/– 7%. The amplification of the load cell signal is carried out by an instrumentation amplifier.

The equation used to relate the applied force and the thread resistance is:

$$R = \frac{\rho}{A * L_I} \left(\frac{F}{Y * A} + 1 \right) \tag{1}$$

Where:

- R = thread resistance
- ρ = thread material resistivity
- A = thread cross section area
- L_I = initial thread length
- F = strength applied to the metal surface
- Y = Young's module

Position Sensor: Linear transducers measure position variations through electrical resistance fluctuations, and usually by means of a moving contact that travels along with a resistive material as indicated in Fig. 1. The equivalent circuit is similar to a potentiometer, whose moving part changes the resistance that varies according to its position. The linear position transducer has a stable signal, so no conditioning is necessary. The measured resistance can be translated in terms of length when it is treated as voltage, which later is digitized by a data acquisition device and computed processed by the HMI.

Data Acquisition Board: In order to acquire data from the position sensor and the load cell a National Instruments MyDAQ data acquisition [5] device is used. In order to reduce noise during data acquisition, signals from the sensors were conditioned.

2.2 Software Design

For this prototype, BASCOM Software Compiler was used to program the ATmega16 microcontroller and, a graphic interface was developed in LabVIEW [4] to build the HMI that supervises, controls, and acquires data from the UTM T5000.

Microcontroller Program: The program has been developed in the microcontroller ATMega16 which acquires data from the motor encoder, generates PWM signals to control the DC motor, drives the keyboard of the front panel, detect the state of ends of stroke, buttons and switches; sends and receives data from the PC and allows visualization of current situation of the testing machine.

The program is arranged with subroutines, providing an efficient structure divided into tasks defined according to the requirements of the system. Figure 4 shows a flowchart with the program implemented. The main loop initiates asking if the working mode will be Local from the machine interface, or from the computer. Once the working mode is selected, the parameters are entered, and the separation between jaws is verified in order to initiate the test. Once the machine is calibrated the test begins and data is collected, allowing to re-initiate the machine to perform a new test.

HMI Design: The interface has been designed on LabVIEW graphical programming platform. For this purpose signals that intervene in the tensile and compression tests are monitoring and controlling by this software. Besides, the HMI is a tool that allows the operator to communicate with the UTM T5000 obtaining numerical and graphic results from different tests.

3   Implementation and Results

Implementation results of the UTM T5000 automation is presented in Fig. 5. The analog panel was replaced by a Digital panel which contain the liquid cristal display, a keyboard, an emergency stop button and the communication port to connect the UTM T5000 with the HMI application running in a PC.

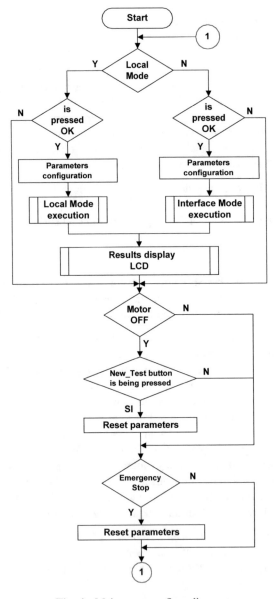

Fig. 4. Main program flow diagram

(a) Original UTM T5000

(b) Upgraded UTM T5000

Fig. 5. In (a) the image shows the original analog panel with manual knobs to control the speed and cell capacity in a range of 0.5 and 10 N. It does not have a results display panel. (b) It has a keyboard for data entry, screen, indicator lights and supports communication with PC to process and display data.

The system allows to test directly from the interface, but first the user must do an automatic calibration. Once completed, the test type must be configured to perform: tensile or compression. The configuration parameters required by each one are listed in Table 1.

Tensile Test. As a functionality proof of the upgraded UTM T5000, a tensile test was made using two different materials, nylon and polyester. The initial characteristics of both thread are presented in Table 2.

Obtained results are presented in the HMI in two screens. One shows the numerical values (see Fig. 6) and, the other one illustrate the graphic functions of strength versus elongation and, stress versus deformation (see Fig. 7).

In the strenght vs. elongation function, it can be observed a maximum peak of 41 N corresponding to the breaking force when the thread had an elongation of 53.438 mm. In the stress vs. deformation function, a maximum deformation of 21.375% along with a break tenacity of 50.58 cN/tex was achieved.

Table 1. User parameters for tensile and compression testing

Parameter	Unit
Tensile Test	
Initial thread length	mm
Speed	mm/min
Thread title[a]	tex[b]
Compression Test	
Normal maximum position	1 mm of minimum separation
Speed test	mm/min
Number of cycles	n

[a]Thread title: Measure of the linear density of a thread.
[b]Tex: unit of linear density, equal to the mass in grams of 1000 m of fiber or thread

Table 2. Initial characteristics of thread

Thread type	Linear Low Density	Initial length	Test speed
Nylon	79.58 tex	49.4 mm	30 mm/min
Polyester	81.11 tex	250 mm	30 mm/min

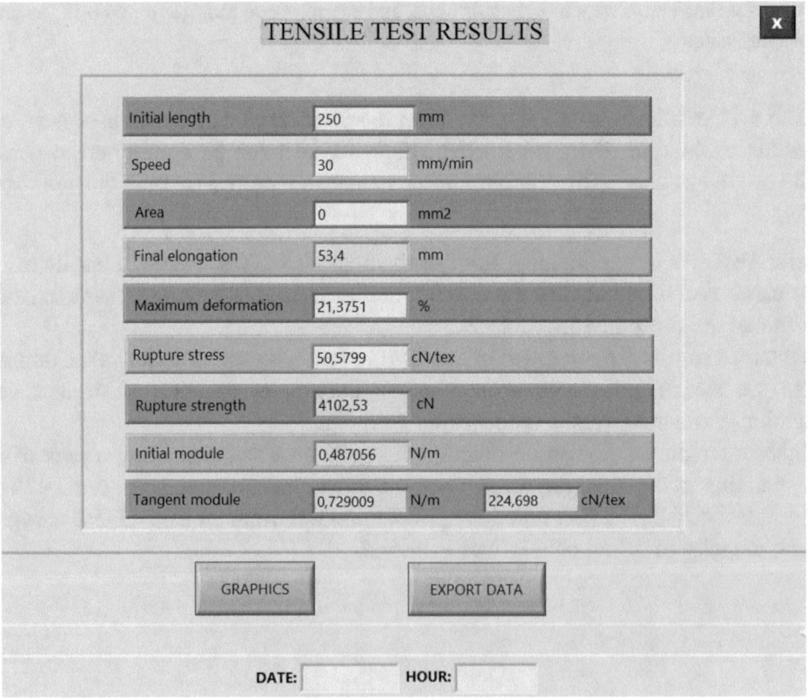

Fig. 6. HMI - Tensile test numerical results

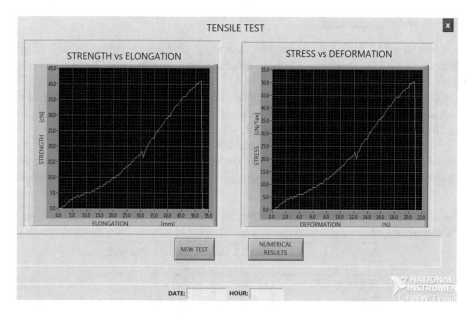

Fig. 7. HMI - Tensile test, graphic of Strength vs Elongation and Stress vs Deformation

Compression Test. In this case a 5 mm thickness synthetic sponge was used. The result of this test can be seen in Fig. 8. In the illustrated case, 3 compression cycles were performed on the material to determine the elongation and maximum force applied in each case.

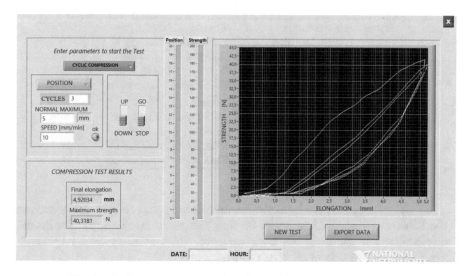

Fig. 8. Cyclic compression test in a 5 mm thickness synthetic sponge

3.1   Result Comparison

In order to probe the reliability of the presented prototype, a comparison with Lloyd Instrument LF UTM with similar characteristics which are detailed in Table 3 has been made.

Using the same type of thread samples, and under the same initial conditions of length and speed, a tensile tests were carried out on the Lloyd Instrument LF plus machine. Results are depicted in Table 4.

With the purpose of calculating the relative error, Lloyd data are considered as real values, and data obtained with the UTM are taken as measured values. The obtained relative error is 2.99% for maximum strength and 2.87% for maximum deformation.

Table 3. Technical specifications comparison between Testing Machine T5000 and Lloyd Instrument LF plus

Assessment	UTM T5000	Lloyd Instrument LF plus
Displacement	1100 mm	750 mm
Speed	10–500 mm/min	0.05–1270 mm/min
Engine DC voltage	100 N	1 kN
Load cell capacity	120 Ω	–
Load cell sensitivity	+/– 7%	<0.5%
Power supply	220 V	220 V

Table 4. Tensile Test Results

Type	Material properties	T5000	Lloyd Instrument LF plus
Nylon	Initial length	49.4 mm	49.4 mm
	End elongation	55.97 mm	54.36 m
	Maximum deformation	113.299%	110.040%
	Break force	44.37 N	43.09 N
	Break tenacity	55.76 cN/tex	54.15 cN/tex
Polyester	Initial length	250 mm	250 mm
	End elongation	53.438 mm	51.948 mm
	Maximum deformation	21.375%	20.779%
	Break force	4102.534 cN	3983.429 cN
	Break tenacity	50.58 cN/tex	49.11 cN/tex

The main source of error is due to the difference in load cell sensitivity between UTM T5000 and the Lloyd Instrument LF plus. The precision in the quantification data also contributes to the error. As a consequence, the parameters obtained with the implemented control system are considered adequate.

4 Discussion

From the experimentally achieved results obtained with the upgraded UTM T5000, it can be highlighted that it allows to perform traction tests and cyclic and normal compression for textile fibers.

A similar work presented in Torres [6] presents a recovered Acco Riehle Universal Testing Machine with a software application which allows to perform tensile and compression test, but only display the strength vs elongation function. In our case is possible to get both strength vs deformation and stress vs deformation functions. Although the work of Torres incorporates the functionality of recording trials with a video camera, this feature can be easily incorporated into our system. In addition, the use of a more precise load cell would improve the performance of our prototype.

5 Conclusions

The system of the universal testing machine has been improved using reengineering providing an important contribution that allows the user obtain the resulting data in a computer and process them later. The new software implemented is efficient, easy to handle, and allows the results to be obtained reliably and quickly, which helps the textile industry to carry out the necessary tests to study the manufactured textile fibers.

The results obtained with the implemented control system are satisfactory. A comparison has been made with another machine of similar characteristics, obtaining an error of 2.99% in maximum force and 2.87% in maximum deformation.

Currently, the 100 N load cell is used, limiting the tests to certain materials. In this case, it is recommended to acquire load cells of greater capacity for materials that require greater strength. Finally, to improve the presented prototype, it can be incorporated a remote communication to send data to a mobile application.

References

1. Asimbaya, D., Topon, B.: Diseño e implementación de un sistema de adquisición de datos para una maquina de ensayos universales para el centro textil de la Escuela Politécnica Nacional, Universidad Politécnica Nacional (2012)
2. Dolez, P.I., Vermeersch, O., Izquierdo, V.: Advanced Characterization and Testing of Textiles. Elsevier Science, San Diego (2017)
3. Huerta, E., Corona, J.E., et al.: Universal testing machine for mechanical properties of thin materials. Rev. Mex. Física. **56**(4), 317–322 (2010)
4. Lazaro, A.: LabVIEW, Programación gráfica para el control de instrumentación (2000)
5. National Instruments. http://www.ni.com/es-cr/shop/select/mydaq-student-data_acquisition-device. Accessed 18 Mar 2019
6. Torres, J.: Repairing and Automation of a Universal Testing Machine, vol 30, no. 2, pp. 171–179 (2009)
7. Zhang, S.H., Yan, M., et al.: Impact of high strain rate deformation on the mechanical behavior, fracture mechanisms and anisotropic response of 2060 Al-Cu- Li alloy. J. Adv. Res. **18**, 19–37 (2019)

Analytical Calculation of the Relations of Profile *w/h* and *s/h* of a Micro-coplanar Stripline with a Dielectric and a Ground Plan for the Static Case

Nayade Vanessa Domenech Polo$^{(\boxtimes)}$ ⓘ,
Wilmer Fabian Albarracin Guarochico ⓘ,
and Yoandry Rivero Padron ⓘ

Universidad Tecnológica Israel, Francisco Pizarro E4-142 y Marieta de
Veintimilla, Quito, Ecuador
{ndomenech, walbarracin, yriverop}@uisrael.edu.ec

Abstract. This paper presents the analytical calculation of the profile relations of a micro-coplanar stripline with a dielectric and a ground plane for the static case. The device consists of a ground plane separated by a dielectric substrate of two conductive strips (tapes) of different lengths. One of the conductors as well as the ground plane is considered to be of infinite lengths. The analysis and synthesis of the device is carried out by means of a physical-mathematical-analytical model based on conformal transformations, on the theory of a complex variable, on the use of electric and magnetic walls, on electromagnetic theory and on a computational numerical calculation. The union of physical, mathematical and computational tools allows the analysis and synthesis of thin film devices and guarantees solutions of the exact analytical type. Three transformations based on the Schwarz-Christoffel transformation were carried out, which made it possible to find a transformation plane to find the profile relations of the Micro-Coplanar Stripline and to obtain the limit for the case of a typical microline.

Keywords: Electrical parameters · Micro-coplanar-stripline · Profile relations

1 Introduction

Radio Frequency Integrated Circuits (RFICs) and Microwave Integrated Circuits (MICs), both hybrid and monolithic, have advanced rapidly over the past two decades. This progress has been achieved not only by the advancement of solid state device technology, but also by the advancement in the study of flat transmission lines.

Among the advances achieved is the development of various methods of analysis for passive structures of radio frequencies, microwaves, millimeter waves and flat transmission lines [1]. These methods have played a very important role in providing electrical parameters of the flat transmission lines for the design of RFICs, as well as in their research and development.

© Springer Nature Switzerland AG 2020
M. Botto-Tobar et al. (Eds.): ICAETT 2019, AISC 1067, pp. 200–210, 2020.
https://doi.org/10.1007/978-3-030-32033-1_19

1.1   Thin Films

At present, the technological trend in the area of telecommunications and very high-speed computers is to model and manufacture thin film devices in flat technology (which uses all the capacity developed so far for the manufacture of integrated circuits). Thus, thin films of superconductors, semiconductors, conductors and insulators (or dielectrics) can be deposited on a substrate or base that can be each of these materials.

The main problem for the manufacture of guided systems in flat technology (integrated circuits) is that, worldwide, it has not been possible to use an analytical physical-mathematical model that can accurately predict the behavior of thin films in terms of the transmission of the electromagnetic field, which does not solve the interference problems presented in these circuits, when they work above 3 GHz.

Among the models used for the analysis of these devices, specifically the flat transmission lines, we have the semi-empirical models, models of approximation to a classic waveguide and matrix relaxation models, among which is mentioned the method of moments of Galerkin [2].

Most of them provide approximate solutions for the static case and in the best of cases they provide solutions with good accuracy but still with certain limitations, among which is the use of many equations as a function of many variables, which implies a great dependence on computational equipment with a fairly high processing capacity, to obtain only some electrical parameters. It is important to note that none of the models mentioned above provide exact solutions to flat transmission lines for time varying fields.

1.2   Micro-coplanar Stripline

The Stripline Micro-Coplanar with a dielectric and a ground plane is a device consisting of two conductors, separated by an S opening, on a dielectric substrate and below these a ground plane. The dielectric substrate is considered in this case the vacuum. The two upper conductors are of different widths; one of the conductive strips has a width of W, while the other is of semi-infinite width and the ground plane is considered of infinite dimensions. The case to study is the static case where losses are not considered.

The micro-coplanar stripline and its cross section are shown in Fig. 1.

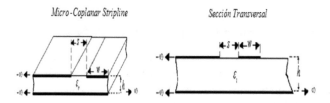

Fig. 1. Micro-coplanar stripline and its cross section.

The method used for the analysis of this device is an analytical physical-mathematical model based on conformal transformations, on the theory of a complex variable, on the use of electrical and magnetic walls, on electromagnetic theory and on a computational numerical calculation, generally used to relax or sediment transcendental equations [2]. With this model it is possible to find a solution of the analytical and exact type, so as to be able to know with great accuracy the behaviour of the electric field and the other electrical parameters of this device, such as the density of surface charge, the electric potential, the characteristic capacitance and the relations of profile w/h and s/h, where s is the separation between the two superior conductors and h is the separation between the two superior conductors and the inferior one.

All this in function of the dimensions of the device, which allows an easy manufacture in the future, placing the model used in an advantageous position with respect to the methods previously mentioned. To know in an exact and analytical way the electrical parameters of these we have the possibility to manufacture circuits and systems RFICs and MICs more exact.

2 Usefulness of Flat Transmission Lines

Currently the technological trend in the area of telecommunications and high-speed computers is to model and manufacture thin film devices with flat technology. Thus, thin films of superconductors, semiconductors, conductors and insulators (or dielectrics) can be deposited on a substrate or base that can be each of these materials [3].

In the specific case of conductors deposited by any method of deposition of the metal on a base substrate, such as a dielectric, we obtain the well-known Microstriplines, Striplines, Coplanar Waveguide (CPW) among others, with all its derivatives or modifications.

The uses of the latter range from the manufacture of components for microwave integrated circuits (hybrid and monolithic) and for flat antennas (antennas of very small dimensions compared to the parabolic antenna, with which it is sought to replace the latter).

2.1 Boundary Conditions for Electric and Magnetic Fields

The boundary conditions for electric and magnetic fields depend on the existing materials on both sides of the interface, these are determined, as is well known, by properly applying the Maxwell equations in an integral manner.

2.2 Boundary Conditions in a Perfect Conductor (Electrical Walls)

Many problems in microwave engineering deal with borders with good conductors (metals), which can often be assumed without loss ($\sigma \to \infty$). In cases of a perfect conductor all field components must be zero within the conductive region. This result can be obtained by considering that a conductor with finite conductivity ($\sigma < \infty$) and that the skin depth, i.e. the depth at which most microwaves penetrate, goes to zero when ($\sigma \to \infty$) [1]. If it is also assumed here that $\underset{Ms=}{\to} 0$, which could be the case if the perfect conductor does not possess magnetic properties $\mu = \mu_0$, then the Maxwell equations are reduced to: $\hat{n}.\bar{D} = \rho_s$, $\hat{n}.\bar{B} = 0$, $\hat{n} \times \bar{E} = 0$ and $\hat{n} \times \bar{D} = \bar{j}_s$. Being ρ_s and $\vec{J}_s$ the surface charge density and the surface current density, respectively, at the interface; $\hat{n}$ it is the unitary pointing vector coming out of the perfect conductor. This boundary is known as an Electric Wall, the components of $\vec{E}$ which vanish on the surface of the conduit.

2.3 Boundary Conditions in a Perfect Conductor (Magnetic Walls)

The boundary condition for a Magnetic Wall occurs when the tangential components $\vec{H}$ must fade away. Such a boundary does not really exist in practice, but can be approximated by a corrugated surface, or in certain problems of flat transmission lines [1]. The Maxwell equations are written as: $\hat{n}.\bar{D} = 0$, $\hat{n}.\bar{B} = 0$, $\hat{n} \times \bar{E} = -\bar{M}_s$ and $\hat{n} \times \bar{H} = 0$, where $\hat{n}$ it is the unit vector of Pointing coming out of the region of the Magnetic Wall.

3 Schwarz-Christoffel Transformation

The most commonly used conformal transformation in the analysis of planar transmission lines is the Schwarz-Christoffel transformation, which transforms the axis w_x (real axis in the plane W) and the upper semi-plane of the plane, W, in the perimeter of a closed polygon and its interior in the plane Z respectively and vice versa [1] (Fig. 2).

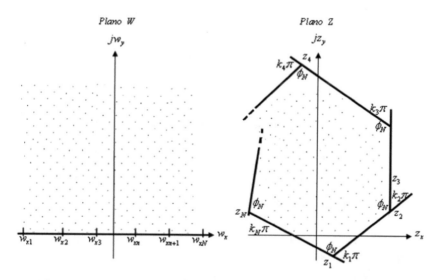

Fig. 2. Schawrz-Christoffel transformation from the W plane to the Z plane.

This is represented by the following equation:

$$z = A \int_0^w \prod_{n=1}^N (w - w_{xn})^{-k_i} dw + B \tag{1}$$

where A and B are complex constants representing amplifications (expansion or contraction) and translations (when the point $w_{xn} = 0$, does not correspond to the point $z_N = 0$) of the polygon in the plane Z [4]. Being $k_i = 0, \frac{1}{2}, 1, \frac{3}{2} o 2$.

3.1   Calculation of the Relations of Profile w/h Y s/h

The following conditions are assumed:

- The strips are perfect conductors ($\sigma \to \infty$).
- The conductive strips are infinitesimally flat.
- Two of the conductive strips have a semi-infinite length (flat to ground and upper left conductor).
- The conductive strips have semi-infinite depth.
- The interfaces between the dielectric and the air, except where the strips reside, represent a perfect magnetic surface.

Now doing an analysis of the lines of electric and magnetic flux the magnetic and electric walls are located, which are represented by the dotted lines and the thick lines respectively, with all these considerations and to facilitate their analysis the device was conveniently divided in three zones shown in detail in Fig. 3.

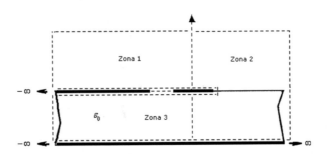

Fig. 3. Zone division of the Stripline Micro-coplanar.

The electrical and magnetic walls (conductors and dotted lines respectively) allow us to obtain three closed polygons. In each of them the Schwarz-Christoffel transformation is applied in the analysis, as will be shown below.

Zone 1 Analysis and Synthesis
By performing the Schwarz-Christoffel transformation to the closed polygon of zone 1, the transformation plane w in figure is obtained (Fig. 4).

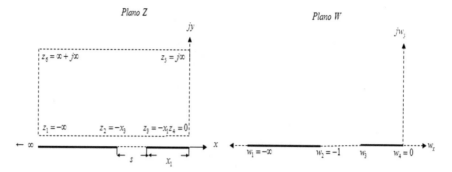

Fig. 4. Physical plane Z and transformation plane W, zone 1.

Analyzing this transformed plane, the dimensions shown in Table 1 are obtained.

Table 1. Dimensions zone 1.

x
$x_1 = -2jA_{w'}\sqrt{1 - w_3'}$
$x_3 = -2jA_{w'}$

Analysis and Synthesis of Zone 2

In the same way a closed polygon is obtained for zone 2. By performing a procedure similar to that performed for zone 1, the polygon for this case is shown in Fig. 5. The same procedure as [2] is used for this zone.

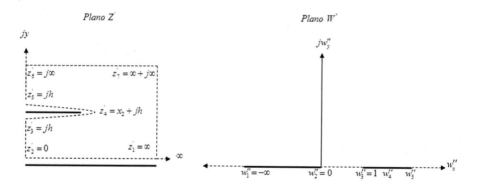

Fig. 5. Plane Z' y plane W', zone 2.

The following table shows the relations obtained for Zone 2 of the constants x_2 and h respectively (Table 2).

Table 2. Zone 2 dimensions.

Constantes: Zona 2
$x_2 = -2A_{w''}t_5Z(\varphi'/\alpha')$
$h = -\dfrac{A_{w''}t_5\pi}{K(\alpha')}$

Relating the expressions obtained in zones 1 and 2, you have to:

$$\frac{x_2}{h} = \frac{2Z\left(\frac{\varphi'}{\alpha'}\right)K(\alpha')}{\pi} \tag{2}$$

Analysis and Synthesis of Zone 3

By applying the Schwarz-Christoffel transformation to the closed polygon in zone 3, the transformation plan is obtained W'', Fig. 6.

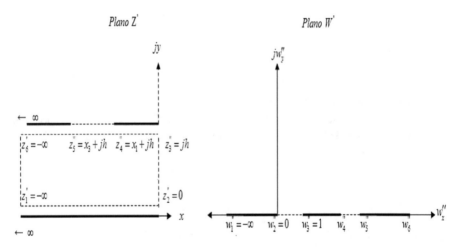

Fig. 6. Plane Z' and transformed plane W", zone 3

Now using a bilinear transformation to the transformed plane w''' (similar to the previous one), to take the point $w_5'''\to\infty$ and join the point w_6''' with w_1''' obtaining the next transformed plane w''' (Fig. 7).

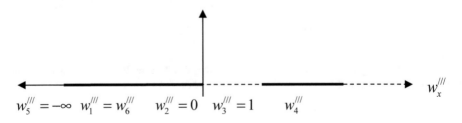

Fig. 7. Transformed plane W''', zone 3.

$$Plano\ W'''$$

$$jw_y'''$$

The following table shows the dimensions found for zone 3 (Table 3).

Table 3. Zone 3 dimensions.

Constants: Zone 3

$$h = \frac{A_3\pi}{\sqrt{\left(|w_1'''|^2 + |w_1'''|\right)}}$$

$$x_1 = \frac{A_3}{\sqrt{|w_1'''|^2 + |w_1'''|}} \ln\left(\frac{(2|w_1'''|+1)-2\sqrt{\left(|w_1'''|^2 + |w_1'''|\right)}}{2\sqrt{\left(|w_1'''|^2 + |w_1'''|\right)\left(1 - (2|w_1'''|+1)w_4''' + \left(|w_1'''|^2 + |w_1'''|\right)(w_4''')^2\right)} -2\left(|w_1'''|^2 + |w_1'''|\right)w_4''' + (2|w_1'''|+1)}\right)$$

The relationship x_1/h of Zone 3 is given by the following is given by the following expression:

$$\frac{x_1}{h} = \frac{\ln\left(\dfrac{(2|w_1'''|+1)-2\sqrt{\left(|w_1'''|^2 + |w_1'''|\right)}}{\sqrt[2]{\left(|w_1'''|^2 + |w_1'''|\right)\left(1 - (2|w_1'''|+1)w_4''' + \left(|w_1'''|^2 + |w_1'''|\right)(w_4''')^2\right)} -2\left(|w_1'''|^2 + |w_1'''|\right)w_4''' + (2|w_1'''|+1)}\right)}{\pi} \tag{3}$$

Finally, you determine the profile relationships w/h and s/h using the relationships (2) and (3) found above. This relationship is given by:

$$\frac{w}{h} = \frac{x_1 + x_2}{h} \quad y \quad \frac{s}{h} = \frac{x_3 - x_1}{h} \tag{4}$$

$$\frac{w}{h} = \frac{\ln\left(\dfrac{(2|w_1'''|+1)-2\sqrt{\left(|w_1'''|^2 + |w_1'''|\right)}}{\sqrt[2]{\left(|w_1'''|^2 + |w_1'''|\right)\left(1 - (2|w_1'''|+1)w_4''' + \left(|w_1'''|^2 + |w_1'''|\right)(w_4''')^2\right)} -2\left(|w_1'''|^2 + |w_1'''|\right)w_4''' + (2|w_1'''|+1)}\right) + 2Z(\varphi'/\alpha')K(\alpha')}{\pi}$$

$$\tag{5}$$

$$\frac{s}{h} = \frac{\ln\left(\dfrac{(2|w_1'''|+1)-2\sqrt{\left(|w_1'''|^2+|w_1'''|\right)}}{\sqrt[2]{\left(|w_1'''|^2+|w_1'''|\right)\left(1-(2|w_1'''|+1)w_4'''+\left(|w_1''|^2+|w_1'''|\right)(w_4''')^2\right)} -2\left(|w_1'''|^2+|w_1'''|\right)w_4'''+(2|w_1'''|+1)}\right) + \left(\dfrac{1}{\sqrt{1-w_3'}}-1\right)}{\pi}$$

(6)

4 Checking the Results Obtained

The expression obtained for the profile ratio w/gives congruent results since, when making the limit when $|W_1'''| \to \infty$ we have to $|W_1'''^2| = |W_1'''|$, comparando este resultado con otros trabajos realizados [2], comparing this result with other works carried out [2], obtaining a profile ratio w/h for a typical microline. This proves, once again, that the model provides analytical and exact solutions to any type of device placing the model in the first alternative when analyzing thin film devices.

Below will be shown the $|W_1'''| \to \infty$ limit of the profile ratio obtained for a Stripline Micro-Coplanar with a dielectric and a ground plane. At the limit the device is a typical Microstripline.

$$\frac{w}{h} = \frac{2Z(\varphi'/\alpha')K\alpha'}{\pi}$$

(7)

5 Conclusions

Luego de la revisión bibliográfica correspondiente al tema, se procedió hacer un análisis detallado de las paredes eléctricas y magnéticas de una Micro-Coplanar Stripline con un dieléctrico y un plano a tierra, para el caso estático. Obtuvieron las ecuaciones que representan las relaciones de perfil de una Micro-Coplanar Stripline, se obtuvieron las relaciones de perfil s/h y w/h, en las cuales s representa la abertura entre el conductor finito y el semi-infinito, w es la dimensión del conductor finito y h es la separación entre los planos superiores e inferior respectivamente.

Se compara los resultados obtenidos, para una Micro-Coplanar Stripline con un dieléctrico y un plano a tierra con la Microstripline, al hacer el límite cuando $w_1''' \to \infty$ se encontró la solución para la microlínea típica mencionada, como era de esperarse.

After the bibliographic review corresponding to the subject, a detailed analysis of the electric and magnetic walls of a Stripline Micro-Coplanar with a dielectric and a ground plane was carried out, for the static case. The equations representing the profile relations of a Stripline Micro-Coplanar were obtained, the profile relations s/h and w/h were obtained, in which it represents the aperture between the finite conductor and the semi-infinite, w is the dimension of the finite conductor and h is the separation between the upper and lower planes respectively.

The results obtained, for a Stripline Micro-Coplanar with a dielectric and a ground plane with the Microstripline, are compared by making the limit when the solution was found for the typical microline mentioned, as expected.

By knowing exactly and analytically the dimensions of the flat transmission lines we have the possibility to manufacture more accurate circuits and RFICs and MICs systems.

This model also gives solution to other planar devices providing advantages with respect to other models already mentioned, since in the model a strong mathematical model is made, in a beginning, in terms of functions and integral elliptic, obtaining, in the end, solutions in function of few variables. With this, we propose and give solutions of the analytical type and we leave the soft work to the computer, being able to use personal computers. While the other models leave the hard work to computers with high processing capacity giving approximate solutions depending on many variables.

References

1. Nguyen, C.: Análisis Methods for RF, Microwave, and Millimeter-Wave Planar Transmission Line Structures, 2nd edn. Wiley, New York (2001)
2. Galo, A.: Dispositivos de Líneas Planares de Alta Frecuencia. IVIC (1981)
3. Galo, A.: Unpublished results. (Actualmente conforman una serie de trabajos realizados por el profesor Agusto Galo en el LIT)
4. Plonsey, R., Collin, R.: Principles and Applications of Electromagnetic Fields, 4th edn. McGraw-Hill Book Company, New York (1979)
5. Di Paolo, F.: Networks and Devices Using Planar Transmisión Lines, 1st edn. CRC Press LLC (2000)
6. Rainee, S.: Coplanar Waveguide Circuits, Components, and Systems, 1st edn. Wiley, New York (2001)
7. Misra, D.: Radio-Frecuencia and Comunication Circuits: Análisis and Design, 1st edn. Wiley, New York (2001)
8. Collin, R.E.: Antennas and Radiowave Propagation, 1st edn. McGraw-Hill Book Company, New York (1985)
9. Abramowitz, M., Stegun, I.: Handbook of Mathematical Functions, 5th edn. Dover Publications, New York (1965)
10. Bahl, I., Bhartia, P.: Microwave Solid State Circuit Design, 2nd edn. Wiley, New York (2004)
11. Callarotti, R., Galo, A.: IEEE Trans. MMT **16**, 1021–1027 (1984)
12. Collin, R.E.: Foundations for Microwave Engineering, 2nd edn. IEEE Press, Piscataway (2001)
13. Galo, A.: CIEN **1**, 93–104 (1993)
14. Galo, A.: Nuevo modelo de dispersión para una microcinta abierta utilizando el modo de propagación LSE. CIEN., dic. 2004, vol. 12, no. 4, pp. 290–297. ISSN 1315-2076
15. Gradshteyn, R.: Table of Integrals Series and Products. Jeffrey, A. (ed.) 5th edn. (2002)
16. Golio, M.: The RF and Microwave Handbook, 1st edn. CRC Press LLC (2001)

17. Henrici, P.: Applied and Computational Complex Analysis, 1st edn. Wiley, New York (1974)
18. Pozar, D.: Microwave Engineering, 2nd edn. Wiley, New York (1998)
19. Reinmut, K.: Handbook of Microwave Integrated Circuits, 1st edn. Artech House, Inc., Boston (1987)
20. Wadell, B.: Transmission Line Design Handbook, 1st edn. Artech House, Inc., Boston (1991)
21. Wunsch, D.: Complex Variables with Applications, 3rd edn. Pearson Addison Wesley, Boston (2005)

e-Business

Organizational Effectiveness and Tools of eCollaboration Antecedents: An Empirical Exam in Ecuador's Automotive Sector

Vicente Merchán[1,2]([⊠]) [iD] and Danny Zambrano[1] [iD]

[1] Universidad de las Fuerzas Armadas ESPE, Sangolquí, Ecuador
{vrmerchan, dizambrano}@espe.edu.ec
[2] Otavalo University, Otavalo, Ecuador
vmerchan@uotavalo.edu.ec

Abstract. Day by day, the automotive sector in Ecuador is looking for ways to incorporate electronic collaboration tools (eCollaboration) into their planning processes. There is a belief, in general, that good organizational performance is due to investments in eCollaboration technologies. Under this point of view, there is little local empirical research that supports assumptions about the factors that determine effectiveness in the IT organization performance. As consequence, this study analyzes the criteria of a framework that has been provided by researchers in eCollaboration. This grid based on time/space collaboration, was developed at the beginning of the 1990s and classified as valid up to today. A descriptive, correlational and regression research was carried out to test the hypothesis based on 2016 data provided by 47 companies in the automotive sector. The results finally show that Income and Continuous Task have a simple marginally significant relationship and that Remote Interactions have the most important influence within the group of variables. In addition, there is an important interpretation of the main findings that are incorporated to the mentioned data through conclusions and recommendations for future work.

Keywords: Association · eCollaboration · Effectiveness · Time/Space

1 Introduction

Since the beginning of the 21st century, collaboration and teamwork, more than ever, have become important issues that have given impetus to the global economy. The heads of companies are increasingly interested in employees collaborating with each other, because there is a positive impact on collaboration, and in their environment and profitability [1]. In this regard, there is little empirical research available that supports assumptions about what factors ensure effectiveness in the performance of an Information Technology (IT) organization. The objective of this study is to fill out this gap.

Laudon and Laudon [2] define collaboration as "working with others to achieve shared and explicit goals. It focuses on tasks or missions and usually takes place in a company or another type of organization, and between one company and another;" in addition, they reflect on its importance for the following reasons: (1) Changing the nature of work; (2) Growth of professional work; (3) Changing the organization of the

© Springer Nature Switzerland AG 2020
M. Botto-Tobar et al. (Eds.): ICAETT 2019, AISC 1067, pp. 213–225, 2020.
https://doi.org/10.1007/978-3-030-32033-1_20

company; (4) Changing the scope of the company; (5) Emphasis on innovation; and, (6) Changing the culture of work and business. In the same line of collaboration, Kock and Nosek [3] define eCollaboration as "the collaboration between individuals involved in a common task that uses electronic technologies." Finally, Bouras, Giannaka and Tsiatsos [4] define eCollaboration as "Collaboration, which is carried out without face-to-face interaction between individuals or members of virtual teams committed to a common task using Information and Communication Technologies ICT." These technologies refer to eCollaboration technologies, of which we can say that they include not only the communication medium created by the technology, but also the particular characteristics that have been designed to support eCollaboration technology which can have a strong effect on how technology is currently being used by a group of people to achieve a certain collaborative task. That is to say, these technologies are characterized by the diversity of tools and uses, which, unlike the classical ones, affect communication and management processes.

In the same order of importance, over time, research has revealed that investments in eCollaboration technologies produce organizational improvements with interesting returns with respect to the amount of investment and, with the greatest benefits for sales functions, marketing and research and development [5–7].

We can presume that business growth has been characterized in an important way by the use of collaboration technologies; however, your choice may also cause a problem that impacts your finances. A framework that has been provided by erudite researchers in eCollaboration is the time/space collaboration matrix that was developed in the early 1990s [8] and whose classification criteria are still valid [9]. Therefore, we address the research question:

PI: To what extent does the number of eCollaboration tools, by qualifying criteria, influence the effective performance of an IT organization in the automotive sector?

If the number of eCollaboration tools affects IT organizations supported by IT from the economic point of view, then it is important to know what kind of technological tools could improve income and to what extent. So, we intend to answer this question by proposing and empirically examining a model that explains the effectiveness of the eCollaboration criteria (see Fig. 1).

In the 2015 report [10], many companies investigated in the automotive sector were categorized as having a better economic position for 2014 due to their perceived income, providing an interesting research space that is made in this document.

The present work has been structured. In Sect. 2, the theoretical background and the study hypothesis are presented. In Sect. 3, the research methodology is presented. In Sect. 4, the results obtained are analyzed. In Sect. 5, the discussions are presented and, in Sect. 6, the conclusions of the study are presented.

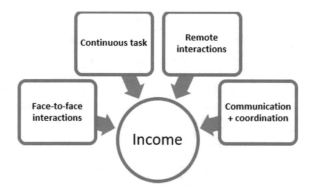

Fig. 1. Research model. Source: author's contribution.

2 Background

Hernández [11] defines collaborative group work as the technology used to communicate, cooperate, coordinate, solve problems, compete and negotiate, in order to
facilitate group work, through computer networks and the services inherent to the same.
On the other hand, Ortega [12] mentions that the term Computer-Supported Cooper-
ative Work (CSCW) was introduced by Grief and Cashman as "a way to describe how
computer technology can help users work together in groups;" in addition, he adds:
"[…] it is the scientific discipline that describes how to develop groupware applications, also having as an objective the theoretical and practical study of how people
work in cooperation and how groupware affects group behavior."

Sosa, Zarco and Postiglioni [13] define "groupware is the software and hardware
that supports and helps group work."

According to Wikipedia [14], collaborative software can be divided into three
categories: collaboration-communication tools, conference tools and collaborative or
group management tools.

With so many available electronic collaboration tools, it is difficult to identify those
that solve specific business problems. However, there is a framework, based on the
time/space collaboration matrix, which has been useful in locating tools and focusing
on the problem in a concrete way [8].

The technological tools and the services inherent in them, such as: electronic mail,
wikis, blogs, electronic conversations, hypertexts, collaborative work discussion
groups, among others; they are categorized in two dimensions of the collaboration
problem: time and space. For example, time is an obstacle to collaboration on a global
scale when groups of people work from different countries with extreme schedules.
Similarly, the space inhibits collaboration in multinational or even national and
regional companies [2].

Using the matrix time/space framework shown in Table 1, it will be more appropriate to identify the eCollaboration tools.

Table 1. Matrix time/space tools of eCollaboration [8].

	Same time (synchronous)	Different time (asynchronous)
Same place (collocation)	*Face-to-face interactions*	*Continuous task*
	Presentation support	Computers shared by different users
	Software for presentations	Use of general automated applications
	Audiovisual equipment	
	Presentation room	
Different place (remote)	*Remote interactions*	*Communication + coordination*
	Videoconferences	Email
	Instant messaging	Wikis
	Computer networks	Blogs
	Multi-user systems	Asynchronous conferences

For Bubenko and Ellis (1991, cited in [13]) a groupware serves to "increase the effectiveness of work in three key levels that support group interaction: communication, collaboration and coordination." The concept shared by Saadoun (1997, cited in [15]) points out that one of the benefits of groupware is to reduce bureaucracy and make the hierarchical structure of the organization "flat" in terms of collaboration, communication, team spirit and reinforces the human interactions in the fulfillment of objectives.

For E. Dyson (cited in [16]) there is a taxonomy based on three variables of the groupware that is the main object of attention:

- Focused on the people: It locally manages the work of each people within a group.
- Focused on the document: The system ensures the management of the tasks assigned to a document.
- Focused on the process: The system controls the conclusion of activities.

Coinciding with the previous taxonomy, although there may be many more, one stands out which is expressed in the appendix of the Delta project of the European Community 7002 [17]. The same one is classified to the main objective of the group's activities:

- Oriented to forms
- Process oriented
- Oriented to communication structure
- Oriented to conversational models

Being these elements those that allow propose a research that seeks to know the relationship between the number of eCollaboration tools, by qualifying criteria (face-to-face interactions, continuous task, remote interactions and Communication + Coordination), and the income received by an IT organization in the automotive sector, according to data provided by Vistazo magazine and reliable sources.

3 Methodology

The research carried out is a descriptive, correlational and regression type; that using the defined framework, seeks to test the research model of Fig. 1. Therefore, the methodology was structured in two steps: (1) Selection of automotive companies to be analyzed; and, (2) Definition of the survey to collect data [18]. Finally, the data is identified and proceeds with the analysis of correlation and regression between variables; subsequently, conclusions and future work in the area are established.

3.1 Selection of Companies in the Automotive Sector

This step constitutes the selection of companies in the automotive sector categorized as having the best economic position in 2014; and, their activity is the commercialization or automotive assembly. As illustrated in the special edition No. 1153 of September 25, 2015 of the magazine Vistazo of Ecuador [10]; the information observed was delivered by the same company, the superintendence of companies and/or the internal revenue service. However, the income that was used for the analysis corresponds to the year 2016.

The reason for having selected these companies is the assumption that they have eCollaboration tools whose characteristics allow them to maintain a good rhythm of productivity and contribute to the high reported income. Therefore, the technical structure of the companies surveyed is shown in Table 2.

Table 2. Technical structure of the companies responded.

Variable	Feature
Identified	60
Contacted	50
Surveyed	47
Companies without polling	3
Overall completion rate	78.33%

3.2 Survey

The survey developed followed the guidelines of [18] and the theoretical background of the time/space matrix. According to Hernández Sampieri et al. [19], a survey is constituted in a technique of opinion (survey) useful in processes of non-experimental transversal or transeccional descriptive or correlational causal research.

Before operating the survey, an understanding analysis was conducted with two research professors and eighteen undergraduate students, as a result of which the clarity and consistency of the terminology adopted in each question is confirmed.

Survio's website[1] was used to operate the survey, through which it was possible to consolidate the opinion data of IT leaders. Then, with IBM SPSS Statistics 22 [20] and R language [21], the data was processed for the respective regression analysis.

---

[1] (www.survio.com/survey/d/examinacion)

4   Results

4.1   Descriptive Analysis

Table 3 shows that 47 companies were surveyed, of which 40 (85.11%) correspond to automotive marketing companies and 7 (14.89%) to the automotive industry.

Table 3. Companies of the automotive sector.

Activity	Registered	Recorded
Industry	8 (13.33%)	7 (14.89%)
Commercial	50 (83.33%)	40 (85.11%)
Services	2 (03.33%)	0 (00.00%)
TOTAL	60 (100.00%)	47 (100.00%)

Table 4. Surveyed by gender.

Gender	Answers	Percentage
Male	41	82.23%
Female	6	12.77%
TOTAL	47	100.00%

In Table 4 it is observed that 41 (87.23%) people represent the male gender and 6 (12.77%) people the female gender.

Figure 2 shows that 46 (97.87%) companies believe that they have e-mail tools and audiovisual equipment; 45 (95.74%) companies have instant messaging tools; 44 (93.62%) companies use general-purpose automated applications; 38 (80.85%) have software for presentations; 37 (78.72%) companies have tools in the category of computer networks and videoconferences; 36 (76.60%) companies in the software category for presentations; 35 (74.47%) companies in the category of group calendars and presentation support; 33 (70.21%) companies in the category of multi-user systems; 25 (53.19%) companies in the category of asynchronous conferences; 19 (40.43%) companies in the category of computers shared by different users; 17 (36.17%) companies in the category of workflow and systems for group decision making; 11 (23.40%) companies in the category of discussion forums; 9 (19,15%) companies in the category of blogs; and, 1 (2.13%) company in the category of wikis.

The previous figure shows that companies in the sector have implemented an important number of eCollaboration tools.

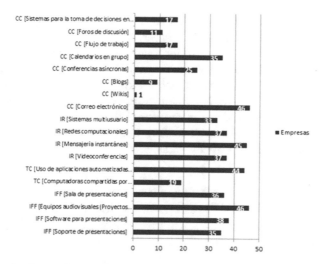

Fig. 2. Companies by eCollaboration tool. Source: author's contribution.

4.2   Correlation and Regression Analysis

The Kendall Tau Test is used because the variable Income is Non-Parametric according to the Shapiro-Wilk normality test. In addition, the data set is small and many scores are at the same level [21].

Based on the opinion expressed by each of the IT leaders of the 47 companies surveyed, the aim is to determine the influence of the number of eCollaboration tools available on the economic income generated by companies in the automotive sector.

The proposed multiple regression conceptual model is shown in Eq. (1).

$$Y_t = \beta_0 + \beta_1 X_1 + \beta_2 X_2 + \beta_3 X_3 + \beta_4 X_4 + \varepsilon \tag{1}$$

The value of β_j corresponds to the constant linear coefficients of independent variables. Y_t corresponds to the Economic Income of 2016. X_1, X_2, X_3 and X_4 correspond to the Face-to-Face Interaction Tools, Continuous Task, Remote Interactions and Communication Tools + coordination, respectively. Finally, ε corresponds to the estimation error.

In Table 5, as can be seen, the significance values obtained allow to accept the research test.

Table 5. Test of normality Shapiro-Wilk.

Variable	Test statistic	Significance
Income	0.605	8.957e-10
Face-to-face interactions	0.927	0.005957
Continuous task	0.826	6.486e-06
Remote interactions	0.930	0.007959
Communication + coordination	0.898	0.0006187

In Table 6, as illustrated, a significance value of less than 0.05 is proof of the existence of a close linear relationship between the variables. If we look at the Income column, we see that there is a marginally significant simple relation between Income and Continuous Task. In addition, there is a close correlation between the face-to-face interaction and remote interaction variables, significant relationships between communication + coordination and both face-to-face interaction, continuous task and remote interactions; which makes the variance of the estimators of the coefficients of the three variables (Face-to-Face Interaction, Continuous Task and Remote Interactions) is high if included as predictor variables.

Table 6. Correlation.

		Income	Face-to-face interactions	Continuous task	Remote interactions	Communication + coordination
Correlation	Income	1.000				
	Face-to-face interactions	.104	1.000			
	Continuous task	.287	.288	1.000		
	Remote interactions	.205	.488	.284	1.000	
	Communication + coordination	.221	.302	.407	.415	1.000
Significance (a tail)	Income	.				
	Face-to-face interactions	.242				
	Continuous task	.025	.025	.		
	Remote interactions	.083	.000	.026	.	
	Communication + coordination	.068	.019	.002	.002	.

Table 7 shows the summary of the model, where the R-square value of 0.2494 is observed, which means that the variations in the independent variables explain a 24.94% of the variation of income for the sample of selected companies.

Table 7. Summary of the model.

Concept	Value
Multiple correlation coefficient	0.499
Coefficient of determination R^2	0.249
R^2 adjusted	0.178
Typical error	88.27
Observations	47

Strategically, as the typical error decreases, the R squared increases, which constitutes the coefficient of determination between the variables.

Table 8 shows ANOVA, where it is observed that the critical distribution value of F is 2.61 and the calculated distribution value of F is 3.489. Since 3.489 is in the null hypothesis rejection zone, it is concluded that the influence of the number of eCollaboration tools available on the economic income generated by the selected automotive companies is different. That is, each variable has the ability to explain the variation that income can suffer.

Table 8. ANOVA.

	Degrees of freedom	Squares	Half quadratic	F	Critical of F
Regression	4	108754.562	27188.640	3.48	0.015
Residues	42	327259.862	7791.901		
Total	46	436014.425			

The column with the heading "Coefficients" of Table 9 gives the Eq. (2) of multiple regression.

$$Y_t = 12.393 + 11.312X_1 + 35.398X_2 - 35.181X_3 + 16.689X_4 + \varepsilon \qquad (2)$$

Table 9 shows that critical value of the test statistic follows the distribution t whose two-tailed test value is ± 2.021. The calculated values t of the independent variables are found in both regions of the hypothesis. The calculated t ratio is -2.937 and a calculated significance level lower than the critical level of 0.05 for the Remote Interactions. This value is in the rejection region. In conclusion, the regression coefficient for the variable Remote Interactions is not equal to zero, so it is a significant predictor for Income. The calculated t ratio is 1.174 for Face-to-Face Interactions, 1.778 for Continuous Task, and 1.634 for Communication + Coordination. These three variables are less than critical t and are in the region of non-rejection of null hypothesis, consequently, they are not significant predictors and can be eliminated from this analysis if required. In addition, the values of significance obtained corroborate with the reasons t. Therefore, not all the values obtained can be generalized to all companies in the sector. In other words, this analysis would not support the argument that revenues will increase if the number of tools is increased at the level of Face-to-Face Interactions, Continuous Task and Communication + Coordination. This conclusion is only true in a model that includes these four prediction variables.

Table 9. Regression coefficients.

	Coefficients	Typical error	Coefficients Beta	t	Sig.
Constant	12.393	42.433		0.292	0.772
Face-to-face interactions	11.312	9.6326	0.183	1.174	0.247
Continuous task	35.398	19.910	0.266	1.778	0.083
Remote interactions	−35.181	11.980	−0.476	− 2.937	0.005
Communication + coordination	16.689	10.213	0.255	1.634	0.110

Figure 3 shows that a high positive atypical case is observed when the number of Face-to-Face Interactions is approximately 3. In addition, it is observed that most of the tools are concentrated between 2 and 4 and can produce an effect on the coefficient of the slope of the line of regression. Otherwise, no exceptional pattern is observed.

Figure 4 shows that no exceptional pattern is observed, except for the high positive atypical case when the Continuous Task is 3. In addition, it is observed that the concentration of tools is between 1 and 2, mainly.

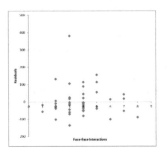
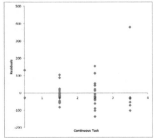

Fig. 3. Residuals in relation to Face-to-Face Interactions.

Fig. 4. Residuals in relation to Continuous Task.

Figure 5 shows that the residue pattern does not suggest an alternative to the linear relationship, except for the high positive atypical case when the number of Remote Interactions is 0. In addition, there is a concentration of tools between 1 and 4. Finally, in Fig. 6 the residues are shown in relation to Communication + Coordination. Again, no exceptional pattern is observed, except for the high positive atypical case when the Communication + Coordination is 4 and, a large concentration of tools between 2 and 4.

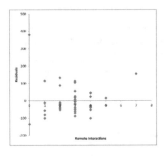
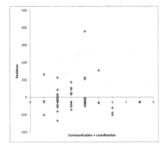

Fig. 5. Residuals in relation to Remote Interactions.

Fig. 6. Residuals in relation to Communication + Coordination.

5 Discussion

The study examines the tools of eCollaboration that are implemented in 47 companies of the automotive sector of Ecuador, and their influence on income for the year 2016, through a reference framework matrix time/space that has been operationalized in the leaders of IT of the companies surveyed. The research instrument reveals demographic and classificatory information. In the matrix of regression coefficients, the estimators of the coefficients and their typical errors are indicated. The t statistics of the coefficients of face-to-face interactions, continuous task and communication + coordination, in an absolute way, do not influence as predictors in the income perceived by the IT organizations. However, due to the marginal values obtained, the continuous task coefficient can be considered as a variable that has some influence as a predictor, but its effect is not as strong as that of an absolutely predictive variable. With this we see that there is a margin to extract predictor variables. The variable of remote interactions stands out as a significant negative predictor for revenue. This implies that employees or managers who are in different parts of the planet collaborate in the solution or attention of problems at the same time making their participation extraordinary with the least number of efficient tools that influence the annual income of the companies. In this way, the companies studied should reasonably consider the implementation of an adequate number of Remote Interactions tools as strategic support in the activities carried out by their personnel because the hypothesis that the number of eCollaboration tools does not influence the income of an IT organization can't be rejected.

6 Conclusion

The study determines the impact on the adoption of a certain number of eCollaboration tools by IT organizations in the automotive sector of Ecuador and that influence the income received in 2016 through the time/space matrix. The results are conclusive in the sense that not all the variables contribute significantly in the prediction model and only the variable of remote interactions and, to some extent, the continuous task variable, influence the income of the automotive companies included in this study. Therefore, based on the regression analysis carried out, it is concluded that there is no absolute proof that the number of eCollaboration tools, by classifying criteria as a whole, influences the income received by the IT organizations in the automotive sector. This conclusion is only true in a model that includes these four variables. Note also the value generated by the variable remote interactions in the sector sample of study. Finally, future studies could analyze other sectoral activities, as well as, types of variables such as profit, profitability by sales or number of employees and measure the impact of eCollaboration tools in all formally registered sector companies; being able to contribute to the strengthening of the predictive value of the multiple regression model.

References

1. Fernández-Cardador, P., Agudo-Peregrina, Á.F., Hernández-García, Á.: Antecedents of collaborative behavior in companies: an analysis of the use of corporate blogs. Dirección y Organización **48**, 5–10 (2012)
2. Laudon, KC., Laudon, J.P.: Sistemas de Información Gerencial, Décima ed., Cruz Castillo, L.M. (ed.), p. 736. Pearson educación, México (2012)
3. Kock, N., Nosek, J.: Expanding the boundaries of e-collaboration. IEEE Trans. Prof. Commun. **48**(1), 1–9 (2005)
4. Bouras, C., Giannaka, E., Tsiatsos, T.: e-Collaboration Concepts Systems and Applications. In: E-Collaboration: Concepts, Methodologies, Tools, and Applications, p. 1890. IGI Global (2009)
5. Frost & Sullivan, Meetings Around the World II: Charting the Course of Advanced Collaboration, Verizon & Cisco (2016). http://www.verizonenterprise.com/resources/whitepapers/wp_meetings-around-the-world-ii_en_xg.pdf. Accessed 25 Nov 2016
6. Meroño, A., Sabater, R.: Análisis de los efectos del empleo de tecnologías colaborativas: el caso del e-mail. Murcia (2005)
7. Aral, S., Brynjolfsson, E., Van Alstyne, M.: Productivity Effects of Information Diffusion in Networks (2007)
8. Johansen, R., Sibbet, D., Benson, S., Martin, A., Mittman, R., Saffo, P.: Leading Business Teams: How Teams Can Use Technology and Group Process Tools to Enhance Performance (1991)
9. Schauer, B., Zeiller, M., Riedl, D.: Reviewing the e-collaboration marketplace: a survey of electronic collaboration systems. In: Proceedings of the 2010 International Conference on e-Business (ICE-B) (2010)
10. Editores Nacionales S.A. «500 mayores empresas del Ecuador» Vistazo, no. 1154 (2015)
11. Hernández Arias, A.: «La tecnología de trabajo colaborativo en el contexto universitario» . www.ucla.edu.ve/dac. Accessed 21 July 2017
12. Ortega, M., Bravo, J.: «Trabajo cooperativo con ordenador» Universidad de Castilla-La Mancha. http://aipo.es/libro/pdf/13Cooper.pdf. Accessed 17 May 2017
13. Sosa, M., Zarco, R., Postiglioni, A.: «Modelando aspectos de grupo en entornos colaborativos para» Revista de Informática Educativa y Medios Audiovisuales, vol. 3, no. 7, pp. 22–31 (2006)
14. Groupware, «wikipedia» Enciclopedia libre. https://es.wikipedia.org/wiki/Software_colaborativo#Definición. Accessed 18 Oct 2017
15. Salinas, S.A.: «Software para trabajo colaborativo y bibliotecas» http://eprints.rclis.org/14721/1/Software_para_Trabajo_Colaborativo_y_Bibliotecas1.pdf. Accessed 12 Mar 2017
16. Saadoun, M.: El proyecto groupware: de las técnicas de dirección a la elección de la aplicación groupware. gestión 2000, Barcelona (1997)
17. Ortega, M., Bravo, J., Prieto, M., De Lara, J.: Groupware y Educación (1997)
18. Pfleeger, S.L., Kitchenham, B.A.: Principles of survey research part 1: turning lemons into lemonade. ACM SIGSOFT Softw. Eng. Notes **26**(6), 18–20 (2001)
19. Hernández Sampieri, R., Fernández Collado, C., Baptista Lucio, P.: Metodología de la Investigación, Quinta edición ed., Mcgraw-Hill/ Interamericana Editores S.A. de C.V., México D.F., pp. 299–350 (2010)

20. IBM. «Guía breve de IBM SPSS Statistics 22» (2013). ftp://public.dhe.ibm.com/software/analytics/spss/documentation/statistics/22.0/es/client/Manuals/IBM_SPSS_Statistics_Brief_Guide.pdf. Accessed 11 Nov 2015
21. Ferre Jaén, M.E.: «Modelos de regresión» . www.um.es/ae/FEIR/40/#correlacion-lineal-simple. Accessed 10 Dec 2017

Communication

Evaluation of the Implementation of Li-Fi in the Communes of Santiago

Álvaro González[1] , Juan C. Castillo[2] , Raúl Carrasco[3](✉) ,
Carolina Lagos[4] , Eduardo Viera[5] , Leonardo Banguera[6] ,
and Gabriel Astudillo[7,8]

[1] Departamento de Ingeniería Industrial, Universidad de Santiago de Chile,
Santiago, Chile
[2] Universidad Tecnológica de Chile INACAP, Rancagua, Chile
[3] Facultad de Ingeniería, Ciencia y Tecnología, Universidad Bernardo O'Higgins,
Santiago, Chile
raul.carrasco.a@usach.cl
[4] Pontificia Universidad Católica de Valparaíso, Valparaíso, Chile
[5] Departamento de Ingeniería Eléctrica, Universidad de Santiago de Chile,
Santiago, Chile
[6] Facultad de Ingeniería Industrial, Universidad de Guayaquil, Guayaquil, Ecuador
[7] Escuela de Ingeniería Civil Informática, Universidad de Valparaíso,
Valparaíso, Chile
[8] Departamento de Ingeniería Informática, Universidad de Santiago de Chile,
Santiago, Chile

Abstract. This work seeks to implement the functioning of a network communication system, this is done through visible light in the urban sectors of the metropolitan region. While it should be known that, Li-Fi (Light Fidelity) is a wireless communication system that uses visible light as a means of data propagation, becoming an alternative to wireless radio frequency systems. The multicriteria method called Elimination et Choix Traduisant la Réalité (ELECTRE). An extended version of ELECTRE I method will be used to obtain a reliable result, since it is possible to obtain a decision with several alternatives, deriving a single solution from the analysis. It must be taken into account that the election process is about the correct place to implement Li-Fi connections, taking into account as a dependent variable places with a better standard of living in the Metropolitan Region, and random variables will be about areas of interest such as; physical, climatological, among others. Likewise, it must be borne in mind that the advantages, disadvantages and hierarchy are evaluated in order of preference.

Keywords: ELECTRE I · Li-Fi · VLC

1 Introduction

There are several Multicriteria decision making methods used mostly as Analytic Hierarchy Process (AHP) [1–3], the Analytic Network Process (ANP) [4–6],

© Springer Nature Switzerland AG 2020
M. Botto-Tobar et al. (Eds.): ICAETT 2019, AISC 1067, pp. 229–237, 2020.
https://doi.org/10.1007/978-3-030-32033-1_21

ANP-DEMATEL (Decision Making Trial and Evaluation Laboratory) [7,8], ELECTRE I [9,10] among others. ELECTRE methods are based on the concepts of external relations, concordances and disagreements. ELECTRE I is designed to solve complex problems, which help decision making, with a small set of preferable alternatives to finally choose the most preferable one among them [11].

The form of connection to get internet connection has been changing from connection with cables to wireless formats, which are even mobile [12,13]. The Wi-Fi (Wireless Fidelity) is a type of connections that has been maintained over time, with several changes or updates. Recently, we have come to find more efficient and effective ways to obtain a connection to the network considering a new type of Li-Fi link. It is a wireless network link that uses a light source instead of radio frequency to perform data transfer also called VLC (Visible Light Communication) [14].

The assembly by visible light consists of a means of propagation of data in the frequency ranges [450–900 nm], therefore VLC is classified as a set of wireless communication technologies. That operates in conjunction with a solid-state lighting system (SSL, Solid State Lighting). In other words, ordinary LED lights will be used with the aim of functioning as a Router, that is, as a modulating emitter [15]. The paper is organized as follows: The second section describes the ecosystem of the metropolitan region, in the third section the description of the problem to be solved appears, in the fourth the analysis of results and finally the Conclusions.

2   Ecosystem of the Metropolitan Region

The product of the Census executed in the year 2017, reveals that the Chilean population registered is of a total of 17,574,003 people. Considering that of them, 8,601,989 (48.9%) are men and 8,972,014 (51.1%), women. The number of registered households was of 6,499,355, considering that 6,486,533 (99.8%) belong to private homes and 12,822 (0.2%), to collectives or others. The above reflects that between 2002 (last census prior to 2017) and 2017, the average annual growth rate of the population was 1.06%. In as much, that the object of study the Metropolitan Region counts on the greater amount of population, with 7.112.808 people (40.5% of the total) [16].

The Metropolitan Region has 52 communes, the majority being urban and only 18 rural. In turn, the Metropolitan Region is constituted by six provinces; Santiago, Cordillera, Talagante, Maipo, Chacabuco and Melipilla.

Describing the geography as data that can be considered in the variables is the relief, within the Metropolitan Region there are three compositions: the Cordillera de los Andes, the Intermediate Depression and the Cordillera de la Costa [17].

The Cordillera de los Andes, in this region, is high and robust, and the volcanic activity appears in turn with the presence of volcanic cones that give the highest heights of this mountain range. Among them we can name the Tupungato

(6,570 m above sea level), Marmolejo (6,108 m above sea level), Nevados del Plomo (6,050 m above sea level), Nevado de Los Piuquenes (6,017 m above sea level) and San José (5,856 m above sea level).

In as much it is precise to emphasize the description of the original composition of the geography of the Metropolitan Region first it is had; The Intermediate Depression, called in turn, as the Santiago basin has its corresponding limits according to the cardinal points: in the northern zone, the Chacabuco cord is located; to the east, it is the Cordillera de los Andes; to the south, the narrowness of Paine is located, and while finally to the west, is the Cordillera de la Costa.

Although it is said that the Santiago basin has an approximate length of 80 km in the north-south route, and 35 km in width, considering the east-west direction. As a general average considering mainly the urban area, it has an average height of 520 (meters above sea level). Meanwhile, related to the Cordillera de la Costa, it is the boundary between the Metropolitan Region and the Valparaso Region, and its geographical genotype is presented as an extended relief that limits the west with the Santiago basin. It is widely said that the region includes heights that exceed 2,000 (meters above sea level), among the most outstanding; the cord of the hills of El Roble, La Campana, Vizcachas, El Roble Alto and the high places of Chicauma and Lipangue. Meanwhile, the main hydrographic source of the Metropolitan Region is the Maipo river, which is of mixed origin, because its waters originate because of winter precipitation and melting. In its physical structure, it is formed by a basin that drains about 15,380 km^2 and its average flow is 92,3 m^3/s.

The climate has a volatile formation of precipitations, that in his generality happen in the stations of autumn and winter, with an annual average of 384 millimetres. In both the warmer seasons such as summer is a drier environment, with temperatures that in some situations are greater than 30 °C (degrees Celsius). Taking into account average data, the average annual temperature is close to 14 °C, taking into account, in turn, that it contains a winter average of 9 °C, and a mean in summer of 22.7 °C [18].

3   Description of the Problem to Be Solved

Trying to look at new technologies and advances, looking for sustainability, we need to provide a closer look to generate a future implementation for this study. Related to the assembly of new techniques to execute connections of networks, estimating or rather considering variables and according to how it could behave in different atmospheres. The Metropolitan Region is an essential part of the country, concentrating more people, the economy and the development of the country, while the following study will take characteristics of specific areas within this region, among which, among others, the variables related to physical characteristics (geographical, climatological, etc.). Thanks to this study of multiple variables a final result will be obtained that will provide the best location among different sectors to the lesson within the region already mentioned [19].

3.1   Use of the Method

– Criteria
  - C1 = Solar Radiation $Kws/m^2/day$
  - C2 = Wind Speed m/s
  - C3 = Height m
  - C4 = Emission of airborne particles $\mu g/m^2 N$
  - C5 = Air Temperature °C
  - C6 = Accumulated Precipitation mm
– Location
  - A1 = Las Condes
  - A2 = Providencia
  - A3 = Lo Barnechea
  - A4 = Santiago
  - A5 = La Florida

To use the ELECTRE method, the first step is to generate the matrix with the applicable criteria and arguments [10], in this case they will be the locations, this was achieved through information collection [20] and then it was summarized, as shown in Table 1 [21].

Table 1. Matrix criteria

	C1	C2	C3	C4	C5	C6
A1	6.25	1.2	681	21	14.6	21.0
A2	6.22	1.1	598	24	14.8	24.0
A3	6.69	1.2	829	20	13.9	20.0
A4	6.27	1.1	534	26	14.9	26.0
A5	6.33	1.0	578	25	14.6	29.2

There are two types of matrix compared in ELECTRE's concordance and discordance. In the case of this study, we will first obtain the agreement matrix, to generate this type of matrix, the rows must be superimposed on each other. An overview will be given of how this matrix was obtained in this case, the weight evaluation is considered, as these should be added and the algebraic composition would be added (w = 2 + 6 + 4 + 2 + 3 + 1) = 18, It should be noted that these weights are the valuation or importance that is considered to each variable.

As an exercise or realization of how to develop said matrix, row A1 is examined with respect to (A2, A3, A4, A5). To be able to specify this comparison, the difference of the arguments between the rows is developed, for example: if we take the matrix A1, A2 and the first serious argument (6.25 − 6.22 = 0.3) which is a variable positive therefore there is no sub-dominance towards A1 and the weight is considered zero bone 0. In the case that exists as it is between A1, A3 (6.25 − 6.69 = −0.44) ie a negative result where A3 dominates and the weight considered for this variable is taken in this specific case would be 2, so it will remain successively until the column C1 is completed, and so on, all the columns of the matrix are completed when this matrix is obtained with the full weights. These weights will be added obtaining a result of each row example in row A1, A2 will be $(2 + 3 + 1 = 6)$ and at the end of evaluating all the rows respect to A1 being this 0, then a new column is generated, the which will be row A1 of the new matrix agreement and this must be done with the respective analysis (A2, A3, A4, A5). Finally, when these are completed and will generate the matrix shown in Table 2:

Table 2. Matrix concordance

	A1	A2	A3	A4	A5
A1	0	6	6	8	5
A2	12	0	2	2	5
A3	6	6	0	8	8
A4	4	2	6	0	6
A5	4	6	6	6	0

The average value of the new antecedents is $p = 4.56$, p is the new constant in which the concordance matrix is evaluated to obtain the final evaluation table. We will take $p = 5$, since p is an ordinal variable. Having the p estimate, the final dominance between the variables will be developed according to their agreement. The conclusive concordance matrix will give the dominance by rows, to obtain it, the values are evaluated, if these are $p \geq 5$, having as result the development shown in Table 3:

Table 3. Matrix final concordance

Alternative	Dominance	Number
A1	2	A3, A5
A2	1	A1
A3	4	A1, A2, A3, A4
A4	4	A1, A2, A3, A5
A5	1	A1

4   Analysis of Results

As stated earlier it is assessed two matrices, so now matrix discordance is generated, for it will normalize the initial matrix. When analysing this matrix, it is observed in a generic way that there is a difference between some variables, thus normalizing the scale can be decreased. In the next study, a reduction scale to 10 will be used, to generate it, they act as positive variables, and the higher a greater range it will be. Only each column should be divided by its largest value and multiplied by 10, the normalized matrix of Table 4 will be generated [22].

Table 4. Normalized matrix

	C1	C2	C3	C4	C5	C6
A1	9.3400	10	8.21470	8.0800	9.8000	7.1920
A2	9.3000	9.2000	7.21350	9.2300	9.9300	8.2190
A3	10	10	10	7.6900	9.3300	6.8490
A4	9.3700	9.2000	6.4415	10	10	8.9040
A5	9.4600	8.3000	6.9723	9.6200	9.8000	10

To obtain the nonconformity matrix, we must look for the difference between the rows as in the case of the matrix agreement column C1 as a reference, "example: it will be; $9.34 - 10 = -0.66$", so on. Obtaining result as explained above and like the case as the matrix agreement is formed. When the matrix is formed, the row is analysed and the largest negative number must be found and this will remain as an absolute value, which will generate a column that will finally form the new row according to the analysed variable. In the end, the results of all these operations are summarized in Table 5 [23].

Table 5. Discordance matrix

	A1	A2	A3	A4	A5
A1	0	1.447	6.000	6.022	3.231
A2	8.308	0	6.000	4.462	7.692
A3	9.913	6.010	0	6.022	7.300
A4	6.486	1.447	2.009	0	3.231
A5	9.221	2.928	2.477	2.735	0

Like the concordance matrix it must be evaluated according to a constant found through the average and in this case, it is called q. The average value is $q = 1.388$, this will be obtained by adding all the numbers of the divided matrix by the sum of the numbers without counting the 0, we will take $q = 1$. To obtain the values $q < 1$ will be searched, but by column, opposite to that of the concordance matrix, visualized in Table 6 [24].

Continuing with the ELECTRE I method, a level of hierarchy must be generated that gives a level of importance to the arguments, thus locating the best

Table 6. Final discordance matrix

Alternative	Dominance	Number
A1	1	A3
A2	2	A4, A5
A3	0	0
A4	2	A2, A5
A5	0	0

option, then it is made known in the summary matrix, this is made up of columns that deliver the attributes, both matrix concordance as nonconformity, which is observed in A° and are the alternatives that occur in the analysis.

While dominance per row has the purpose of teaching the dominance of the matrix agreement, while dominance per column is that which reveals the results of the nonconformity matrix, the column that delivers the dominance differential between the two matrices is also represented. Then ending with Ranking, which represents the ranking of positions, displayed in Table 7.

Table 7. Summary

A°	Dominance per row	Dominance per column	Δ Dominance	ΔN°	Ranking
A1	A2, A3, A4	A3	A2, A4	2	3°
A2	A1	A4, A5	A4, A5	−2	5°
A3	A1, A2, A3, A4	0	A1, A2, A3, A4	4	1°
A4	A3, A5	A2, A5	A2, A3	1	4°
A5	A2, A3, A4	0	A2, A3, A4	3	2°

5   Conclusions

In terms of the results already generated, it can be said that the location A3 represented by the commune of Lo Barnechea has a great dominion over the rest, being the option considered in this variable of analysis of physical and climatological type. Within the result that the analysis yielded, there is a clear domain, giving a unique solution without controversies, this result was thanks to the fact that in this study there are well-defined differences between the variables. With this it is observed that a precise adjustment is found to then carry out the implementation of this type of connection Li-Fi, which also has as a plus to be an additional source of electricity. While the analysis can be mentioned that the second position expressed with the variable A5 (Florida) has dominance in some aspects in reference to physical criteria, but does not exceed all the benefits that can be delivered by the implementation in the commune of what Barnechea. The third position can be an option, since it indicates a positive domain, but it is not significant in front of the results delivered by the other two, being unimportant. While the other two variables can not be considered since their results are not optimal, not even considerable. The study in general tried to find the best option according to the weight of each given criterion, considering discrete variables and see what influences more, proposing the Multicriterio methodology. It is important to know that the metropolitan region is affected by pollution and microclimates that influence decision-making; therefore, the connection factors will depend both on the behavior of the geographical and climatological variables, that is why they were taken as a criterion in this study, and thus find the best location to implement this connection system. As a synthesis within the communes of the metropolitan region, Lo Barnechea is the most optimal stable place according to the present study, so in the near future to be able to specify the execution of the use of Li-Fi, to give greater confidence of its use and legitimacy [25].

Acknowledgments. The authors acknowledge the funding for the investigation to FVF Ingeniería y Consultoría Ltda.

References

1. Saaty, T.L.: The Analytic Hierarchy Process: Planning, Priority Setting, Resource Allocation. McGraw-Hill, New York (1980)
2. Mac-Ginty, R., Carrasco, R., Oddershede, A.M., Vargas, M.: Strategic foresight using an analytic hierarchy process: environmental impact assessment of the electric grid in 2025. Int. J. Anal. Hierarchy Process **5**(2), 186–199 (2014). https://doi.org/10.13033/ijahp.v5i2.195
3. Lagos, C., Fuertes, G., Carrasco, R., Gutierrez, S., Vargas, M., Rodrigues, R.: Facing the data analysis complexity for the energetic efficiency management at great copper mining, in Codelco Chuquicamata, Chile. In: 2016 IEEE International Conference on Automatica (ICA-ACCA), pp. 1–6. IEEE, Curicó, Chile, October 2016. https://doi.org/10.1109/ICA-ACCA.2016.7778482
4. Saaty, T.L.: Time dependent decision-making; dynamic priorities in the AHP/ANP: generalizing from points to functions and from real to complex variables. Math. Comput. Model. **46**(7–8), 860–891 (2007). https://doi.org/10.1016/j.mcm.2007.03.028
5. Blair, A.R., Mandelker, G.N., Saaty, T.L., Whitaker, R.: Forecasting the resurgence of the U.S. economy in 2010: an expert judgment approach. Socio Econ. Plann. Sci. **44**(3), 114–121 (2010). https://doi.org/10.1016/j.seps.2010.03.002
6. Durán, C., Carrasco, R., Sepúlveda, J.M.: Model of decision for the management of technology and risk in a port community. Decis. Sci. Lett. **7**(3), 211–224 (2018). https://doi.org/10.5267/j.dsl.2017.10.002
7. Pamučar, D., Mihajlović, M., Obradović, R., Atanasković, P.: Novel approach to group multi-criteria decision making based on interval rough numbers: hybrid DEMATEL-ANP-MAIRCA model. Expert Syst. Appl. **88**, 58–80 (2017). https://doi.org/10.1016/j.eswa.2017.06.037

8. Durán, C., Sepulveda, J., Carrasco, R.: Determination of technological risk influences in a port system using DEMATEL. Decis. Sci. Lett. **7**(1), 1–12 (2018). https://doi.org/10.5267/j.dsl.2017.5.002

9. Oral, M., Kettani, O.: Modeling outranking process as a mathematical programming problem. Comput. Oper. Res. **17**(4), 411–423 (1990). https://doi.org/10.1016/0305-0548(90)90019-4

10. Gonzalez, A., Carrasco, R., Soto, I., Alfaro, M., Fuentealba, D.: Location evaluation for implementation of VLC, in Chile. In: 2018 7th International Conference on Industrial Technology and Management (ICITM), pp. 275–278. IEEE, Oxford, March 2018. https://doi.org/10.1109/ICITM.2018.8333960

11. Çal, S., Balaman, e.Y.: A novel outranking based multi criteria group decision making methodology integrating ELECTRE and VIKOR under intuitionistic fuzzy environment. Expert Syst. Appl. **119**, 36–50 (2019). https://doi.org/10.1016/j.eswa.2018.10.039

12. Lorenzo, B.: Estudio del Estado del Arte de los sistemas de comunicaciones por luz visible (VLC). Trabajo fin de grado: Grado en ingeniería de las tecnologías de telecomunicación, Universidad de Sevilla (2016)

13. Haruyama, S.: Visible light communications. In: 36th European Conference and Exhibition on Optical Communication, pp. 1–22. IEEE, Torino (2010). https://doi.org/10.1109/ECOC.2010.5621174

14. Tarannum, S.: Data transmission through smart illumination via visible light communication technology. Int. J. Tech. Res. Appl. **4**(2), 136–141 (2016)

15. Sathiya, T., Divya, E., Raja, S.: Visible light communication for wireless data transmission. Int. J. Innovative Res. Electr. Electron. Instrum. Control Eng. **2**(2), 1084–1088 (2014)

16. Unidad de Estadísticas del Medio Ambiente: Medio ambiente: Informe anual 2016. Medio ambiente, Instituto Nacional de Estadisticas, Santiago, Chile (2016)

17. Romero, H., Irarrázaval, F., Opazo, D., Salgado, M., Smith, P.: Climas urbanos y contaminación atmosférica en Santiago de Chile. EURE **36**(109), 35–62 (2010)

18. Soto, M.V., Castro, C.P., Rodolfi, G., Maerker, M., Padilla, R.: Procesos geodinámicos actuales en ambiente de media y baja montaña: Borde meridional de la cuenca del río Maipo, Región Metropolitana de Santiago. Revista de Geografía Norte Grande **35**, 77–95 (2006). https://doi.org/10.4067/S0718-34022006000100006

19. Seremi de Salud Región Metropolitana: Aire: Santiago de Chile indices de calidad del aire (2018)

20. Centro de Ciencia del Clima y la Resiliencia (CR)2: Explorador Climatico (CR)2 (2019)

21. Ministerio de Energía: Explorador Solar (2019)

22. Romero, C.: Análisis de las Decisiones Multicriterio, Publicaciones de Ingeniería de Sistemas, vol. 14. Isdefe, Madrid, España, primera edn (1996)

23. Ruiz, J.: Métodos de decisión multicriterio electre y topsis aplicados a la elección de un dispositivo móvil. Proyecto fin de carrera ingeniería industrial, Universidad de Sevilla, Sevilla, España (2015)

24. Justel, D., Pérez, E., Vidal, R., Gallo, A., Val, E.: Estudio de métodos de selección de conceptos. In: XI Congreso Internacional de Ingeniería de Proyectos, vol. 1, pp. 561–572. Lugo, España (2007)

25. Figueira, J., Mousseau, V., Roy, B.: Electre methods. In: Figueroa, J., Greco, S., Ehrogott, M. (eds.) Multiple Criteria Decision Analysis: State of the Art Surveys, International Series in Operations Research & Management Science, vol. 78, chap. 4, pp. 133–153. Springer, New York (2005)

Crowdsensing and MQTT Protocol: A Real-Time Solution for the Prompt Localization of Kidnapped People

Ana Zambrano[1], M. Eduardo Ortiz[1]([✉]), Marcelo Zambrano Vizuete[2], and Xavier Calderón[1]

[1] Departamento de Telecomunicaciones y Redes de Información, Escuela Politécnica Nacional, 170525 Quito, Ecuador
{ana.zambrnao,eduardo.ortiz, xavier.calderon}@epn.edu.ec
[2] Carrera de Ingeniería en Telecomunicaciones, Universidad Técnica del Norte, 100110 Ibarra, Ecuador
omzambrano@utn.edu.ec

Abstract. One of the biggest problems that Ecuadorian society and Latin American have to confront in general is insecurity, a condition under which anyone can become a victim. This proposal offers a first step towards linking Ecuador within a new model of security management for smart cities. It takes advantage of new Information and Communications Technologies (ICT), including the *Internet of Things* (IoT), and *Cloud Computing*. The main thesis is to develop a real-time communications system that would reduce the time needed to rescue kidnapping victims. The local community would be notified of any kidnapping attempts via their *smartphones* using an opportunistic application and improving the *crowdsensing*. This network uses the *Message Queue Telemetry Transport protocol* (MQTT), which is ideal for IoT applications in which energy consumption is an important factor. The proposed architecture has improved previous works obtaining the reduction of rescue time, achieving 30,3 min in advance to them.

Keywords: Internet of Things · Smart city · Smartphones · Real-time system · Distributed system · Message Queue Telemetry Transport

1 Introduction

A community's rights and needs change according to its location, its culture, and other variables. For example, a person's right to walk freely on the public walkways, or let their children play in the park, is becoming in Latin America more of a need for rather than a right to security. According to the *Centro Internacional para Niños Desaparecidos* (CINDE, International Center for Missing Children), more than 1.8 million children worldwide are victims of sexual exploitation. Although fortunately, the rate for young children is decreasing in Latin America, the problem now falls on girls and young women in the 11–22 years old group, according to the founder of *Asociación de Niños Robados y Desaparecidos* (Association of Stolen and Missing Children) [1].

© Springer Nature Switzerland AG 2020
M. Botto-Tobar et al. (Eds.): ICAETT 2019, AISC 1067, pp. 238–247, 2020.
https://doi.org/10.1007/978-3-030-32033-1_22

Following these trends, Ecuador's situation is not encouraging; according to statistics from Ecuador's Ministry of the Interior, during 2015 there were 217 kidnapping cases. However, 179 of them the victim were rescued alive, and the rest did not survive. Of the survivors, 27% were rescued on the same day as the kidnapping, while 73% were recovered within 7 days of their disappearance [2]. It should note that many of these cases were solved by the *Dirección Nacional de Delitos Contra la Vida, Muertes Violentas, Desapariciones, Extorsión, y Secuestros* (DINASED, National Board of Crime Against Living Persons, Violent Death, Disappearances, Extortion, and Kidnappings). But what does it really mean if rescue time changes from days to hours? It needs to take into account the psychological trauma that kidnapping victims suffer, which is proportional to the time held against their will. Currently, search efforts are based on logistical and bureaucratic steps that take too long time. Moreover, the local community is not involved in the process, nor does it cooperate with search and rescue agencies. This is a community that could have relevant information about the neighborhood (Situational Awareness), the kidnapping victim, the kidnappers, etc. This proposal offers a real-time, cloud-based innovative solution that exploits technology and resources available to the community, i.e., smartphones, along with the *Message Queue Telemetry Transport* (MQTT) protocol for IoT and M2 M applications, in order to improve kidnapping rescue times.

This article first lays out a clear explanation of the problem. Next, it reviews the State of the Art projects related to the situation. This review is followed by a detailed description of the architecture and protocols required. The fourth section outlines some of the preliminary results obtained. Finally, it shows conclusions, and the future work is presented.

2   The State of the Art

Various projects have been designed to address the problem of kidnapping. However, the majority of them have been perceived as unorthodox because they did not use technological tools to achieve their goals, as [3]. Among the most relevant is one from the non-profit organization, **Civil Rights Defenders**, dedicated to Natalia Estemirova and human rights work. They introduced the *Natalia Project bracelet*, described in [4]. The project distributes bracelets to aid workers that sends alerts through social networking sites. The bracelet sends a mobile signal about an attack or kidnapping while transmitting the victim's GPS location. Nevertheless, the biggest problem with this project is the huge financial investment needed to distribute bracelets to every person. Another project is the US DOJ's *Child Abduction Alert System, Amber Alert* [5]. This system is a broadcast network that sends multimedia alerts about any kidnapped people. The statistics of the alert *Amber* [6], reveal that the average time to a send child's report is 1 h, and at least 1,5 h for the recovery of the missing person. It has been extended to 18 countries, but these have huge budgets too. As is expected, the success of this project relies on a high level of political and citizen support [5].

Another example is the *TrackChild* project [7] in India, whose main objective is to reunite missing children with their families through the use of computer technology, preventing these children fall into the networks of sex trafficking or prostitution. In India more than 500,000 children disappear every year [8, 9]; the Government has created a web application that connects with a large database with information on missing children. This application allows uploading the photo and through facial recognition performs a search in the aforementioned repository. The application has been used in hospitals, orphanages, and other institutions in India. However, this post-kidnapping solution, contrary to the present system, that refers to a system developed in real-time where the first notification is the priority. For Ecuador, this new proposal would be the first program to address the kidnapping issue, introducing new communications standards and, at the same time, including society as part of the solution.

3   Technical Architecture

This proposal incorporates a real-time system capable of reducing the rescue time of a kidnapped person that integrates security forces and the community (public transportation services, police, and individuals) as part of the solution. Figure 1 details the technical architecture of the system. Referring to this Fig. 1, when a vulnerable person (child, senior citizen, or person with special needs) is kidnapped, it is essential for her or his victim's custodian to immediately notify the security forces (police). They, in turn, use *SECURITY_MOB_APP* to communicate in real time to the "local community" within a "virtual perimeter" (to determinate in the next section) in order to resolve the event. This communication will disseminate the characteristics of the missing person; for example, in the case of a child, their physical traits, the color of their clothing, and including improve the situational awareness with multimedia content such as photographs, video, and audio recordings. Therefore, the community has a vital role in the diffusion of information via smartphone; the smartphone is an optimal device for this role, capable of transmitting and processing data after being customized with an application like *COMMUNITY_MOB_APP*, known as an *opportunistic application* [10]. Thus, the smartphone is key to the success of this proposal. The more devices are in the system, the greater the efficiency that can be achieved. This is known as *crowdsensing* [11].

The process alerts the community immediately, keeping it up to data on the attacker. Once a person who has received an alert observes something/someone that conforms to the detailed characteristics transmitted (such as a child with a yellow shirt, blue shorts, no hair, white skin, etc.), they notify to the pertinent agents using the same *COMMUNITY_MOB_APP* application that has identified the event. This allows agents to proceed with identification and rescue before all traces of the victim are lost, as well as transmit data about the possible suspect. In this way, the community is integrated into the search and rescue process of a missing person, which has the effect of increasing situational awareness. Consequently, they can improve the decision-making as they prevent a kidnapper from leaving the city since the first notification is intended to be as soon as possible.

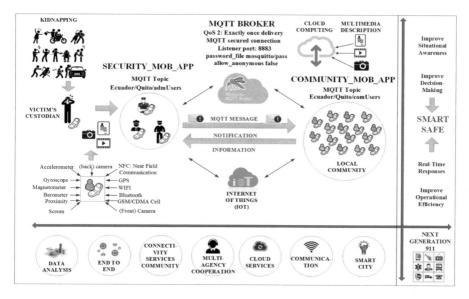

Fig. 1. Technical architecture of the system.

For this, it has been used avant-garde technologies such as Cloud Computing, which allows the system to be deployed to be reachable by any smartphone at any time and place. In addition, this technology allows for great scalability, since it is necessary to consider what the application is destined for society, therefore, an increase in the number of users is expected, considerably, limiting the system's deployment.

3.1   Geographical Limit and Software Tools

An important issue that should be evaluated is the range of the "virtual parameter" of notifications. In other words, it is the maximum range in kilometers for the first warning notification. Quito covers an area of 370,39 km^2, where the extension of the city covers approximately 50 km long and 8 km wide; and to avoid the commotion throughout the country, the city has been geographically limited in a virtual area of 400 km^2 with the objective of limiting the scope of the notification only to users with potential collaboration.

Using this criteria and the characteristic of hierarchical topics of the MQTT protocol, the diffusion of sending notifications can be limited geographically. For example, for Quito city, the communication has been divided into several sub-channels named: Ecuador/Quito/admUsers, Ecuador/Quito/comUsers for sending messages to administrators and the community respectively. The messages circulate through this way, and later, are processed in the mobile application to show notifications to the corresponding user. In Listing 1 you can see the code used in the mobile application to discriminate the different topics MQTT and display the information to the end user.

```
1   if (topic.equals("Ecuador/Quito/admUsers")) {
2   mensajeRecibido = new String(message.getPayload());
3   try {
4           JSONObject obj = new JSONObject(mensajeRecibido);
5           mensajeNotificacion =
6   obj.getString("mensajeNotificacion");
7           cedulaCaso = obj.getString("cedulaCaso");
8           notificacion(mensajeNotificacion, cedulaCaso);
9       } catch (Exception e) {
10      }
11  }
```

Listing 1. Algorithm for receiving MQTT messages.

The architecture uses the MQTT protocol [12] that is often used in SmartCities applications [13], thanks to its many advantages. MQTT is a lightweight publish/subscribe transport that allows users to interact with one or more users [14]. MQTT is ideal for mobile applications, because it consumes little power, distributes data and consumes a minimum of storage in the host device. This protocol is implemented with MQTT Broker Eclipse Mosquitto v3.1. Amazon Web Services has also been selected as the leading provider [15], as it offers a wide range of proven advantages as a provider of cloud computing. The platform is for the Ubuntu Linux 16.04 LTS server, which is widely available. A MySQL 5.5 database for information management. Finally, the mobile application was implemented using Android Studio IDE together with a REST service structure. As described in Fig. 2.

A real-time system is one that accepts any type of data and responds to inputs and stimuli within a specific and finite period. For example, a seismic detection system [16] should respond in 4 s. In this proposal, the data must be processed and the community must be notified within a time limit of 1 h, shorter than in previous studies such as [17], whose time limit is 1 to 3 h; statistics show that 74% of children die 3 h after their abduction [18]. The notifications are information relevant to the user, as shown in Fig. 3.

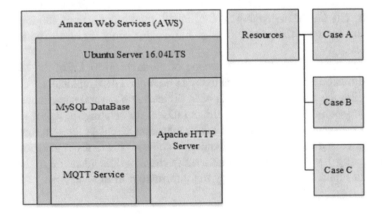

Fig. 2. Design of architecture.

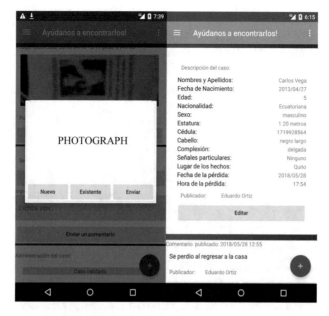

Fig. 3. Notification of a missing person.

4 Operational Testing and Results

The present notification system of disappeared persons was tested in Quito, Capital of Ecuador; and, given the uncertainty to determine the existence of a kidnapping, the validation of the proposal has been proposed through the simulation of ten cases of disappearance. The results obtained correspond to the integration tests carried out in the scenario and characteristics detailed in Fig. 1. For the simulation of cases of kidnapping, it has counted on the help of security agents, during a period of 30 days and the collaboration of 50 people from the community.

The notification system has been evaluated in the performance fields, these data are important because of the real-time approach of the system. Technical aspects of the developed mobile application have been analyzed, because if these parameters present inappropriate results, these may be possible causes for the removal of the mobile application, decreasing the *crowdsensing* [19]. It was tested the application functions under different scenarios, such as using different mobile devices. It shows that the applications consumed a minimal amount of energy, comparable to that used by WhatsApp, the most popular instant messaging application, which also uses MQTT for its operations [20]. As you can see in Fig. 4.

Table 1 shows the verification results of the average time to upload/download files of images and multimedia video for a case, in different types of communication media such as WIFI, 3G, and LTE. To carry out the tests, 50 people participated with the applications installed in the test devices; these people played the role of the security forces and community users. On the other hand, Table 2 shows the delay in sending messages through the **SECURITY_MOB_APP** application, and in the reception of messages using the **COMMUNITY_MOB_APP** and the MQTT server.

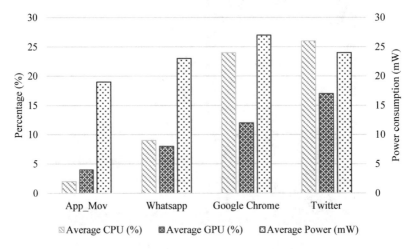

Fig. 4. Comparison with known social applications.

Table 1. Time upload/download case of disappearance.

TIME (seconds)	WIFI		3G		LTE	
	Img[a]	Vid[b]	Img[a]	Vid[b]	Img[a]	Vid[b]
Upload	4,90	23,8	18,40	34,50	0,46	0,39
Download	1,91	6,87	2,34	26,27	0,12	0,10

[a]2 MB Image
[b]10 MB Video

Table 2. MQTT connectivity times.

MQTT update time	5 min			15 min		
	3G	WIFI	LTE	3G	WIFI	LTE
Battery consumption per hour (mW)	0,15	0,02	0,10	0,02	0,03	0,01

Table 3. Results of experimental disappearance cases.

Núm.	Community participation	Solution distance (km)	Solution time (Min)
Avg.	37%	24,5	30,3

Consequently, the average time for the publication of the disappearance case was a few minutes, showing that the administrator users intervened quickly in the validation of the aforementioned cases. As seen in Table 3, the average time for the solution of the case was 30,3 min, even though; it depends on the number of community users because increasing this number also increases the probability of finding a missing person. These 30,3 min is a result that surpasses other projects seen in the state of the art. In addition, the percentage of community participation was 37%. Besides, the evidence revealed that the average recovery distance was 24,5 km, indicating that the cases were resolved within the same city, as it is an effect of the early response of the community that has prevented the disappeared from leaving the city, verifying the established range "virtual perimeter". Certainly, a virtue of this system is that it has the possibility of adapting to other cities in Latin America that have the same social problem either to a greater or a lesser degree of severity of the problem, adjusting to their respective "virtual perimeter".

5   Conclusions and Future Work

The statistics regarding personal security in various Latin American cities are becoming alarming. Therefore, governments are searching for economic and technologically efficient solutions provided by new information and communications technologies that incorporate direct community involvement. In Ecuador for example, 500 reports of missing persons per month are recorded on average, which provokes in the population a state of constant fear with respect to the social security provided by the Government. Despite the efforts of the National Police to curb this problem, the disappearances continue, showing their need for help in dealing with this issue. This article presents a system for the location and alerting of missing persons, improves rescue times. Moreover, it should emphasize that implementing this project will directly and positively influence the productive matrix of Ecuador since insecurity corrupts and obstructs any country's growth under any conditions.

The proposal presents a real-time notification system that reuses the technology deployed in society as smartphones to propose a solution that integrates the community in the area of recovery of missing persons, this means taking advantage of sensors at zero cost. The results of the simulated cases showed that it is possible to reduce the recovery time of 1,5 h in the case of Amber from the USA. at 30,3 min evidencing that a system has been developed capable of optimizing the search and recovery times of a disappeared person within the city of Quito. In addition, the developed mobile application has presented a good performance in the fields of battery consumption; light storage, because it does not occupy significant installation space and, a little delay in the connection with the data server.

This study emphasizes the use of MQTT as a communications protocol and cloud-based data management through Amazon Web Services. MQTT is a protocol based on a publish/subscribe model for the Internet of Things (IoT), Machine to Machine technology (M2M) with the following characteristics: efficient one-to-one and many-to-many message transmission, low-energy consumption, and lightweight transport. Therefore, it is ideal for real-time systems. MQTT incorporates an internal hierarchy that facilitates orderly maintenance of every missing person case. The AWS platform is flexible enough to meet user requirements, including during a major catastrophe, i.e. as the amount of information and user requests increases for the system, the hardware resources of the server will be dynamically adapted to the same.

Finally, for the future work, the implementation of the hierarchical topic's characteristic of the MQTT protocol has been planned for the segmentation of the communication channel for different entities of the country such as hospitals, shelters, among others. In addition, the implementation of MQTT-Bridge is planned to add several notification servers and thus avoid a point of failure within the notification system. In this way it is possible to replicate the notification mechanism to several servers, improving the availability in case of suffering an incident.

Acknowledgments. Authors gratefully acknowledge the financial support provided by the Escuela Politécnica Nacional, for the development of the project PII-DETRI-2019-02 "Smart Safe Cities: Un Sistema Genérico en Tiempo Real de Ágil Notificación de Rescate de Personas Desaparecidas Utilizando Comunicación IoT".

References

1. Navas, M.E.: El drama de los niños desaparecidos de América Latina. https://www.bbc.com/mundo/noticias/2013/11/131106_ninios_perdidos_desaparecidos_explotados_america_latina_men
2. Ministerio del Interior: Dinased resolvió el 84.8% de los casos de desaparecidos, este año. https://www.ministeriointerior.gob.ec/dinased-resolvio-el-84-8-de-los-casos-de-desaparecidos-este-ano/
3. Gatti, G.: El lenguaje de las víctimas: silencios (ruidosos) y parodias (serias) para hablar (sin hacerlo) de la desaparición forzada de personas. Universitas Humanística (2011)
4. Isaacson, B.: "Natalia Project" Bracelet Aims To Protect Human Rights Defenders. https://www.huffingtonpost.com/2013/04/08/natalia-project-bracelet-human-rights-defenders_n_3024220.html
5. U.S. Department of Justice: AMBER Alert - America's Missing: Broadcast Emergency Response. https://www.amberalert.gov/
6. U.S. Department of Justice: AMBER Alert – Statistics. https://www.amberalert.gov/statistics.htm
7. Redacción BBC Mundo: La tecnología que ayudó a encontrar a 3.000 niños en tan solo 4 días en un país donde cada año desaparecen medio millón. https://www.bbc.com/mundo/noticias-44068357, (2018)
8. Infosecurity News: Infosecurity News - 14 años. http://www.infosecurityvip.com/newsletter/productos_may18.html
9. Bhattacharyya, R.: Sociologies of India's missing children. Asian Soc. Work Policy Rev. **11**, 90–101 (2017)
10. Chon, Y., Lane, N.D., Li, F., Cha, H., Zhao, F.: Automatically characterizing places with opportunistic crowdsensing using smartphones. In: Proceedings of the 2012 ACM Conference on Ubiquitous Computing, pp. 481–490. ACM, New York (2012)
11. Ganti, R.K., Ye, F., Lei, H.: Mobile crowdsensing: current state and future challenges. IEEE Commun. Mag. **49**(11), 32–39 (2011)
12. Official Website MQTT: MQTT. http://mqtt.org/

13. Atmoko, R.A., Riantini, R., Hasin, M.K.: IoT real time data acquisition using MQTT protocol. J. Phys: Conf. Ser. **853**, 012003 (2017)
14. Luzuriaga, J.E., Cano, J.C., Calafate, C., Manzoni, P., Perez, M., Boronat, P.: Handling mobility in IoT applications using the MQTT protocol. In: 2015 Internet Technologies and Applications (ITA), pp. 245–250 (2015)
15. Panahi, M.S., Woods, P., Thwaites, H.: Designing and developing a location-based mobile tourism application by using cloud-based platform. In: 2013 International Conference on Technology, Informatics, Management, Engineering and Environment, pp. 151–156 (2013)
16. Zambrano, A., Pérez, I., Palau, C., Domingo, M.: Sistema distribuido de detección de sistemas usando una red de sensores inalámbrica para alerta temprana. Presented at the Revista Iberoamericana Automática e Informática (2015)
17. Sakaki, T., Okazaki, M., Matsuo, Y.: Tweet analysis for real-time event detection and earthquake reporting system development. IEEE Trans. Knowl. Data Eng. **25**, 919–931 (2013)
18. U.S. Department of Justice: Case Management for Missing Children Homicide Investigation (1997). https://www.ncjrs.gov/pdffiles1/pr/201316.pdf
19. Liu, C., Zeng, F., Li, W.: Synergistic based social incentive mechanism in mobile crowdsensing. In: Chellappan, S., Cheng, W., Li, W. (eds.) Wireless Algorithms, Systems, and Applications, pp. 767–772. Springer, Cham (2018)
20. Thosar, A., Nathi, R.: Air quality parameter measurements system using MQTT protocol for IoT communication over GSM/GPRS technology. In: Balas, V.E., Sharma, N., Chakrabarti, A. (eds.) Data Management, Analytics and Innovation, pp. 421–433. Springer, Singapore (2019)

An Adapted Indoor Propagation Model for Colonial Buildings

Danny Curipoma Hernández, Patricia Ludeña-González[✉][ID],
Francisco Sandoval[ID], Carlos Macas Malán, and Marco Morocho-Yaguana

Universidad Técnica Particular de Loja, 11-01-608 Loja, Ecuador
{dfcuripoma,pjludena,fasandoval,camacas1,mvmorocho}@utpl.edu.ec
https://www.utpl.edu.ec

Abstract. In Latin America, there is a large number of colonial buildings that, given their great historical and cultural value, are preserved and used for the development of different activities that demand wireless communication network services. The internal structure of these buildings consists basically of rammed earth walls or mud-bricks, locally named "*tapia*" and adobe respectively. Most popular propagation models are based on measurements developed in modern structures and materials, for this reason, it is necessary to establish a propagation model that allows the correct design of wireless networks in colonial buildings. The main objective of this work is to propose an adaptation of a typical propagation model considering the characteristics of indoor environments for buildings of colonial architecture, defining loss coefficients based on empirical data obtained in a measurement campaign for the 2.4 GHz band. The adaptation mechanism applied is the method of least squares. A comparison with the Log-distance, One Slope, Motley-Keenan, ITU model and the measured data is given. The measurement campaign is developed in three scenarios, where a specific measurement methodology and various types of obstacles are considered. Then, the proposed model is validated with a low average error in two control scenarios and therefore it could be applied in this type of buildings.

Keywords: Indoor propagation model · Motley-Keenan model ·
Log-distance model · ITU model · One Slope model · Colonial
buildings · Least squares method · Mud wall · Rammed earth wall

1 Introduction

The study of the propagation of electromagnetic waves in indoor environments is a key factor for the design of wireless communications systems, for example, wireless local area networks (WLANs), wireless telephony, mobile telephony networks, and any other system based on Radio Frequency communication. In general, the objective of modeling wave propagation is to determine the probability of a satisfactory performance of a wireless communication system or any other system that depends on the propagation of electromagnetic waves [21].

© Springer Nature Switzerland AG 2020
M. Botto-Tobar et al. (Eds.): ICAETT 2019, AISC 1067, pp. 248–258, 2020.
https://doi.org/10.1007/978-3-030-32033-1_23

Basically, a propagation model allows predicting the average intensity of a signal received at a given distance from the transmitter. Propagation models can be classified as empirical and deterministic [12]. Deterministic models simulate the propagation of electromagnetic waves based on the Maxwell equations and taking in count the propagation environments, for this reasons they are considered complex and difficult to apply [17]. On the other hand, empirical models are based on the statistical use of results obtained from the tests carried out on model scenarios [8], so its application and replication is simpler than with the deterministic models. However, the effective accuracy of empirical models depends on the similarity between the model scenario and the site of interest to be analyzed. In our knowledge, there is no model that considers colonial structure and materials.

A lot of cities in Latin America have been developed around colonial buildings. Some of them are considered world heritage and they attract the attention of hundreds of historians, researchers, and tourists. Nowadays, in Ecuador, there are preservation and conservation policies in order to protect a large number of buildings of colonial architecture. Specifically in Loja city, where this work was developed, there are 592 cataloged colonial buildings [1], most of them date from the nineteenth and early twentieth centuries [6]; and preserve the classic architectural features of the colonial period: a central interior courtyard as an articulating element and rooms around it (see Fig. 1). Many of these properties continue to be used for public or private activities, such as trade centers, houses, office buildings, museums, or as headquarters of government institutions. Therefore, they demand telecommunications services and, given that these structures cannot be physically altered, wireless communication networks are the most recommended for these environments.

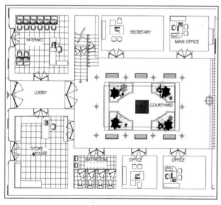
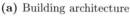

(a) Building architecture

(b) Building view

Fig. 1. A colonial building used for the measurement campaign

The colonial buildings are made of earth-based materials, such as rammed earth walls (REW) and mud-bricks. REW are based on an old construction technique that consists of compacting earth using high pressure to make it solid, and are locally named *"tapia"*. Mud-bricks or adobe are made in open timber frames. The earth used for the manufacture of both building materials consists of a mixture of clay, silt, and sand. Also, it usually contains straw and small stones for greater resistance.

Pauli et al. [18] and Afrizal et al. [2] show that, compared to other materials, construction materials made of earth present a high level of attenuation to high-frequency electromagnetic radiation. For example, Minke et al. [16] determine that at the frequency of 2 GHz a wall made with mud-bricks of 24 cm thick can attenuate a signal up to 24 dB; while a wall built with limestone and the same thickness attenuates it in only 7 dB. The use of earth as a building material is widespread around the world. It is estimated that approximately 20% of the world's population lives or works in buildings made of earth [20]. Although the proposed model is based on measurements of buildings in Ecuador, the selection of the analyzed scenarios and the methodology used will allow applying this model to buildings of similar characteristics and materials in other countries.

In the literature, there are some research works [14,15,22] that propose adapted indoor propagation models for the 2.4 GHz band, where some constructive characteristics are considered such as the type of walls, their thickness, and the building materials used. The works in [9,23] use a linear regression technique known as the least squares method for model fitting, while [10] uses non-linear regression techniques. On the other hand, in [7,17] an intensive measurement campaign is carried out in buildings on a university campus, where they expose methodologies to measure the attenuation of a signal in different frequency bands, obstacles, and scenarios. This work uses the methodology proposed by [7].

The objective of this paper is to propose an indoor propagation model adapted to the constructive characteristics of colonial buildings, which is based on the Motley-Keenan propagation model. The adjustment of model is made through an intensive measurement campaign which allows obtaining attenuation coefficients for different obstacles at 2.4 GHz band. In order to achieve this goal, this work is divided into four phases. The first one is to carry out a measurement campaign in which the necessary data for adapting the model is obtained. Next, a propagation model is chosen and then adapted. The adaptation mechanism used is the method of least squares. Finally, the proposed propagation model is validated in a second measurement campaign, where the results show that the average error between measured and modeled losses is low.

The rest of this paper is structured as follows. Section 2 describes some indoor propagation models and their parameters. Section 3 presents the scenarios where the measurement campaign is developed, describes the equipment and methodology used to carry it out and the validation process of the proposed model. In Sect. 4, the proposed propagation model is described. The results are shown in Sect. 5 and the conclusions of this work are presented in Sect. 6.

2 Indoor Propagation Models

Indoor propagation depends on reflection, diffraction and dispersion. These effects, either individually or collectively, contribute to the degradation of a signal [21]. Four popular indoor deterministic models such as the One Slope, Log-Distance, Motley-Keenan, and ITU model are described in the following sections. Prestigious organizations and researchers have defined a series of loss coefficients for each of these models through propagation measurements carried out in different indoor environments and transmission frequencies. Such is the case with the International Telecommunications Union for ITU model [11], the European Cooperation in Science and Technology for One Slope and Motley-Keenan model [5], and with Andersen, Rappaport and Yoshida for the Log-distance model [3]. Therefore, they are considered as key models and are widely used.

2.1 One-Slope Model

The One-slope is one of the simplest propagation models. It assumes that the path loss in dB depends logarithmically on the distance between the transmitter and receiver [24]. If $PL(d_0)$ is the path loss at the reference distance taken as free space loss at $d_0 = 1\,\mathrm{m}$, the path loss PL is given by

$$PL[dB] = PL(d_0) + 10n\log(d), \tag{1}$$

where n is the loss exponent and d is the separation distance between the transmitter and receiver in meters. Values of the loss exponent at different indoor scenarios are given in [5].

2.2 Log-Distance Model

It follows the distance power law (Eq. 2), where the average power received from a signal decreases logarithmically with distance [19]. It can be applied in a wide range of environments.

$$PL[dB] = PL(d_0) + 10L_{LD} \cdot \log(d/d_0) + X_\sigma. \tag{2}$$

L_{LD} represents the path loss distance exponent, and X_σ is a Gaussian random variable with zero mean and standard deviation σ. Some typical values for the path loss exponent and standard deviation at different indoor propagation cases can be reviewed in [21].

2.3 Motley-Keenan Model

The Motley-Keenan model assumes an additional loss introduced by the number of walls between the transmitter and the receiver [14]. If k_{w_i} represents the number of penetrated walls of type i and n_{w_i} is the loss coefficient for a wall of type i, the Motley-Keenan model is given by:

$$PL[dB] = PL(d_0) + 20\log(d) + k_{w_i}n_{w_i}. \tag{3}$$

Some values for the parameters previously described are presented in [5].

2.4   ITU Indoor Path Loss

In this site-general propagation model, the power loss coefficients due to distance implicitly take into account the transmission through walls and other loss factors that can be encountered inside a building. It is defined in Eq. 4, where f is the frequency in MHz, L_{ITU} is the distance power loss coefficient, which have an implicit tolerance for transmission through walls and across other type of obstacles, as well as other loss mechanisms that are likely to be found within a floor of a building [4]. Some loss coefficients for different environments are given at [11]. Meanwhile, L_f represents the floor penetration loss factor and n_f the number of floors between the transmitter and receiver, which is considered as 0 in this work.

$$PL[dB] = 20\log(d)f + L_{ITU} \cdot \log(d) + L_f(n_f) - 28 \qquad (4)$$

3   Methodology

3.1   Scenarios

Five colonial buildings of the city of Loja-Ecuador were chosen, whose predominant construction material is REW. Three of them were selected to develop the measurement camping in order to obtain the necessary data to formulate the proposed propagation model. The two remaining were used to the validation phase. In the first group of buildings, we can find REW up to 92 cm thick, while in the second group there are REW up to 100 cm thick. In general, these buildings have common architectural features for colonial buildings, such as those described in Sect. 1.

3.2   Measurement Methodology

The system used to execute the measurement campaign consists of four elements: a USB-TG124A Tracking Generator (100 KHz–12.4 GHz), a Signal Hound USB-SA124B Spectrum Analyzer (100 KHz–12.4 GHz), and two grid antennas HG2415G with a gain of 15 dBi. The transmitter and receiver antennas are fitted on metal supports at a height of 1.30 m above ground level and configured in vertical polarization. The transmission power is set to -12 dBm at the frequency of 2.437 GHz and the receiver is configured in the band of 2.4 GHz at the central frequency of 2.437 GHz.

Three sets of measurements are made in each building on different days and hours, in order to consider all the factors that may influence the data collection. Two types of obstacles are considered: single REW and two REW (multi-wall obstacle). In the case of single REW, the measurements are made on 30 cm, 65–69 cm, 89–92 cm thick walls. The transmission and reception antennas are located at an initial separation distance of 150 cm. Then, the receiving antenna moves in a straight line every 25 cm and the measurement data is taken in each new position until the scenario is finished. For the case of two REW, measurements are made on two walls of 63, 65, and 66 cm thick each. The transmitting and receiving antennas are placed 50 cm apart from each wall, then, the receiver is displaced in a straight line and the measurement is made each 25 cm.

3.3 Validation of the Proposed Model

The validation tests are carried out in two colonial buildings. Measurements are developed for four cases, two are for single REW with a thickness of 76 cm and 100 cm, and two for multi-REW in two combinations: 45–53 cm thick and 47–90 cm thick. The error calculation method chosen is the absolute error, which consists of obtaining the difference between the data obtained in the measurements and the data predicted by the adapted model. Based on [13], to validate the proposed propagation model, this work considers as an acceptable value an average absolute error equal to or less than 3 dB.

4 Proposed Model

Based on the data obtained in the measurement campaign, one of the propagation models described in Sect. 2 is chosen and adapted. First, the slope of the curve generated by each propagation model is determined, which is compared with the slope of the curve of the measured data. The selected model is the one whose slope is closest to the slope of the measured attenuation.

Table 1. Slope of each propagation model and measured attenuation data

Obstacle: REW [cm]	ITU	Log-Distance	One Slope	Motley-Keenan	Measured attenuation
30	4.63	4.63	8.03	3.09	2.20
65 - 69	4.32	4.32	7.48	2.88	3.18
89 - 92	4.32	4.32	7.48	2.88	2.88
66 (with concrete)	4.63	4.63	8.03	3.09	3.09

Table 1 shows the slope of each propagation model and the slope of the measured attenuation for REW of different thickness. As we can see, the slopes of the Motley-Keenan propagation model are the closest to the slopes of the measured data, and therefore it is the model chosen to be adapted.

In order to adapt the Motley-Keenan model, we need to calculate the reception power predicted by the model (P_{model}) at the distance d_i, that is given by the load balance equation:

$$P_{\text{model}}[\text{dBm}] = P_{\text{Tx}} + G_A - \text{PL} - L_s, \tag{5}$$

where P_{Tx} is the transmission power, G_A the gain given by the transmission and reception antennas (sum), and L_s represents the sum of the system losses (cables, connectors). If for simplicity, we consider only one penetrated wall of type i by each experience, i.e., $k_{w_i} = 1$, the path losses, PL, in the Eq. 5 can be calculate by the Eq. 3. Then, the reception power predicted by the model is:

$$P_{\text{model}}[\text{dBm}] = P_{Tx} + G_A - \text{PL}(d_0) - 20\log(d) - n_{w_i} - L_s. \qquad (6)$$

Following, we use the least squares method (LSM), which allows reducing the error between the measured values, to adapt the model. According to the LSM, the error E as a function of the attenuation coefficient n for a particular obstacle, between the measured and calculated values, is defined as:

$$E(n)[\text{dB}] = \sum_{i=1}^{s}[P_{\text{measured}}(d_i) - P_{\text{model}}(d_i)]^2, \qquad (7)$$

where i and s are the sample and the total number of samples, respectively, and P_{measured} represents the reception power measured at the distance d_i. Thus, to find the attenuation coefficient (n) that minimizes the error, we must derive and equal to zero the error $E(n)$, i.e., $\frac{\partial E(n)}{\partial n} = 0$. Table 2(a) shows the calculated attenuation coefficients (n_{w_i}) for REW of different thickness.

Table 2. Coefficients of the adapted model

(a) Attenuation coef. for single-REW   **(b)** Correction factor for multi-REW

REW thickness [cm]	n_{w_i} [dB]	Two REW thickness [cm]	F_m[dB]
30	7.96	63	3.82
65 - 69	23.02	65	0.68
89 - 92	35.16	66	0.64
66 (with concrete)	26.79	Average value	1.71

Table 2(b) presents an average correction factor (F_m) for multi-REW obstacles, which was obtained by applying the same adaptation process described above considering two REW of different thickness.

Finally, the proposed model is presented in Eq. 8, where $\text{PL}(d_0)$ is replaced by the average free space losses measured in each scenario, and the correction factor (F_m) for multi-REW obstacles is added.

$$\text{PL}[\text{dB}] = 42.83 + 20\log(d) + k_{w_i}n_{w_i} + F_m \qquad (8)$$

5   Results

This section shows the results obtained in the validation of the proposed propagation model. A comparative analysis is performed based on the average absolute error calculated between the attenuation predicted by the model and the measured attenuation.

The next two figures compare the measured attenuation (MA) and the calculated attenuation through the proposed propagation model (PMA) for four measurement cases: case 1 (100 cm thick REW), case 2 (76 cm thick REW), case 3 (two REW of 45 cm and 53 cm thick), and case 4 (two REW of 47 cm and 90 cm thick). Fig. 2 corresponds to cases 1 and case 2, while Fig. 3 corresponds to case 3 and case 4.

According to Fig. 2, the attenuation for case 1 and case 2 increases on average 0.94 dB and 0.91 dB respectively each 25 cm. The proposed propagation model has an average absolute error of 1.16 dB and 1.35 dB for case 1 and case 2, respectively. The variations that can be seen in the measured attenuation curves, specially for case 1 between 2 m and 3 m, are essentially due to multipath fading and to the changes that these indoor environments may have experienced in their characteristics during the measurement campaign. These changing environments induce reflection, diffraction and dispersion, which cause the signal levels in the receiver to vary in different points of the measurement.

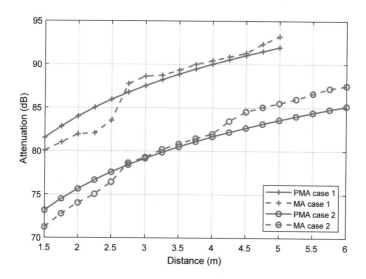

Fig. 2. Measured attenuation (MA) and proposed model attenuation (PMA) for case 1 (100 cm thick REW) and case 2 (76 cm thick REW).

In Fig. 3, the attenuation increases 0.59 dB and 0.29 dB each 25 cm for case 3 and case 4 respectively; and the average error of the proposed model for case 3 and case 4 is 2.03 dB and 1.47 dB respectively. In these cases, the measurements are after the second wall in order to evaluate the effect of this wall, for this reason, the distances are larger than in cases 1 and 2.

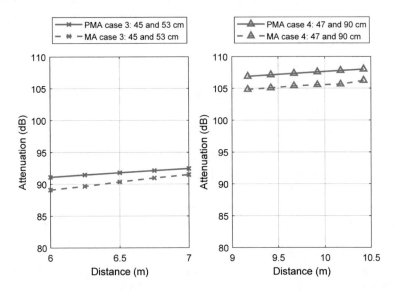

Fig. 3. Measured attenuation (MA) and proposed model attenuation (PMA) for case 3 (two 45 and 53 cm thick REW) and case 4 (two 47 and 90 cm thick REW).

6   Conclusions

This paper presents an adaptation of the Motley-Keenan indoor propagation model for the 2.4 GHz band, considering the constructive characteristics of colonial buildings. We define new attenuation coefficients values (n_{w_i}) for single REW of different thickness and a correction factor (F_m) for multi-REW. These values were obtained combining an intensive measuring campaign and a mathematical adaptation method, which were then validated in a second measurement campaign. The average absolute error for single REW is 1.16 dB and 1.35 dB for a thickness of 100 cm and 76 cm, respectively. In addition, for multi-REW it obtains errors of 2.03 dB and 1.47 dB for 45–53 cm and 47–90 cm combinations. In all cases the proposed model is validated, therefore we conclude that this adapted propagation model is suitable for indoor environments of colonial buildings with similar characteristics.

Future research lines could use deterministic and semi-deterministic methods, i.e. Maxwell equations, for the extraction of loss coefficients, that allow a more accurate modeling of propagation losses in new environments with special constructive characteristics.

References

1. Instituto Nacional de Patrimonio Cultural: Sistema de Información Cultural Ecuatoriano SIPCE. http://patrimoniocultural.gob.ec/sistema-de-informacion-del-patrimonio-cultural-ecuatoriano-sipce/. Accessed 20 Jan 2019
2. Afrizal, A., Khadiko, N., Sulistya, F., Septyani, V., Lubis, A.: Electromagnetic wave attenuation in lossy dielectric. Appl. Mech. Mater. **815**, 180–182 (2015)
3. Andersen, J., Rappaport, T., Yoshida, S.: Propagation measurements and models for wireless communications channels. IEEE Commun. Mag. **33**, 42–49 (1995)
4. Barclay, L.: Propagation of Radiowaves. The Institution of Electrical Engineers, London (2003)
5. COST: COST Action 231: Digital Mobile Radio Towards Future Generation Systems. European Commission, Directorate-General Telecommunications, Information Society, Information Market, and Exploitation of Research, Luxembourg (1999)
6. Cueva, K.: Flexibilidad de uso y nuevo uso de las viviendas republicanas en Ecuador. El caso de la ciudad de Loja, Academia XXI (2016)
7. Erreyes, A., Poma, B.: Optimización de un Modelo de Pérdidas de Propagación en Ambientes Internos. Caso Campus UTPL. Bachelor's thesis, Universidad Técnica Particular de Loja (2016)
8. García, N.: Modelo de cobertura en redes inalámbricas basado en radiosidad por refinamiento progresivo. Ph.D. thesis, Universidad de Oviedo (2006)
9. Granados, O., Santiago, F.C.: Caracterización de la Propagación en Interiores para una Red WLAN Basada en IEEE 802.11. Master's thesis, Escuela Superior de Ingeniera Mecánica y Eléctrica Unidad Zacatenco (2013)
10. Hernández, D.: Modelo para Predicción de Pérdidas de Propagación en Redes WLAN, Bandas No Licenciadas (2.4 GHz y 5.8 GHz, en Ambientes Interiores). Master's thesis, Escuela Superior Politécnica del Litoral (2016)
11. ITU-R: Datos de propagación y métodos de predicción para la planificación de sistemas de radiocomunicaciones de interiores y redes de radiocomunicaciones de área local en la gama de frecuencias de 900 MHz a 100 GHz. Technical report, ITU (2017)
12. Julio, G., Omar, R.: Desempeño de Modelos de Propagación en Comunicación Móvil para la zona de Caldas parte 1: Modelos para áreas urbanas. X Congreso Internacional de Telecomunicaciones (2002)
13. Lee, W.C.Y.: Wireless and Cellular Telecommunications, 3rd edn. McGraw-Hill, New York (2006)
14. Lima, A., Menezes, L.: Motley-Keenan model adjusted to the thickness of the wall. In: SBMO/IEEE MTT-S International Microwave and Optoelectronics Conference, pp. 180–182 (2005)
15. Mardeni, R., Solahuddin, Y.: Path loss model development for indoor signal loss prediction at 2.4 GHz 802.11n network. In: International Conference on Microwave and Millimeter Wave Technology (ICMMT) (2012)
16. Minke, G.: Building with Earth. Birkhäuser, Kassel (2006)
17. Naranjo, C.: Caracterización práctica del canal estrecho en las bandas de 2.4 GHz, 3.3 GHz y 5.5 GHz para interiores. Master's thesis, Universidad Autónoma de Madrid (2013)
18. Pauli, P., Moldan, D.: Reduzierung hochfrequenter Strahlung - Baustoffe und Abschirmmaterialien. AnBUS e.V, Fürth (2015)

19. Rappaport, T.: Wireless Communications Principles and Practice, 2nd edn. Prentice Hall, Hoboken (2002)
20. Rocha, M.: CEB implementation Report. Technical report, Vale das Lobas (2017)
21. Seybold, J.: Introduction to RF Propagation, 1st edn. Wiley, Hoboken (2005)
22. Shen, X., Xu, K., Sun, X., Wu, J., Lin, J.: Optimized indoor wireless propagation model in WiFi-RoF network architecture for RSS-based localization in the Internet of Things. In: International Topical Meeting on Microwave Photonics jointly held with the 2011 Asia-Pacific Microwave Photonics Conference (2011)
23. Tummala, D.: Indoor Propagation Modeling at 2.4 GHz for IEEE 802.11 Networks. Master's thesis, University of North Texas (2005)
24. Xie, M.: Indoor radio propagation modeling for system performance prediction. Ph.D. thesis, Institut National des Sciences Appliquées de Lyon (2013)

Hand Prosthesis with Nitinol: Shape Memory Alloy

René E. Cortijo$^{(\boxtimes)}$ ⓘ, Millard Escalona ⓘ, and Victor Laverde

Universidad Israel, Francisco Pizarro E4-142, Quito, Ecuador
{recortijo,mescalona,vlaverde}@uisrael.edu.ec

Abstract. This article presents the result of an investigation, to develop a hand-held prosthesis with Nitinol springs, which is a shape memory alloy (SMA). The purpose was to develop an automated prosthesis capable of generating 5 hand movements, similar to those of a human being. It shows the advantages of using SMA to replace the muscular action of the hand compared to other actuators such as motors, pneumatic or mechanical elements and provide the ease of being remotely controlled from a mobile device. An investigation is carried out on the alloys with form memory and is emphasized in the use of Nitinol springs for the development of the prosthesis for its excellent capacity of form memory, for its good relation strength-weight and easiness of activation of the material by inducing through it an electric current. It also presents the results of the implementation of the prosthesis within a structure designed in 3D and the tests carried out that validate the fulfillment to the satisfaction of the five movements of the hand fingers proposed.

Keywords: Nitinol · Austenite · Martensite · Shape memory

1 Introduction

The use of automation and robotic technologies in the development of upper limb prostheses for people who have suffered the loss of these limbs, have shown relevant results in recent decades [1]. Hand prostheses have been developed with actuators based on servo-controlled electric motors or stepper motors, pneumatic, mechanical and hydraulic elements that present significant operating limitations due to weight, noise, friction as well as the complexity of handling and configuration [2].

The prosthetic systems with myoelectric elements present, unlike the previous ones, better degrees of rehabilitation with greater force and speed of grip, however they also present limitations due to the weight, cost and requirements of external sources to provide power [3]. Shape memory alloys (SMA) are very strong, with significantly reduced size and weight [2]. Research shows that motion generators based on these intelligent materials can reduce their magnitude by two or three times, retaining the same output ratio [1].

This project seeks to implement a hand prosthesis based on Nitinol which is a combination of Nickel and Titanium with shape memory property, discovered by William Beuheler in the laboratories of the United States Navy and which is currently presented in two primary categories: Super Elastic, characterized by a great capacity of

© Springer Nature Switzerland AG 2020
M. Botto-Tobar et al. (Eds.): ICAETT 2019, AISC 1067, pp. 259–270, 2020.
https://doi.org/10.1007/978-3-030-32033-1_24

recovery of tension and Memory of Form characterized by the capacity of recovering a preset form when it is heated above its temperature of transformation [4].

In order to be able to use this element, which is considered to be intelligent, in a hand prosthesis, a form memory learning must be carried out. This is achieved by subjecting the material to mechanical stress while internal temperature variations occur, which are known as the Martensite and Austenite phases. These training processes are thermomechanical treatments of cyclic type to achieve a spontaneous and reversible change, because this effect is not an intrinsic property of SMA [5].

Once the learning process is finished, the shape of the intelligent material will be controlled through a control circuit made up of optocouplers and MOSFET transistors, which will regulate the current through the actuator without generating wear in it. A 16-channel PWM pulse width modulation block will also be used to control the number of Nitinol actuators needed to make the prosthesis, without increasing the physical space where they will be installed.

The control will be in charge of an Arduino Nano V3 module that, together with the PCA 9685 module, will generate the PWM pulses. With them, the mechanical shape of the intelligent material will be controlled. In addition, FSR 402 force sensors will be used to measure the force exerted on objects, and LM35 temperature sensors will be used to measure the heating conditions of the intelligent actuator.

It is intended that the prosthesis can perform defined movements, which will be of five types: open hand, closed hand, individual movement of each finger, cylindrical grip, hook grip and lateral grip.

1.1 State of the Art Robotic Hand Prototypes Using SMA

The research carried out shows important previous work related to the use of intelligent materials in the design of prostheses, such as: "Design and application of an SMA actuator in the control of robotic hands". In this project the design of a methacrylate finger was carried out in which the SMA wires were fixed and functional tests were carried out with different types of mechanical arrangements, control systems and a cooling system for the alloy [6]. Despite the fact that the beginning of this study was considered a success in its initial objectives, the tests carried out concluded that the use of PID (Integral Proportional Derivative) systems was ineffective due to the non-linearity of the SMA materials.

UNAM students detail another of the works carried out with memory form alloys in the article "Robótica y Prótesis Inteligentes", where mention is made of the study carried out for the design, simulation and implementation of a robotic hand prosthesis with the use of memory form alloys during 2004. The project was declared unsuccessful because the tests and analyses carried out showed that the SMA wires received the necessary electrical pulse, but the intelligent material was not able to generate any type of movement and presented problems of excessive heating [7].

The Saarland University in Germany was responsible for carrying out the project "Bionic hand based on bundled Nitinol wires activated by electricity", which is one of the best projects carried out. The research team inserted Nitinol wires into a plastic arm to work as small muscles that can pull on individual fingers to open and close.

According to the reports, they managed to move each phalanx individually by controlling the intelligent cables through a single semiconductor chip [8].

Charles Pfeiffer and Kathryn DeLaurentis, from the Engineering Department of the University of New Jersey developed the project: "Shape Memory Alloy Actuated Robot Prostheses: Initial Experiments", in which they implemented an articulated arm based on human anatomy using artificial muscles with SMA. Two prototypes were preset that emulate the structure of the arm and fingers of the hand. They concluded that it is possible to generate large movements and that their use is possible for the design and implementation of prostheses [1].

With the exposed information and taking into account that no evidence of SMA prosthesis development was found in Ecuador, the present project has as its purpose the creation of a robotic hand through the use of SMA springs, to evaluate the functioning of intelligent materials and be able to have a basic prototype on which to start researches for the development of new technologies.

2 Fundamentals of Shape Memory Alloys. Nitinol

The properties of form memory and super-elasticity are associated with a phase transformation into solid state, the thermo-elastic Martensitic transformation. The term Austenite, or beta phase (β), is used for the mother phase, stable at high temperatures, and the term Martensite, for the phase that comes from Austenite by a Martensitic transformation. These phases can be consulted in Massalski's book, "Binary Alloy Phase Diagrams" [9].

Form Memory Alloy (SMA) transformation temperatures are used to characterize the material. These temperatures are Ms and Mf (Martensite start and Martensite finish). The foundation of this principle is based on a phase transition between an Austenite and a Martensite structure [9].

Austenite is a form of cubic arrangement of the crystalline structure of solids formed by iron and carbon atoms. On the other hand, Martensite is a crystalline phase in ferrous alloys that is produced from a transformation of phases without infiltration of foreign particles, at a speed close to that of sound in the material. These physical properties mean that these alloys have a unique response to physical and temperature changes, which is why they have been classified as intelligent materials that are frequently used to generate forces, movements or even store energy [2].

The phase change of Austenite and Martensite forms a hysteresis cycle or loop, characterized by 4 characteristic or transformation temperatures that are: start temperature of the Martensite phase (Ms) end temperature of the Martensite phase (Mf, start temperature of the Austenite phase (As), end temperature of the Austenite phase (Af). Super-elasticity in an SMA occurs in a temperature range starting at the final temperature of the Austenite Af phase up to the maximum temperature for transformation of Martensite by stress Md [2, 10].

2.1   Use of Nitinol as a Shape Memory Alloy

Nitinol is a memory shaped alloy of 55%–56% Nickel and 44%–45% Titanium with almost the same molar concentrations. Electrically, it behaves like a resistance that varies according to its length and diameter. It is presented in two stable phases: one, at high temperature, which is Austenite and the other, at low temperature which is Martensite, as mentioned above and which are responsible for the super-elasticity of the material. Table 1 summarizes the main mechanical properties of this material [4].

Table 1. Mechanical properties of NiTi alloys

Properties	Austenite	Martensite
Maximum tensile strength (MPa)	800–1500	103–1100
Elastic limit (MPa)	100–800	50–300
Elastic module (GPa)	70–110	21–69

2.2   Advantages and Disvantages of Nitinol Compared to Other Alloys

This material has a greater memory capacity in proportion of 8% in relation to other materials with shape memory. Its electrical resistivity is greater in comparison with other alloys allowing greater ease in the activation of the material when an electric current is induced, it is more resistant to environmental factors such as corrosion and thermally it is more stable which allows a greater temperature range for the change of phases. However, the cost of Nitinol is higher than other alloys and is more difficult to mechanize, and the hysteresis cycle is very pronounced, which causes large deformations that prevent maintaining a control proportional to temperature variations [4, 11].

2.3   Shape Memory Control in Nitinol

According to revised information from DYNALLOY INC, world leader in the manufacture of shape memory alloys based on Nitinol, it has been found that there is a relationship between the force produced in the material, between the required current and the diameter of its structure. The greater the area of the cross section of Nitinol, the greater the force that will be produced without depending on its length; but the greater the area, the greater the current will be necessary, allowing the material to reach its state change temperature [4].

Since the discovery of Nitinol in 1963 at the Naval Ordinance Laboratories-NOL, which gave rise to the acronym Nitinol for alloy, different types of systems to control shape memory have been proposed, ranging from complex PID (Proportional Integral Derivative) systems, position control systems as a function of material temperature, position control systems as a function of the resistive variation of the intelligent material, and electronic control systems using PWM (Pulse Width Modulation) signals. The latter is the most accepted due to the ease of implementing electronic circuits that generate PWM signals and the ease of handling its position by applying pulses with a predefined width.

2.4   Calculations of the Nitinol Springs Use in the Prosthesis

Nitinol springs are a form of physical presentation of this material that undergoes a process that prevents the new form adopted from returning to its wire form. In this way, it will have the same properties as the SMA and, to the same extent; it will adopt the properties of the springs, which allows the material to produce linear movements in two directions. An important parameter in Nitinol springs is the SR stretch ratio, which makes the difference with the wires of this material and is directly proportional to any length L and inversely proportional to the length of the spring in its original solid form SL, according to the following equation:

$$SR = \frac{L}{SL} \tag{1}$$

This characteristic can be present in both the Martensite and Austenite phases [4]. For the calculations, some important technical considerations of the material were taken into account such as recovery force, given by the weight of the joint and the friction force in the joint joints: $F_d = m * g$ where F_d is the weight of each joint. The frictional force in the joint is given by: $F_r = \mu * N$, where $\mu = 0,6$ represents the coefficient of friction for its plastic state and N the normal force on the joints. In addition, the characteristics of the Nitinol spring used were taken into account, such as: diameter 750 μm, deformation force 6 N, length variation of 14 cm for 6 N, restoring force 8,4 N (F_{niti}), working temperature range in austenite phase 75 °C–85 °C, outer diameter of the coil 6 mm, contraction of 1,8 s with a current of 2 A [4]. Therefore, the calculations performed are as follows:

$$F_{niti}8,4N > F_d + F_r$$
$$F_d = m * g, F_r = \mu * N = \mu * m * g \tag{2}$$

$$F_{total} = (1 + \mu)m * g \tag{3}$$

Where m = 88 g, average mass of a finger and:

$$F_{total} = (1 + 0,6)0,088Kg * 9,8\frac{m}{s^2} = 1,37N$$

As $F_{niti} = 8,4N > F_{total} = 1,37N$, it is concluded that the spring of 750 μm has the necessary force to generate hand movements. In addition, with the data we can obtain the coefficient of elasticity:

$$F = K * \Delta x, \quad K = \frac{F}{\Delta x} = \frac{6N}{0,14m} = 42,85N/m \tag{4}$$

3   Design of the Robotic Prosthesis with Nitinol Springs

The SMA operate in a relationship of temperature and motion, which can be controlled by the flow of current through the alloy; therefore it was decided to implement a circuit that manages the flow of energy through a PWM signal generated by a PCA9685 module [10]. This module is an integrated electronic circuit, which allows generating 16 PWM output channels with a resolution of 12 bits and a single communication channel I_2C to a microcontroller and programmable frequency within the range of 40 Hz to 100 Hz with a useful cycle, which can vary between 0% and 100% [12].

All handheld functions are controlled from a smartphone device, which is why an Android application was developed. With a Bluetooth connection, orders are sent to execute the desired types of movements. Once the device is initialized, the Arduino controller performs a temperature and force measurement until the command is received from the mobile device. When the Arduino receives the data sent from the smartphone through the Bluetooth HC-05 module connected to its serial port, it performs the verification and chooses the type of movement selected and then sends a command to the PWM channel, which will be used from the PCA 9685 module, by means of a connection I_2C. The selected channel produces the PWM signal that is sent to the Nitinol intelligent material control circuit.

3.1   Electronic Current Circuit

For the design of the power circuit 4N25 optocouplers were used, which can work with voltages of up to 30 V to ensure proper isolation of the digital power circuits, because the energy source used for the prosthesis is 11 V and high amperage. The design for the optocoupler is shown in Fig. 1 where the value of R1 = 330 Ω, which is very easily calculated by assuming a voltage drop of the diode of 1,7 V. The power circuit was designed with a transistor MOSFET IRFZ44 N that has a fast commutation ON/OFF, a low drainage resistance of 15 mΩ, maximum drainage current of 49 A and a maximum dissipation power of 110 W, which is according to the received PWM signal.

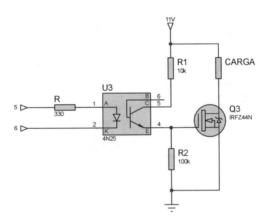

Fig. 1. MOSFET IRFZ44N polarization circuit

As long as there is no current circulating through the polarization circuit formed by the transistor of the optocoupler and resistors R4 and R5, the MOSFET transistor behaves like an open circuit between the source and drain terminals, which prevents current, circulating through the Nitinol spring. If the optocoupler becomes saturated, a current is delivered to the MOSFET gate, which causes the MOSFET to behave like a closed circuit and, in this way; the Nitinol spring starts its contraction process according to the established memory. The design results are shown below.

The SMA will work in ranges of up to 3 A, which in turn will be the maximum current circulating through the transistor and the power dissipated in the form of heat is 52.5 mW, which is calculated using the Eq. (5).

$$P = Rds * IDs^2 \tag{5}$$

3.2 Dimensioning and Design of the Energy Source

The consumption of the circuit is 3,86 W plus a tolerance of 10%, which gives a power of 4,24 W, from which we started for the calculation of the battery considering the most critical case of autonomy for 20 h, depth of discharge, average temperature of the battery of 25 °C and battery voltage 11 V. According to the calculations with Eq. 6, the LIPO 11.1 V/10000 mA lithium battery was chosen

$$Capacidad = \frac{(4,24W * 20h) + (4,24 * 20h) * 30\%}{11 volts} = 10,03Ah \tag{6}$$

Table 2 below details the current consumption and dissipation powers of all the components used in the project, with which the calculations necessary to correctly dimension the energy source to be used were carried out.

The LIPO lithium battery of 11 V and 10 A/h coupled with a variable voltage regulator (LM317T) and capacitors of 10 and 100 μF, can deliver an output voltage in the range of 1.5 V to 11 V, controlled through a potentiometer μF. The LM 317T delivers a maximum current of 1.5 A, so it was necessary to use a transistor bypass formed by a resistance of 1 Ω/10 W through which the MOSFET PNP TIP 147

Table 2. Total energy consumption of the robotic hand prototype

DEVICE	Current consumption	Supply voltage	Power W
Arduino Nano Module	15 mA	5	0,075
PCA9685 Module	35 mA	5	0,175
HC05 Bluetooth Module	50 mA	3,3	0,165
Lm35 Sensor	60 uA	5	0,0003
FSR 402 Sensor	0,35 mA	5	0,00175
4n25	60 mA	1,15	0,069
Nitinol	2 A	0,08 ohm	(0,32 * 5)
Voltage Regulator	160,41 mA	11.1	1,78

transistor is energized through which will circulate a maximum current of 10 A emitter collector to be delivered to the regulator output as shown in Fig. 2. A second power supply stage of 5 V based on LM7805 was designed to provide power to the control stage and sensors.

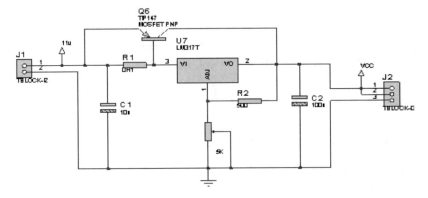

Fig. 2. Power supply

3.3   Mechanical Design of the Hand Structure

The joints that make up the prototype of the hand prosthesis are made from a 3D model obtained in Thingiverse [13]. The template was modified according to the needs of movements through the 3D Builder program. This assistant was used to design the elements of the hand and forearm structure, which has bases where the electronic cards will be co-located. Once the printed circuits were placed in the forearm base structure, a protection was installed that separated the internal part from the external part. Figure 3 shows the final design of the mechanical structures of the prototype.

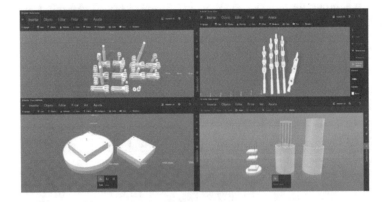

Fig. 3. Structures of the fingers and forearm

3.4   Programming for the Control of the Prosthesis

For the control of the prosthesis from a mobile device was designed an application in App inventor version 2, which is a free platform for Android. The programming initially consisted of configuring a startup screen using the screen1 function. Afterwards, the application screen was programmed, from which the movements of the designed hand will be controlled. For this phase the Button commands were used, which allow you to create Label buttons. They allow creating List Picker texts to produce Horizontal Layout lists that are used in the separations, Bluetooth that activates the Bluetooth function and Clock to generate counters. In addition, we proceeded with the programming of the selection of the type of movement for which there are two buttons in the mobile application, which allow to activate or de-have the movement of the memory alloys in the following way.

Two stages were defined for programming the Arduino module: sensors and control. In the sensor stage, the variables to be measured were considered, such as force, temperature and voltage, which will be read by the microcontroller and then conditioned for quick reading by the user.

As for the control stage, the serial communication enablement was defined to perform the information exchange between the smartphone device and the Arduino. Once the variables and libraries are started in the controller, the program expects to receive a signal by serial communication and if it receives any information, it compares the value with the data stored in a variable and if these coincide, the functions corresponding to each movement are started.

3.5   Assembling the Hand Prosthesis

The electronic circuits described above and the Arduino module were implemented in two sections of a 3D-designed plastic structure that simulates a person's forearm. It was taken into account for the design, leaving the necessary free space between the components and the location of the modules, because a certain amount of heat is generated and, therefore, it is essential to place dissipaters. Figure 4 shows the implemented structure of the hand prosthesis made with the placement of the outer covers that protect the electronic circuits and the Nitinol springs.

3.6   Results and Discussion

According to the information provided by the manufacturer, Nitinol springs can shrink in less than 1 s after being subjected to a current of 2 A [4]. During the tests it was found that the contraction depended - to a large extent - on the applied pulse, but the effective current applied by the PWM pulse was less than the constant supply current required, which caused the increase of the contraction time of the Nitinol spring.

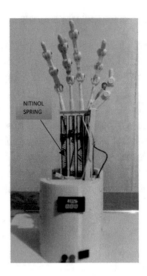

Fig. 4. Prototype of the assembled hand prosthesis

The tests performed, varying the pulse and frequency are shown in Table 3 below.

Table 3. Functional tests

ARTICULATION	FREQUENCY (Hz)	PULSE WIDTH	PULSE WIDTH %	CONTRACTION TIME (sec)	DISPLACEMENT (mm)
Index	2	1024	25%	5	20
Index	2	2048	50%	5	20
Index	2	4000	97.7%	5	20
Index	2	1024	25%	10	20
Index	2	2048	50%	10	20
Index	2	4000	97.7%	10	20
Index	100	1024	25%	5	20
Index	100	2048	50%	5	20
Index	100	4000	97.7%	5	20
Index	100	1024	25%	10	20
Index	100	2048	50%	10	20
Index	100	4000	97.7%	10	20
Index	350	1024	25%	5	20
Index	350	2048	50%	5	20
Index	350	4000	97.7%	5	20
Index	350	1024	25%	10	20
Index	350	2048	50%	10	20
Index	350	4000	98%	10	20

The results of these tests conclude that the best ratio of displacement of the Nitinol springs as a function of pulse width variation was achieved at a frequency of 2 Hz and that the variation of pulse width allowed to decrease the contraction time of the alloy, thus achieving greater speed in generating movements. However, we also found that the increase in the velocity of spring movements causes a higher current consumption with the corresponding increase in temperature, which in turn implies that the cooling period of Nitinol to reach the Martensite phase is longer.

In the test stages, we had damage to the Nitinol springs due to the adaptation of a new unwanted shape memory, as a consequence of the stresses produced in the material while applying a pulse width of 3000 to 4000 beads of 4096 (12 bits). Therefore a pulse width with a maximum flank of 2500 was used, which avoids damaging the Nitinol springs during the transition from Austenite to Martensite phase.

The research found unresolved problems that must be worked on in the future in a second stage of development. The first is that the power supply integrated by the LIPO battery and the regulation circuits increases the size and weight of the structure of the prosthesis, damaging the portability of the same.

The experience gained in the use of Nitinol allowed us to prove that great forces can be generated with them however, there are limitations to cause smooth and precise movements, so we suggest in a future development, using a combination of spring and Nitinol wires to solve this difficulty.

We must also improve the generation of movements by means of the PWM signal, because we detected inconveniences when we wanted to provoke the simultaneous movement of the five fingers. It was possible to establish the proposed movements in the robotic hand, but not with the same degree of similarity as a real hand due to the physical limitations of the upper structure of the prosthesis, therefore it is also important to improve the structural design and use new techniques to reduce the dimensions of the electronic circuits in order to ensure a lower weight and overall size. On the other hand, the development of a computational algorithm allowed a mobile device to communicate with Arduino modules via wireless communication, which resulted in an achievement in terms of the flexibility of the prosthesis to integrate with different current electronic technologies, providing scalability for continuous improvements in the future.

4   Conclusions

The design met the objective proposed for a first stage, which consisted of generating five types of movements in a hand prosthesis without the use of motors.

A control system was designed using the Arduino module, electronic circuits with optcouplers and MOSFET transistors that allowed the movements defined for the hand to be controlled using PWM signals according to the proposed scope.

It was taken into account, based on experimentation, the relationship between increased speed of movements, current consumption and temperature increase; however, the cooling system of Nitinol must be perfected, in order to achieve faster cooling process and, consequently, transitions between the changes of form.

It is recommended in the future to modify the design of the base structure of the prototype in order to place more springs in parallel to increase the force in each hand joint and to use myoelectric muscle sensors that will allow voluntary hand movements, controlled by the user without the need to use external devices.

References

1. Pfeiffer, C., DeLaurentis, K., Mavroidis, C.: Shape memory alloy actuated robot protheses: initial experiments. IEEE International Conference on Robotics and Automation, vol. 3 (1999)
2. Gómez, A., Restrepo, C.: Cables Musculares. EIA, no. 4, 103–111 (2005)
3. Loaiza, J., Arzola, N.: Evolución y Tendencias en el desarrollo de Prótesis de Mano. DYNA **78**(169), 191–200 (2011)
4. DYNALLOY, INC, "DYNALLOY Makers of Dynamic Alloys" 7 January 2018. http://www.dynalloy.com/tech_data_wire.php
5. Lahoz, R., Puértolas, A.: Entrenamiento y memoria de Forma de Doble Camino. In: VIII Congreso Nacional de Propiedades Mecánicas de Sólidos, pp. 185–194 (2002)
6. Villoslada Peciña, Á.: Diseño y aplicación de un actuador SMA en. Universidad Carlos III, Madrid (2010)
7. Dorador, J., Ríos, P., Flores, I., Jurez, A.: Robótica y Prótesis Inteligentes. Revista Digital Universitaria, vol. 6, no. 1, 18 enero 2004
8. ASM International, "SMST, Shape Memory & Superelastic Technologies." ASM, 05 October 2015. www.asminternational.org/web/smst/newswire/-/journal_content/56/10180/25825248/NEWS
9. Massalski, T., Okamoto, H., Subramanian, L., Kacprzak, L.: Binary Alloy Phase Diagrams, 2nd edn. ASM International, Cleveland (1990)
10. Moreno Martínez, G.: Control de posición usando Shape Memory Alloy SMA. Universidad Politécnica de Cartagena, Cartagena (2014)
11. Salinas Chavez, J.A., Caracterización de aleaciones NiTi y NiTiCu elaboradas por fusión inducida al vacío y por inducción de plasma. UANL, Nuevo León (2011)
12. Naylamp Mechatronics, "Naylamp Mechatronics," 6 December 2016. http://www.naylampmechatronics.com/blog/41_Tutorial-M%C3%B3dulo-Controlador-de-servos-PCA9685.html
13. Thingiverse, "Thingiverse," MakerBot Industries, LLC (2019). https://www.thingiverse.com/

Dielectric Permittivity Measurement of Kraft Paper by the Resonant Ring Method

Luis A. Morocho, Leonidas B. Peralta, Luis F. Guerrero-Vásquez$^{(\boxtimes)}$,
Jorge O. Ordoñez-Ordoñez, Juan P. Bermeo,
and Paul A. Chasi-Pesantez

Universidad Politécnica Salesiana, 010105 Cuenca, Ecuador
{lmorochoml, lperaltap}@est.ups.edu.ec,
{lguerrero, jordonezo, jbermeo, pchasi}@ups.edu.ec

Abstract. Dielectric permittivity is one of the most important parameters in the field of telecommunications, since it indicates an interaction value between the substrate and an electric field. Lately the paper has had a great acceptance for the development of new technologies, therefore, the permittivity of the Kraft paper was measured to be used as a substrate, in telecommunications applications. For the measurement of the dielectric permittivity of Kraft paper, a resonant ring structure was designed. The design of the ring requires a frequency of operation or resonance, moreover, to an impedance coupling between the measuring equipment and the resonator. The structure of the resonant ring was simulated obtaining the coefficient of direct transmission (S21), validating the design for its construction. S21 was acquired by the NI-PXI instrument in a practical way; this parameter allows the calculation of the effective dielectric permittivity and later the relative. Finally, the process revealed a value of 1.40, relative dielectric permittivity for Kraft paper.

Keywords: Resonant ring · Dielectric permitivity constant · Substrate

1 Introducción

The technological proposal of the last years, drives to develop portable and flexible devices with a low cost, motivating to innovate in the materials used for the construction of said devices [1]. Specifically, in telecommunications area there is a tendency to venture into the use of new types of materials, and among the most used for radio frequency applications, some types of flexible polymers, textiles, and paper are considered, the latter being the most economical material to produce [2]. This means that paper has been considered as an excellent substrate for the development of new technologies during the last decades, especially focused on microwave applications. To achieve this goal, it is essential to analyze the electrical properties of this material, which will allow us to know more precisely its possible benefits.

In telecommunications, expressly in the radio frequency area, there is a strong technological deployment based on the development of Microstrip devices, whose construction is mainly based on chemical processes, which are carried out through the use of acids and toxic compounds, causing a significant impact on the environment and

© Springer Nature Switzerland AG 2020
M. Botto-Tobar et al. (Eds.): ICAETT 2019, AISC 1067, pp. 271–279, 2020.
https://doi.org/10.1007/978-3-030-32033-1_25

in the health of people [3]. However, to remedy the environmental impact generated by the construction of these devices, the paper could be used as a possible solution, since this is a highly recyclable material, furthermore have a low cost of production compared to other substrates. [4]. Moreover, Kraft paper, being made of wood pulp has a high level of resistance to tearing and traction, these features make it a good candidate used in the development of devices that require a combination of strength and flexibility [5].

In this article, both the dielectric constant of Kraft paper and the experimental method used to obtain it are analyzed. In the first instance, the measurement of the effective dielectric constant of the Kraft paper was carried out, for which a ring resonant circuit with an attenuation estimate of the wave transfer in a microstrip line was used, this was done because circumference mean of the resonator ring is equal to an integer multiple of the guided wavelength to establish resonance. The ring was designed to operate at a frequency of 3 GHz, with a coupling impedance of 50 Ω for the measuring equipment. The structure of the resonator ring is simulated in ANSYS HFSS obtaining the Direct Transmission Coefficient (S21) and validating the design for its construction. The S21 is acquired by the NI-PXI instrument in a practical way, this parameter allows the calculation of the effective dielectric permittivity and later the relative one.

2   Methodology

Based on the resonant ring method for the measurement of dielectric permittivity, we started from the theory of the same, which allows the design of the ring under the parameters of frequency, impedance and height of the substrate. For the first simulation of the ring in the Kraft paper substrate, an initial value of the dielectric constant of the substrate is required, so it was assigned an average permittivity value of papers used in radiofrequency applications. This value is 1.72 for papers with a grammage of 80 gr/m^2 [6], considering that Kraft paper used for experimentation has a similar grammage [7]. In Fig. 1 the process that was carried out to develop the article is shown.

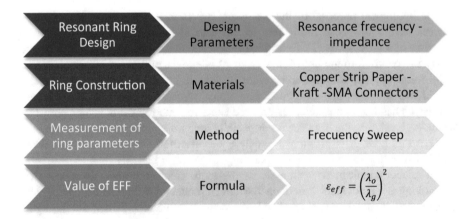

Fig. 1. Block diagram of the methodology used.

The measurements for the ring resonator, were calculated for a frequency of 3 GHz with an impedance of 50 Ω for its feeders, under these parameters the same was built. After this, the resonance frequency of the ring was evaluated by sweeping frequencies from 85 MHz to 6 GHz, providing a graph of the S21 parameter. The resonance frequency given by the S21 is 2.4 GHz, with this value we proceeded to calculate the ε_eff, in addition, the parameters of the feeders. The value of ε_eff allowed the calculation of relative permittivity or dielectric constant ($\varepsilon_r = 1.4$).

With the result $\varepsilon_r = 1.4$ the resonator ring was redesigned, again evaluating the resonance frequency with a result of S_{21} in 3 GHz, followed by the calculation of ε_r ensuring the value of 1.4.

3   Resonator Ring

The resonator ring used is a transmission line resonator, which has a circular geometry, that is excited by a pair of transmission lines [8], as shown in Fig. 2. Using a ring resonator circuit, measurements of the effective dielectric constant of a material can be carried out, using an estimate of the wave transfer attenuation in the Microstrip line [9].

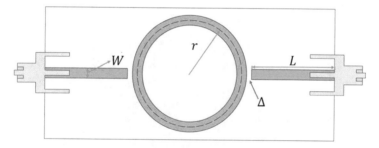

Fig. 2. Diagram of a resonator ring. W (power line width), L (feed line length), r (middle radio), Δ (coupling gap).

When the average circumference of the ring resonator is equal to an integer multiple of a guided wavelength, resonance is established [10]. This can be expressed as:

$$L = 2\pi r \tag{1}$$

For $n = 1, 2, 3, 4. \ldots$

$$L = \frac{1}{2}n\lambda_g \tag{2}$$

Where n is the mode number, λ_g is the guided wavelength, and r is the middle radio of the ring, that is equal to the average of the external and internal radii. From this equation, we can calculate the resonant frequencies for different modes of λ_g [10–12].

For a ring resonator, λ_g, the relation with the frequency is expressed with the following equation:

$$\lambda_g = \frac{\lambda_o}{\sqrt{\varepsilon_{eff}}} \tag{3}$$

Where ε_{eff} is the effective dielectric constant, λ_g is free space wavelength, λ_o is the resonant frequency. The coupling gap (Δ) is the space between the power line and the resonant ring [13, 14], and is given by the next expression:

$$\Delta = 0.41\,h\left(\frac{\varepsilon_{eff}+0.3}{\varepsilon_{eff}-0.285}\right)\left(\frac{\frac{W}{h}+0.262}{\frac{W}{h}+0.818}\right) \tag{4}$$

3.1   Microstrip Power Lines

The Microstrip line is one of the most popular types of flat transmission lines; can be easily integrated with other passive and active components or microwave devices [10, 15].

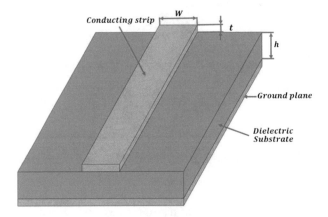

Fig. 3. Microstrip line.

In Fig. 3 a wide W conductor is observed, this is printed on a high dielectric substrate h and relative permittivity ε_r [15, 16], is given by:

$$\frac{W}{h} = \frac{8e^A}{8e^{2A}-2} \quad para \quad \frac{W}{h} < 2 \tag{5}$$

$$Where \quad A = \frac{Z_o}{60}\sqrt{\frac{\varepsilon_r+1}{2}} + \frac{\varepsilon_r-1}{\varepsilon_r+1}\left(0.23+\frac{0.11}{\varepsilon_r}\right) \tag{6}$$

3.2 Design Parameters

Kraft paper has a thickness g = 0.16 mm, then ten sheets of Kraft are stacked, reaching a thickness of h = 1.6 mm for simplicity and feasibility towards the connector SMA. With the already established values of resonance frequency, resistor impedance and substrate thickness, the next step is to use the Eqs. (1), (2), (3), (4), (5), (6), to obtain the construction measures of the resonator ring.

Table 1. Construction Measures of the resonator ring.

Parameter	Measure (mm)
Middle radio (r)	13.9442
Power line width (W)	6.8801
Feed line length (L)	25
Coupling gap (Δ)	0.9211

3.3 Simulation

For the simulation process, the ring was built in the HFSS software, according to the characteristics of the materials to be used, as is the Kraft paper substrate with a height h = 1.6 mm and the copper plate that acts as a conductor with a height of t = 0.04 mm. In addition, was taken into account the measures of the Table 1. Construction Measures of the resonator ring (Fig. 4).

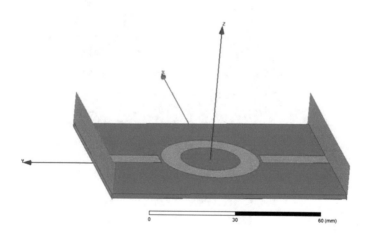

Fig. 4. Ring resonator in the HFSS software.

3.4 Construction and Measurement

For the construction, it was started from a template that is appreciated in Fig. 5, in which the resonant ring was drawn with the exact measurements according to the calculation.

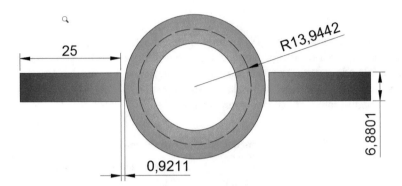

Fig. 5. Ring resonator drawn in AutoCAD (values are in millimeters).

The copper tape was cut according to the shape of the template made of the resonator, using a cutting plotter machine. Finally, we have the gluing of the copper elements, such as the feeders and the ring resonator on Kraft paper. In addition, the SMA connectors were welded on each end of the feeders (Fig. 6).

Fig. 6. Ring resonator in Kraft paper.

The effective dielectric constant is obtained based on the length λ_g of the ring and the measured resonance frequency λ_o of the resonator ring, this last variable was measured by a sweep of frequencies, thus obtaining a graph of the parameter S_{21}.

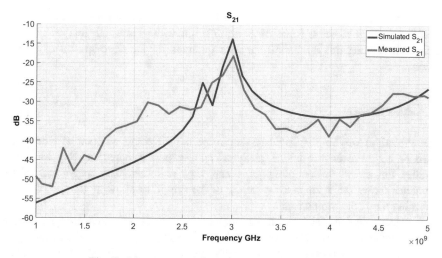

Fig. 7. Measurement of parameter S_{21} in the resonator.

In Fig. 7, the resonance frequency of the ring is shown in the previous figure, from this frequency, its wavelength is calculated and with the ring length the measurement of the effective dielectric constant of the substrate is carried out in this case, the Kraft paper.

$$\sqrt{\varepsilon_{eff}} = \frac{\lambda_o}{\lambda_g} \tag{7}$$

$$\varepsilon_{eff} = \left(\frac{\lambda_o}{\lambda_g}\right)^2$$

Getting the effective dielectric constant of the substrate ε_{eff}, we can calculate the relative dielectric constant ε_r of essential substrate. The ε_r is calculated by clearing this constant from the following equation:

$$\varepsilon_{eff} = \frac{\varepsilon_r + 1}{2} + \frac{\varepsilon_r - 1}{2} * \frac{1}{\sqrt{1 + 12\frac{h}{W}}} \tag{8}$$

Based on the Eqs. (7) and (8) the parameters ε_{eff} and ε_r are calculated. The results obtained in Table 2 are shown.

Table 2. Measurement and calculation of ε_{eff} and ε_r for Kraft paper.

Parameters	
Effective dielectric constant $\left(\varepsilon_{eff}\right)$	1.3027
Relative dielectric constant (ε_r)	1.4000

The previous table shows that $\varepsilon_r = 1.4$ for Kraft paper, this value is similar to that of white paper that has a value of $\varepsilon_r = 1.72$, but with greater advantages of tensile strength indicating this as a substrate with qualities for the development of wireless systems.

4   Conclusions

A new method was implemented to determine the dielectric constant of Kraft paper based on a ring resonator device, analyzed with respect to parameter S_{21}. Model S_{21} includes the effects of coupling gaps and ring radiation. The magnitude of this parameter shows the resonance frequency of the ring, thus allowing the calculation to determine the dielectric constant.

A planar microstrip ring resonator structure was developed. This structure consists of a self-adhesive copper sheet on a stack of Kraft paper rectangles, thus proving the effectiveness of this type of paper as a substrate, opening a channel for future microwave applications of flexible structure, disposable and very low cost. In addition, when using this type of materials, it contributes to the reduction of the environmental impact, which currently exists as a result of the manufacture of PCB printed circuits.

The resonance frequency of the ring has the same value both in the simulation with the HFSS software and in the measurements made in practice, which indicates that the estimation of the dielectric constant by the resonant ring method has a high degree of reliability. All this is based on the equations described in Sect. 3 of the resonant ring, which allowed the design of the ring.

The equations used allow the design of a resonator ring according to the frequency of operation required and depending on the thickness of the Kraft paper base, on which the level of flexibility of the structure depends. The measurements made of the parameter S_{21} made it possible to calculate the ε_r for Kraft paper, resulting in a value of 1.4, the same that was checked according to the resonance frequency of the ring both in the simulation, and in practice.

Kraft paper used as a substrate has a great advantage in terms of flexibility and strength, however, it is not as good as FR4, therefore, it is recommended for future work, to carry out a wider study on different types of paper or polymers that have better characteristics, even as flexible options and waterproof materials.

References

1. Thompson, D.C., et al.: Characterization of Liquid Crystal Polymer (LCP) Material and Transmission Lines on LCP Substrates From 30 to 110 GHz, no. May (2004)
2. Bozzi, M., Moscato, S., Silvestri, L., Delmonte, N., Pasian, M., Perregrini, L.: Innovative SIW components on paper, textile, and 3D-Printed substrates for the internet of things. In: Asia-Pacific Microwave Conference Proceedings, APMC, vol. 1, pp. 1–3 (2016)

3. Moro, R., Bozzi, M., Kim, S., Tentzeris, M.: Novel inkjet-printed substrate integrated waveguide (SIW) structures on low-cost materials for wearable applications. In: 2012 42nd European Microwave Conference, pp. 72–75 (2012)
4. Kim, S., Vyas, R., Georgiadis, A., Collado, A., Tentzeris, M.M.: Inkjet-printed RF energy harvesting and wireless power trasmission devices on paper substrat. In: 2013 European Microwave Conference, pp. 983–986 (2013)
5. Roberto Juacida, P., Sandra Rodriguez, S., Marco Torres, U.: Composición química, obtención de pulpa Kraft y su evaluación papelera en castaño, ciprés y encino. Bosque **23**(1), 125–130 (2002)
6. Mouzouna, Y., Nasraoui, H., Mouhsen, A., El Aoufi, J., Chababi, G.: Miniaturized meander antenna using low cost paper substrate. In: 2016 5th International Conference on Multimedia Computing and Systems, pp. 447–450 (2016)
7. Teschke, K.: Industria del papel y de la pasta de papel
8. Waldron, I.: Ring Resonator Method for Dielectric Permittivity Measurement of Foams. Worcester Polytechnic Institute (2006)
9. Rida, A., Yang, L.: Conductive Inkjet-Printed Antennas on Flexible Low-Cost Paper-Based Substrates for RFID and WSN Applications CONDUCTIVE INKJET PRINTED ANTENNAS ON FLEXIBLE LOW-COST Presented to The Academic Faculty By Amin H. Rida In Partial Fulfillment Of the Requireme, no. July 2014, 2009
10. Shebani, N.: Design curves of microstrip ring resonator design curves of microstrip ring resonator, no. November (2016)
11. Bogner, A., Steiner, C., Walter, S., Kita, J., Hagen, G., Moos, R.: Planar Microstrip Ring Resonators for Microwave-Based Gas Sensing : Design Aspects and Ammonia Sensing (2017)
12. Chen, J., Chen, J.: Tunable Resonances in the Plasmonic Split-Ring Resonator, no. January 2015, 2014
13. Le, T., et al.: A novel strain sensor based on 3D printing technology and 3D antenna design A Novel Strain Sensor Based on 3D Printing Technology and 3D Antenna Design, no. July 2015, 2016
14. Semouchkina, E., Cao, W., Mittra, R.: Modeling of Microwave Ring Resonators using the Finite-difference Time-domain Method (FDTD), vol. 24, no. 6, pp. 392–396 (2000)
15. Balanis, C.A.: Antenna Theory 3rd. (2005)
16. D. M. U. of M. at A. Pozar. Microwave Engineering, 4th edn. (2011)

Implementation of a Wireless Network for an ISP in the Alhajuela Community from Ecuador

Flavio Morales$^{(\boxtimes)}$, Hernan Valencia, and Renato M. Toasa⬤

Universidad Tecnológica Israel, Quito, Ecuador
{fmorales,hvalencia,rtoasa}@uisrael.edu.ec

Abstract. This work was born as a necessity for the reality lived by the sectors far from our country, such as Alhajuela, we emphasize in research and significant contribution with the link to society, which requires thematic problems that can be solved satisfactorily with this type of implementations. The implementation of this wireless network, currently benefits more than 20 households whose population is engaged in agriculture, being outside the benefit and technological development, in many cases due to the high cost of investment in infrastructure, low population density and little economic interest generated by its location, in a rural area of Ecuador. Allowing them access to the Internet and increasingly shortening the existing digital divide in our country, showing reliable results with quantitative values in terms of existing connectivity, which allow to justify a significant growth, which supports universal access, which constitutes a citizen's right. The project started with 25 Mbps and currently works with 50 Mbps, with the projection of continuing to grow, demonstrating the success of this work.

Keywords: Wireless network · Link to society · Internet access · Alhajuela community

1 Introduction

The Internet access service in Ecuador represents one of the services with the highest demand and growth [1], mainly due to the amount of content generated and shared through the network, the development of applications and access to social networks. Denoting a high interest in promoting the increase in the penetration of fixed and mobile broadband services, as indicated in the National Plan of Telecommunications and Information and Communication Technologies 2016–2021 of the Ministry of Telecommunications and Information Society from Ecuador [1].

The National Plan of Telecommunications and Information Technologies of Ecuador aims to be the instrument of planning and management of the telecommunications sector, allowing to achieve greater digital inclusion and competitiveness of the country. The vision projects to locate Ecuador as a regional reference in the year 2021 in terms of connectivity, access and production of Information and Communication Technology (ICT) services, in benefit of the country's economic and social development [1].

© Springer Nature Switzerland AG 2020
M. Botto-Tobar et al. (Eds.): ICAETT 2019, AISC 1067, pp. 280–290, 2020.
https://doi.org/10.1007/978-3-030-32033-1_26

Something important to consider is to promote the use of Infocentros, to achieve two objectives; increase the number of people trained in ICT and ensure connectivity and infrastructure in Infocentros and schoolsn [1]. The trend with respect to fixed Internet access shows that there has been growth in the demand for the service, 11.48% for the fourth quarter of 2018 (10.61% in the fourth quarter of 2017 [2]) Which represents, a growth in the Internet access service; however, connectivity by wireless means does not represent it, on the contrary, it decreases by 0.19%.

In Ecuador, there is a significant number of initiatives and projects supported by ICT as strategic tools, which seek to contribute to improving the quality of life of the population and especially the less favored sectors. The inclusion of radio, video, television, telephone, cell phone, email, Internet, among others, are contributing to the construction of more equitable conditions. Its access constitutes a citizen's right [3]. However, the impact of these initiatives is limited, due to the few spaces for knowledge exchange, collective learning and generation of strategic alliances, added to this the inequality that exists in the country for access to these tools, generating an illiteracy technology at all levels, from children to adults [4].

In order to promote the access and use of ICT in the rural sector, this work was proposed, whose objective is: Implement a wireless network using a Primary Node in the Alhajuela community with Standard Wireless technology 802.11a/n/ac; where it is established that a primary node is the one where the network management centers are connected [5]. In the work [6], it is indicated that the community presents some problems related not only to connectivity, for example its access roads are in poor condition, transport units that do not provide a good service, Infocentros that do not work properly, finally the electric power service and public lighting is deficient. Additional shows the following data, in terms of telecommunications services: 95.11% of households do not have fixed telephony, 72.81% have cell phone, 94.86% do not have a computer and 98.07% do not have Internet service in their homes [6].

It is for this reason that the project proposed in this research is very important, since it will benefit around 20 households in the sector, providing access to the Internet, with a growth projection based on the technological changes of the current world, which have led to society making massive use of the development of Information and Communication Technologies (ICT). Narrowing the call "Digital Divide", making this generate distinction between those who have and do not, access to technology.

In this work we propose, as a first phase, to carry out the survey of the general data of the place, operation of the 802.11a/n/ac wireless technology and the necessary equipment. Then, design the distribution network, based on the results obtained from the research and the establishment of the elements that meet the requirements for the construction of the network in the Alhajuela community. Subsequently, implement two links, both a point to point and a point multipoint to establish communications, all supported by two radio bases. And finally perform performance tests of the service implemented.

The document is organized as follows: Sect. 2 presents the contextualization and state of the art with similar projects implemented in the rural sector of Ecuador, Sect. 3 considers the methodology and materials used where technical aspects of the imple-mented solution are shown, as well as the investment costs, Sect. 4 describes the results obtained in this investigation and finally Sect. 5 presents the conclusions and future work.

2   Contextualization and State of the Art

In rural and urban areas, universal access and the responsible use of ICT are a challenge, since the production of quality information content and effective communication with a development focus is still pending. The trend of individual use of ICT (Internet and cellular) and the unconscious consumption of the abundant information that exists are a barrier to the social use of ICT [7]. The application of ICT and other tools still depend on economic interests and political focus, the quality and coverage of services is not the same in the field, projects and community media are more regulated than encouraged, the costs of the equipment and services are not accessible [8], since these cost are very high.

In 2004, a project was executed as part of InfoDesarrollo [9], focused on communications in Short Wave HF, that allow the intercommunication between cocoa producers in the one province of Ecuador in remote communities. The project consisted of implementing a shortwave communications system plus the integration of digital modems, which allowed to transmit the audio with higher quality, since in the localities at that time there was no conventional and cellular telephone infrastructure, the main objective of the implementation was to allow the cocoa producers to coordinate the trucks that passed collecting the production for each of the communities.

In a study conducted in December 2017, the Internet World Stats page indicates the figures regarding Internet access worldwide and it can be seen that in Latin America, Ecuador is shown as a leader in the penetration rate of Internet access with 81%, above countries in the region such as Argentina (78.6%), Chile (77%), Brazil (65.9%), Venezuela (60%), Colombia (58.1) and Peru (56%) [10]. All this leads us to think that Ecuador has undergone a positive and encouraging change process regarding the use of ICTs and the high impact that this represents for a developing country.

When people in a community are given quick and cheap access to information, they will benefit directly from what the Internet offers. The time and effort saved by having access will result in a smaller value. In the same way, communities connected to the network at high speed have a presence in the global market where transactions happen quickly. People around the world have noticed that Internet access has provided a voice to discuss their problems and everything that is important in their lives, so that neither the telephone nor the television can compete. What until recently seemed an illusion, now becomes reality, and that is built on wireless networks [11].

In [12] it is affirmed that from 2012 to the end of 2014, digital illiteracy decreased, from 21.4% to 14.4% based on ICT training to 177,786 people, carried out in Infocentros Comunitarios and the deployment of an extensive fiber optic network, which reaches 42,758 km coverage throughout the continental country. In the Information Technology Report released at the World Economic Forum, which includes within its "First Pillar: Political and regulatory environment" a sub-index on "Laws related to ICT" and Ecuador is located in 62nd place in a ranking of 139 countries included in the study (p.203). The aforementioned analysis positions Ecuador in the group of countries classified below the average. Regarding infrastructure and connectivity, one of the main approved plans was the Universal Service Plan, which promotes the universalization and massification of telecommunications services. Thanks to the public and private investments made, 386 thousand new subscribers accessed the broadband network, thus contributing to the development of Ecuador. The first step in offering connectivity is to deploy infrastructure to the population, but it is also necessary to ensure that households and companies can use and connect to the deployed network [13].

As it can be evidenced, some implementations have been carried out that solve similar problems to our project, which allows us to determine that this work that involves field research will be useful for the development of the country.

3   Materials and Methodology

The project is based on providing Internet service, which will benefit approximately 20 households that are part of the study community, which will use a wireless technology with Standard 802.11a/n/ac same that will provide a speed minimum of 2 Mbps and maximum exceeds 25 Mbps according to the requirement of the end user [14]. For its realization references and data were taken according to a study of the sector in which many people who need Internet service live, it was noted that in the sector there is no infrastructure implemented by other operators; and because of the need detected, conversations were established with its residents.

Based on surveys carried out, it was determined that most of them did not have access to the Internet service and in the same way to fixed and cellular telephony services, so it was necessary through the implementation of this project to start shortening the existing digital divide, which in certain communities is still maintained.

The type of research methodology used for this project was field-based, because of the collection and analysis of data supported by surveys; as well as a brief diagnosis was raised with some residents of the sector (16 in total). It should be noted that the sample taken does not represent the central objective of the project, but rather constitutes a tool that allows to validate certain aspects such as those shown in Table 1, which complement the study carried out.

The survey had 7 questions. Below is a summary of the surveys carried out on the inhabitants, as a starting point to work on this research:

Table 1. Summary surveys

Questions	Answers		
Do you have a computer in your home?	Yes 8	No 8	
What use do you offer to your computer?	Internet 6	Works 8	Entertainment 5
Do you have access to the Internet in your home?	Yes 7	No 9	
How often do you access at the internet service?	Once 8	Tree times 2	More than 5 times 6

(*continued*)

Table 1. (*continued*)

Questions	Answers		
Do you have a mobile phone?	Yes 16	No 0	
Do you browse Social Networks?	Yes 14	No 2	
Do you know if there are Internet service providers in your community?	Yes 6	No 10	

Table 1 shows the following results: 50% of the inhabitants have a computer in their home, 42% use their computer to perform their tasks, 56% do not have Internet access from their home, 50% enter at least one Once a day to the Internet, 100% have a mobile phone, 88% use social networks and 63% don't know if there are Internet service providers in their community.

3.1   Technical Aspects of the Solution

The project was established in the design and implementation that provides a wireless Internet connection to 20 homes with a transmission speed of 2 Mbps. The technology used is based on the 802.11 a/n/ac with brand equipment whose costs and technological characteristics provided a correct operation.

The IEEE 802.11 standard defines the use of the two lower layers Physics and Data Link of the OSI model (Open Interconnection System), specifying its operating rules in a Wireless Local Area Network (WLAN). 802.11a, works in the 5 GHz band and uses a different modulation scheme called OFDM (Orthogonal Frequency Division Multiplexing), with a maximum speed of 54 Mbps. 802.11n works at a maximum speed of 600 Mbps and can operate in two frequency bands: 2.4 GHz (the one that uses 802.11b and 802.11 g) and 5 GHz (the one that uses 802.11a). Thanks to this, 802.11n is compatible with devices based on all previous editions. IEEE 802.11ac, which is an improvement proposal to the IEEE 802.11n standard, reaches speeds of up to 1.3 Gbps [15].

The financing of the project was provided by the Internet Wireless company, and because it is a testing stage at the beginning, it will have a low cost for the users. Later, in the period of one year, it is expected to reach a projection of 50 homes approximately. The construction of the network implies a short installation time, minimal environmental impact and great scalability, so it will be very beneficial for the community, since it will allow access to ICT.

The equipment used in the implementation was 2 Lhg5 antennas, working at a frequency of 5.8 GHz with a gain of 24.5 dBi, an edge router Rb2011 that will allow to manage the network efficiently, 2 Netbox5 AP/CPE radios/point-to-point double-stranded outdoor with RP-SMA connectors, and a cable cover for protection against moisture. This device is compatible with the new 802.11ac standard that has a higher transmission speed (866 Mbps and 20/40 and 80 MHz channels), this radio needs sector antennas, each has an opening angle of 120°, a CPE device is a high speed, low cost 5 GHz MIMO outdoor wireless device. The unit is equipped with a powerful

600 MHz CPU, 64 MB of RAM, 16dBi dual polarization antenna, POE, power supply and mounting kit, a WiFi 2.4 GHz router which will provide the experience of surfing wirelessly from home, a tower supported by 30-m tensioners and a mast that will serve to have a line of sight from the end user.

By means of a Site Survey [16], the place of implementation for the Node was analyzed, so it was verified that the place had a clean spectrum; that is, frequency channels available without interference, all this through tests carried out with an application on the mobile device called WiFi Analyzer.

3.2 Analysis of Costs and Time Required for Development

It has opted for the best operating recommendations and equipment costs necessary for the proposed solution, the brand. Depending on the time of completion of the project, a man-hours cost was determined (see Eq. 1), the salary considered in the calculations was USD 386, which is a SMBU (unified basic minimum wage) and a number of 40 h worked weekly, giving a total of 160 h per month, with the data establishing the relationship, as shown in Table 2:

$$hours\text{-}man = 386/160 \, hours = USD \, 2,42 \, by \, hour \tag{1}$$

Table 2. Cost by Hour-man

Hours/Day	Days/Week	Weeks	Cost/hour	People	Total (USD)
8	4	8	2,42	3	1858,56

Table 3. Total project cost

Element	Cost (USD)
Budget	11840
Hour/Man	1858,56
TOTAL	13698,56

Table 3 shows the calculation of the total cost of the project, whose amount is adapted to the budget and it is determined that the project is viable.

4 Test and Results

4.1 Analysis and Results

Next, in Fig. 1, the diagram of the physical topology is shown, the physical layout of the equipment, the network and wiring devices is shown; you can see within this topology two types of connections: PTP (point to point) and PMP (point to multipoint).

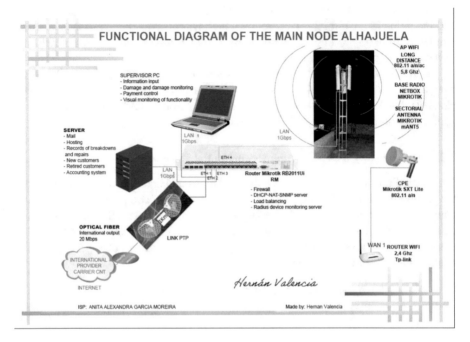

Fig. 1. Physical node topology diagram

Figure 2 shows the trunk PTP link through which the International Internet Service output is available; once it is connected to the edge router, in the tower the main access point is configured as PMP to arrive with the service to the final client.

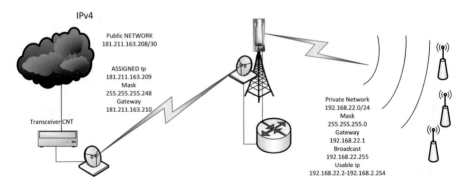

Fig. 2. Logical node topology diagram

The entry to the main router can be done through remote access (Public IP 181.211.163.210 assigned by the provider), which allows to verify the consumption of the network transmission speed, so that the service provided by the carrier can be monitored.

Figure 3 shows the consumption of transmission speed of a daily report, which is updated every 5 min on average, it should be noted that from 02:00 to 06:00 a low consumption is evidenced, since it has been determined that there are no users connected.

"Daily" Graph (5 Minute Average)

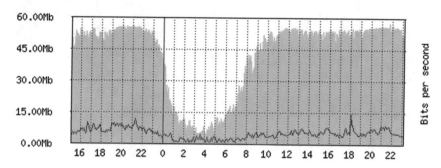

Max In: 57.67Mb; Average In: 43.12Mb; Current In: 52.12Mb;
Max Out: 14.40Mb; Average Out: 4.72Mb; Current Out: 3.82Mb;

Fig. 3. Daily consumption

Figure 4 shows the consumption of transmission speed in a weekly report of 30 min average, with a low consumption according to what is explained in the previous figure.

"Weekly" Graph (30 Minute Average)

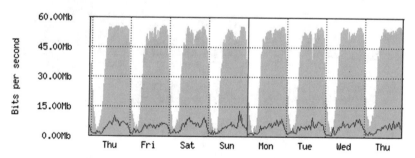

Max In: 55.98Mb; Average In: 38.31Mb; Current In: 54.85Mb;
Max Out: 12.63Mb; Average Out: 4.27Mb; Current Out: 4.51Mb;

Fig. 4. Weekly consumption

Figure 5 shows the consumption of transmission speed in a monthly report of 2 average hours, with a low consumption according to what is explained in Fig. 3.

"Monthly" Graph (2 Hour Average)

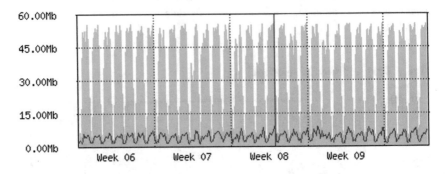

Max In: 55.86Mb; Average In: 37.81Mb; Current In: 55.86Mb;
Max Out: 9.16Mb; Average Out: 4.02Mb; Current Out: 6.99Mb;

Fig. 5. Monthly consumption

Figure 6 shows the consumption of transmission speed in an annual report of 1 average day, it is worth mentioning that in the month of January there was a formal cut that was generated by maintenance of the main fiber optic link.

"Yearly" Graph (1 Day Average)

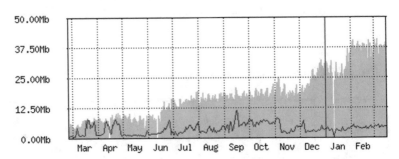

Max In: 40.82Mb; Average In: 18.15Mb; Current In: 39.46Mb;
Max Out: 11.36Mb; Average Out: 3.46Mb; Current Out: 4.35Mb;

Fig. 6. Yearly consumption

With these figures the daily, weekly, monthly and annual monitoring is evidenced, where a consumption of the speed of navigation of the clients is obtained, determining that there were no cuts in less than 1% of the data obtained in Fig. 6, what allows to have a connection reliability in a 99%, see Eq. 1.

$$(R = hy - hly/hy) \tag{1}$$

Where:

- hy = hours per year (8760)
- hly = lost hours per year (24)
- R = reliability = (99.72%)

Finally, Fig. 7 shows the growth of the project, since it started with 20 households and currently there are 127 households connected to the Internet service and with the projection to continue growing.

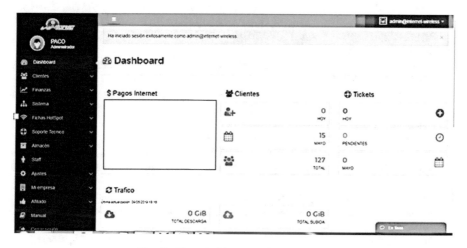

Fig. 7. Households currently connected

5 Conclusions

Two important lines of development such as research and link with society have joined forces to jointly achieve the proposed objectives of this project, providing access to the Internet in a remote community.

With this project it has been possible to see a high impact, in the growth of connectivity, since at the beginning it was expected to benefit around 20 homes and currently it has 127; as well as the growth in the use of bandwidth, 25 Mbps to December 2017 and 50 Mbps to February 2019, which translates into real satisfaction for the Alhajuela community and allows to bridge the digital divide in Ecuador, which supports to universal access, which constitutes a citizen's right.

Allowing access to the Internet through the deployment of infrastructure where it was previously not available, especially in rural areas, makes a difference by benefiting people by providing them with greater opportunities to acquire knowledge through the Internet and making them more productive.

Through a Site Survey, the use of the radio spectrum was analyzed using a mobile application, which determined that it was possible to work in a free spectrum of unlicensed frequencies at 5.8 GHz.

References

1. Ministerio de Telecomunicaciones, "PLAN NACIONAL DE TELECOMUNICACIONES Y TECNOLOGÍAS."
2. Agencia de Regulación y Control de Telecomunicaciones, "Boletin Estadístico."
3. Andrade, D., et al.: INFODESARROLLO: Camino, experiencia y futuro. Corporación Red Infodesarrollo (2016)
4. Area Moreira, M.: Sociedad De La Información Y Analfabetismo Tecnológico: Nuevos Retos Para La Educación De Adultos. Rev. DIÁLOGOS Univ. La Laguna, pp. 6–11 (2002)
5. Agencia de Regulación y Control de Telecomunicaciones, "INSTRUCTIVO DE TRABAJO DE LOS FORMATOS TÉCNICOS PARA EL OTORGAMIENTO DEL TÍTULO HABILITANTE PARA LA PRESTACIÓN DEL SERVICIO DE TELEFONÍA FIJA. Código: IT-DRS-05"
6. GAD Alhajuela, "Programa de Desarrollo y Ordenamiento Territorial de la Parroquia Alhajuela 2015–2019"
7. Selwyn, N.: The use of computer technology in university teaching and learning: a critical perspective. J. Comput. Assist. Learn. **23**(2), 83–94 (2007)
8. Phaal, R., Farrukh, C.J.P., Probert, D.R.: Technology management tools: concept, development and application. Technovation **26**(3), 336–344 (2006)
9. Jiménez Becerra, J.A.: "El papel de las TIC en el desarrollo una mirada desde la construcción social de la tecnología en el caso Ecuatoriano," Íconos Rev. Ciencias Soc. ISSN-e 1390-1249, no. 37, págs. 87–97 (2010)
10. Internet World Stats Group, "South America Internet Usage Stats, Population Statistics and Facebook Reports." https://www.internetworldstats.com/stats15.htm. Accessed 25 March 2019
11. Butler, J., Pietrosemoli, E., Zennaro, M., Fonda, C., Okay, S., Aichele, C., Büttrich, S.: Redes inalámbricas en los países en desarrollo (2013)
12. Moreira, J., Palomares, J.M., Serrano, R., López, J.: Un breve análisis de la brecha digital de acceso en el Ecuador. Zenodo, p. 4 (2017)
13. Ministerio de Telecomunicaciones y de la Sociedad de la Información, "MINTEL promueve el desarrollo del país, a través de la reducción de la brecha digital y la apropiación de las TIC – Ministerio de Telecomunicaciones y de la Sociedad de la Información." https://www.telecomunicaciones.gob.ec/mintel-promueve-el-desarrollo-del-pais-a-traves-de-la-reduccion-de-la-brecha-digital-y-la-apropiacion-de-las-tic/. Accessed 25 March 2019
14. Backes, F.: Transparent bridges for interconnection of IEEE 802 LANs. IEEE Netw. **2**(1), 5–9 (1988)
15. Qiao, D., Choi, S., Shin, K.G.: Goodput analysis and link adaptation for IEEE 802.11a wireless LANs. IEEE Trans. Mob. Comput. **1**(4), 278–292 (2002)
16. Hernández, N., et al.: Continuous space estimation: increasing WiFi-based indoor localization resolution without increasing the site-survey effort. Sensors **17**(12), 147 (2017)

RFID and Wireless Based System
for Restaurant Industry

Andrés Riofrío-Valdivieso$^{(\boxtimes)}$ ⓘ, Francisco Quinga-Socasi ⓘ,
Carlos Bustamante-Orellana ⓘ, and Eddy Andrade-Chamorro ⓘ

Yachay Tech University, Hacienda San José, 100119 Urcuquí, Ecuador
andres.riofrio@yachaytech.edu.ec

Abstract. The convergence of the new information technologies has led to sophisticated ways to improve and automate processes that in the past were done manually. In particular, the inclusion of wireless technologies like mobile applications, RFID and Wi-Fi in the restaurant industry can reduce the levels of entropy in certain processes like order taking and its corresponding dispatch. That is the problem that the proposed restaurant management system called "Fast-Waiter" solves. In this work, we integrate an RFID module, using Arduino, which allows users to pay directly the requested amount using an RFID card. The experimental results show an important reduction in the required time to register and pay an order over conventional methods.

Keywords: RFID · Arduino · Mobile applications · Restaurant industry

1 Introduction

Currently, everybody uses mobile communication devices like smartphones or tablets. One of the wireless communication technologies that those devices use is WI-FI with IEEE 802.11 standards to transmit and receive data [1]. In particular, this work uses the standard 802.11n. The 802.11n standard was released in 2009. This standard employs the OFDM modulation technique. Multiple Input Multiple Output (MIMO) antenna technology is used with this protocol. This technology refers to the ability of 802.11n and similar technologies to coordinate multiple simultaneous radio signals. 802.11n can use two channels of 20 MHz and 40 MHz. The 802.11n standard supports a maximum theoretical network bandwidth up to 300 Mbps. The IEEE802.11n indoor/outdoor ranges are 75 m, and 250 m respectively [1]. The WLAN deployment in this project is through an Access Point AP, in this case, a TP-LINK range extender TL-WA850RE. This device makes possible that the mobile devices, desktop PC's and the server receive the WI-FI signal to work cooperatively while taking advantage of server resources.

On the other hand, Radio Frequency Identification (RFID) is a kind of Automatic Identification (Auto-ID) technology. The term refers to a family of technologies that uses radio waves to automatically identify objects, people or products. The automatic system detection helps to improve the efficiency of this technology. Commonly, a pair of tag and reader is used for identification purposes thanks to a process known as electromagnetic coupling. The communication process is made by a code that is stored in a tag that is attached to a physical object; then, the object transmits this code and the reader can get information from that object [2] (Fig. 1).

© Springer Nature Switzerland AG 2020
M. Botto-Tobar et al. (Eds.): ICAETT 2019, AISC 1067, pp. 291–302, 2020.
https://doi.org/10.1007/978-3-030-32033-1_27

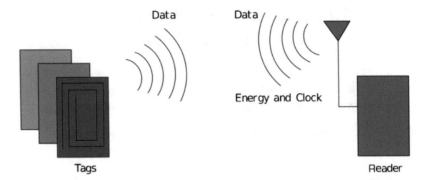

Fig. 1. RFID functioning by a tag and reader pair.

Depending on the RFID application, different frequencies are used: RFID; 433 MHz and 2.45 GHz for active RFID; 25 kHz, 13.56 MHz and 860-930 MHz for passive. The global standardization of RFID is not possible because of the variety of manufactures but a set of organization, like Organization for Standardization (ISO) and the International Electro-technical Commission (IEC) and global, have proposed different standards depending on hardware physics rectification, tag-reader air interface and reader-host command specification [2].

An RFID-RC522 reader module was used for this implementation. It uses 3.3 V as an input voltage and it is controlled through the SPI protocol with a maximum bit rate of 10 Mbps [3]. These characteristics enable compatibility with almost any micro-controller, Arduino or development board. The RC522 uses an advanced modulation and demodulation system for any kind of passive device working over 13.56 MHz. The reader and tags communicate using a protocol defined at ISO/IEC 14443-3 it describes polling for proximity cards entering the field of a reader, the byte format and framing and timing used during the initial phase of communication between reader and tags [4].

In this work, we developed a system to optimize the order taking and processing in the restaurant industry called "Fast-Waiter" by the integration of an android application, a windows platform, and a RFID module (using Arduino) which allows commensals to pay directly the requested amount using an RFID card.

2   Related Work

Until now, a couple of systems that integrate RFID with mobile applications, desktop software or web-based applications have been developed for different purposes. Kiranmayi et al. [5] proposed an RFID-based parking and payment system using Android. Such a system is capable to track a user's car as well as deduct the amount of money in their wallet. When the car enters the parking, the user must scan his RFID card and operation details are sent to the server. This information is also displayed in the admin and user app interfaces. When the car leaves the parking the user scans again his RFID card and if their wallet has enough amount in his application, the user can pay directly the required amount. This project used Android Studio for the Front end, HTML and PHP for the website and SQL Server for the Back end.

Likewise, Shabin et al. [6] presented a system that uses RFID and QR technologies to implement an automated library where a mobile android application is a Front end. Here, the users get connected to the system via Wi-Fi and can search for the location of the desired book. When the user finds the book, they must scan the QR code glued in the book to issue it, and the transaction information is sent to the system database. Each book also has a unique RFID tag associated with it. When the book is taken out the RFID reader gets the tag and, in case the book was not successfully issued an alarm will ring to protect the book from theft.

In a similar way, Shejwal et al. [7] implemented a smart attendance system using RFID, NFC, and Android to minimize the waste of time in registering the attendance of students using traditional methods. In the proposed system, every student has a unique RFID card and the lecturer has an NFC enabled phone with the android application installed on it. When a lecturer starts, the professor must log into the application and set class time and duration for the attendance record. Then the students are requested to scan their RFID card on the teacher's phone. The android application will send the student's information to the server side where the attendance is registered.

3   Proposed Model

A system to optimize the order taking and processing in the restaurant industry called "Fast-Waiter" has been developed. Fast-Waiter uses a desktop Windows' application for the admin side, where the number of tables and available dishes are fixed; and one android application with two working modes depending on the login. In "waiter" mode the application consumes the information specified in the admin side. Here, the waiter registers the order of a selected table. The order information is sent to the admin side to process the corresponding bill. In "chef" mode, the order information is also received and displayed to elaborate the target dish. Communication between Windows and Android application is made via WI-FI. Fast-Waiter mobile application had outstanding participation in the sixth edition of the international tournament of Mobile application called "TuApp" carried out in Lima-Perú in 2018 [8]. In this work, we integrate RFID technology that allows the Fast-Waiter system to directly charge a payment for an order using an RFID card.

4   System Overview

The proposed model system allows users to pay directly a requested amount of money using an RFID card. When a user acquires an RFID card, a new client is created in the database and the client is requested for payment to recharge money to his tag. In a transaction, the user has the option to pay whit his RFID card or with cash. If RFID payment is chosen, the user is requested to scan his RFID card and if he has the required amount then the user account is updated. Otherwise, they can recharge their card or pay with cash. Figure 2 describes the system's flow diagram.

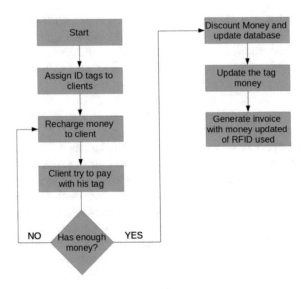

Fig. 2. Flow diagram of developed system.

4.1 Mobile Phone's Wi-Fi Chip Information

The Android application was deployed in an Android-based smartphone, which is equipped with the BCM4334HKUBG Wi-Fi module. This is a single-chip quad-radio device that provides the highest level of integration for wearable, Internet of Things and gateway applications, with integrated dual-band (2.4 GHz/5 GHz) IEEE 802.11 a/b/g/and single-stream IEEE 802.11n MAC/baseband/radio, and Bluetooth 4.0. Besides, it includes integrated power amplifiers and LNAs for the 2.4 GHz and 5 GHz WLAN bands, and an integrated 2.4 GHz T/R switch. This greatly reduces the external part count, PC footprint, and cost of the solution [9].

4.2 Communication

The system is based on the lecture of wireless signals through an RFID module. This module is controlled by the Atmega 328p microcontroller, included in the Arduino Uno board. The RFID reader module is the MFRC522, and the communication between said module and the host microcontroller is a serial communication based on the Inter-Integrated Circuit (I2C) interface. This interface is used to enable the low-cost serial bus interface and a low number of pins for the host. The diagram of communication through I2C can be seen in Fig. 3.

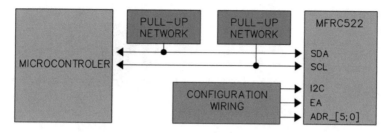

Fig. 3. Communication based on I2C-bus.

The MFRC522 has automatic microcontroller detection that makes this RFID wide compatible. The module supports direct interfacing host using SPI, I2C bus or serial UART interfaces. The automatic microcontroller detection resets its interface and checks the current host interface type automatically after performing power on [10].

The I2C-bus interface is implemented in accordance with the I2C bus inter-face specification of NXP Semiconductors, rev. 2.1, January 2000 [10]. The MFRC522 does not support clock generation or access arbitration interface, as a result, it can only act in slave mode. Therefore, the MFRC522 can act as a slave receiver or as a slave trans-mitter. The MFRC522 implements three modes of function providing different rate transmission: Standard mode, Quick mode, and High-speed mode. The I2C bus data can be transferred at data rates of up to 100 kBd in Standard mode, up to 400 kBd in fast mode or up to 3.4 Mbit/s in high-speed mode [10].

In a similar way, the microcontroller, in this case, the Atmega 328p, must communicate with the computer. The communication implemented is of serial type to avoid difficulties in synchronization and, considering the low complex communication, the parallel communication is impractical. Serial communication is the process of transmitting data one bit at a time over a communication channel in a sequential manner.

The Arduino Uno board supports serial communication through a serial port of type UART. The communication is performed through pins Rx (0) and Tx (1) [11]. The asynchronous receiver-transmitter (UART) port is an integrated circuit used for asynchronous serial communication. UART port allows configuring the data format and transmission speeds.

In the same manner, the communication between the server and network is done through the Wireless adapter of the server. The Wireless Adapter Intel Dual Band Wireless-AC 31-65 is an adapter for wireless communication that supports Wi-Fi and Bluetooth technologies. This adapter that works under 802.11 ac of second generation standard and allows a dual-band communication with speeds up to 433 Mbps [12]. In order to communicate the whole system through Wi-Fi, a private wireless network will be used. This network will be created using the TP-LINK range extender TL-WA850RE. This device supports the IEEE 802.11b, IEEE802.11g, IEEE 802.11n standards and works at a frequency of 2.4 to 2.4835 GHz [13].

Finally, to connect the system, it was necessary to develop a server in XAMPP that allows communication between a database, the mobile app, and the desktop app; considering that all the signals are carried out through a router or signal repeater as Fig. 4 shows. The RFID reader is connected to the desktop app in order to pay the order at the cashier side.

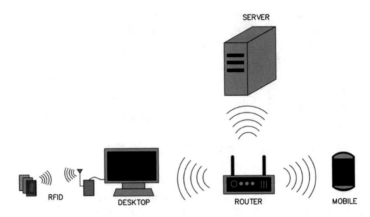

Fig. 4. Connection's structure of the project.

5 Methods

The "Fast Waiter" system was tested by applying an experiment designed with the advice of a Chef specialized in International Cuisine and Bartending who imparted lessons on Installation and Organization of Kitchens.

The experiment consisted in the simulation of order registration and billing procedures with the objective of measuring waiting times for diners. These simulations were carried out in a bar-restaurant with a capacity of 128 diners. The simulation of the order register was done using two waiters, called "waiter 1" and "waiter 2". In this simulation also 3 types of orders were used, called "order 1", "order 2", and "order 3". The simulation of the billing process was made by using only one cashier, called a "cashier".

The simulation of the order registration was done using two registration procedures. The first procedure was the order registration using the traditional method, that is, the handwritten order record. The second procedure was the order registration using the method proposed, that is, using the "Fast Waiter" application. The two registration procedures consisted of taking orders table after table. That is, both waiters took the orders of each of the assigned tables successively and at the end, they went to deliver the orders first to the kitchen and then the bar. The registration time ends when the order arrives at the bar and the distance from the middle of the bar and kitchen to a table is another parameter. The tables assigned to waiters were 1, 14, and 29 for "waiter 1" and 4, 15, and 32 for "waiter 2". Tables 1 and 4 made the request for "order 1" and are located at 21.5 m. Tables 14 and 15 made the request for "order 2" and are located at 13.7 m, and tables 29 and 32 made the request for "order 3" and are located at 1.9 m. The distribution of the tables, along with the tables assigned to each waiter can be observed in Fig. 5.

The consumption charge simulation was made by paying the consumption using the "Fast Waiter" system. The payment through the system consists of using the card with RFID technology to make the payment of the consumption instead of other payment methods.

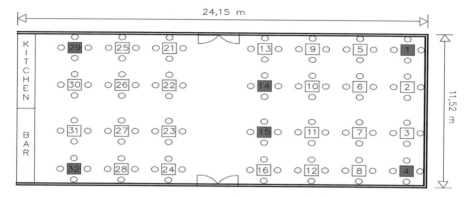

Fig. 5. Schema of the restaurant.

6   Results and Discussion

In Fig. 6 the initial admin screen of the desktop application is shown. Here, the admin can set the number of tables as well as the available dishes and manage the sales. Those functionalities are achieved using the left-hand side buttons.

Fig. 6. Initial admin screen.

In order to pay for an order, in the cashier window of Fig. 7, the key button must be used. This button allows us to obtain the code of the card through the RFID reader and update the client's money. Then, after clicking in the bottom "Pagar Tarjeta" the dishes consumption will be paid.

Fig. 7. View of cashier window.

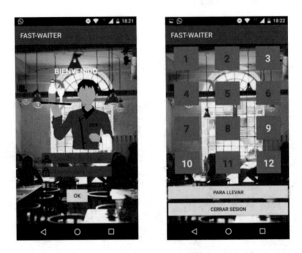

Fig. 8. Initial app screen and table's layout.

The initial Android screen is presented in the left side Fig. 8. Here is where either a waiter or chef can log into the system. Every employee of the restaurant has a unique username and password and the system automatically detects whether a chef or waiter is trying to enter.

The tables of a restaurant are represented as squares with a number inside of it as the right side of Fig. 8 shows. When a table is occupied the number inside of it changes to red, while available tables remain in their original color.

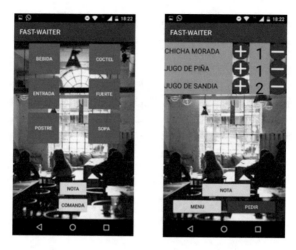

Fig. 9. Food categories and order details

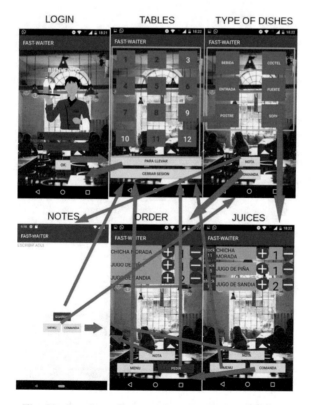

Fig. 10. Interfaces diagram of the Android application.

In the left side of Fig. 9, the food categories that a particular restaurant offers to their customers can be seen. Once a food category has been chosen, the system displays a list of the available dishes, see the right side of Fig. 9. In this interface, the waiter can set the amount of the desired dish.

Finally, Fig. 10 shows a flow diagram of the interfaces in the Android application. The red arrow indicates what screen is displayed when the user presses a particular button. It is important to remark that the number of tables, dishes, and food categories are consumed by the mobile application from the databases in the server.

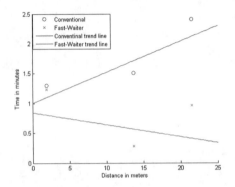

Fig. 11. Registration time for "Water 1" with and without Fast-Waiter.

On the other hand, Fig. 11 shows the resulting registration time for "Waiter 1" after applying the previously described experiment. In this case, when "Waiter 1" doesn't use the proposed system it tends to spend more time to register an order. On the contrary, using "Fast-Waiter", the taking order time tends to decrease as the distance from the kitchen increases.

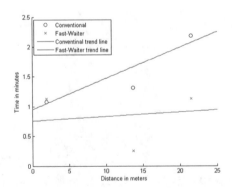

Fig. 12. Registration time for "Water 2" with and without Fast-Waiter.

Likewise, Fig. 12 displays the registration time for the "Waiter 2" after the experiment was performed. This case is slightly different from the above result. Even though the same behavior is observed when not using Fast-Waiter, the order taking time tends to stay constant as the distance increases when the proposed system is used. Nevertheless, using the proposed system always yields to better times while registering an order.

Another advantage of using "Fast-Waiter" is the reduction of wasting time while paying a requested amount of money. However, this analysis is very subjective because the restaurant uses different mechanisms to carry out payments. For example, it could be typewritten bills with cash-based payment, credit card-based payment with automatic bill generation combinations of them. In any case, using "Fast-Waiter" the only waiting time happens when keyboarding an ID number, which is between 10 to 12 s, because the commensal often pays with different credentials.

7 Conclusions

The proposed system is based on communications that are mostly wireless. Communications are used from the reading stage of a card to the interactions of the desktop and mobile applications with the server. The RFID system embedded with the cashier's desktop computer allows the user to pay the dishes consumed through a system RFID, XAMPP server and Wi-Fi of 2.4 GHz. Additionally, the system manages all the information between the apps and the server.

Because the transactions are micro and face to face, the use of RFID for payment is feasible, meaning a very low cost of implementation.

8 Future Work

In order to improve the security of transactions and to replace the use of cards with smartphones, the RFID system will be upgraded to an NFC system with higher standards of security and allowing us to pay by scanning an NFC-enabled smartphone.

References

1. Mohammed, R., Mustafa, A., Osman, A.: A comparison between IEEE 802.11a, b, g, n and ac Standards. IOSR J. Comput. Eng. **17**(5), 26–29 (2015)
2. Chechi, D., Kundu, T., Kaur, P.: The RFID technology and its Applications: a review. Int. J. Electron. Commun. Instrum. Eng. Res. Dev. **2**(3), 109–120 (2012)
3. Jamaluddin, A., Harjunowibowo, D., Rochim, M.A., Mahadmadi, F., Kakanita, H.B., Laksono, P.W.: Implementation of RFID on computer based test (RF-CBT) system. In: Proceedings of the Joint International Conference on Electric Vehicular Technology and Industrial, Mechanical, Electrical and Chemical Engineering (ICEVT & IMECE). IEEE, Surakarta (2016)

4. Woo-Garcia, R.M., Lomeli-Dorantes, U.H., Lopez-Huerta, F., Herrera-May, A.L., Martinez-Castillo, J.: Design and implementation of a system access control by RFID. In: 2016 IEEE International Engineering Summit, II Cumbre Internacional de las Ingenierias (IE-Summit). IEEE, Boca del Rio (2016)
5. Kiranmayi, D., Nasaramma, K., Lakshmi, M.B.: RFID based parking and payment system using android. Int. J. Comput. Sci. Inf. Technol. **8**(3), 338–341 (2017)
6. Rafi, S., Mohan, A., Sankar, A.: Android based advanced book locator and library manager. Int. J. Adv. Res. Sci. Eng. **6**(4), 562–565 (2017)
7. Shejwal, B., Walke, A., Solanki, R., Pandit, A., Shelar, P.: Smart attendance system using radio frequency identification (RFID) and android. J. Eng. **10**, 44–48 (2018)
8. TuApp. https://tuapp.org/. Accessed 29 March 2019
9. Cypress. https://www.cypress.com/file/299506/download. Accessed 29 March 2019
10. NXP Semiconductors: MFRC522 Standard performance MIFARE and NTAG fronted (2016)
11. Kurniawan, A.: Arduino Uno: A Hands-On Guide for Beginner, 1st edn. PE Press, Riverside (2015)
12. Intel: Intel Dual Band Wireless-AC 3265 (2015)
13. TP-LINK: 300Mbps Universal Wi-Fi Range Extender: TL-WA850RE

Design and Implementation of a Panic Alert System to Notify Theft Events

Miguel Baquero Tello, Marcos Gavela Moreno, Iván Sánchez Salazar,
Nathaly Orozco Garzón[✉], and Henry Carvajal Mora[✉]

Faculty of Engineering and Applied Sciences (FICA), Telecommunications
Engineering, Universidad de Las Américas (UDLA), Quito, Ecuador
{nathaly.orozco,henry.carvajal}@udla.edu.ec

Abstract. Unfortunately, crime rates around the world increase every year. In particular, robberies in the streets are the events with the highest rate of occurrence. In addition, these types of events are not notified fast enough. Hence, in some cases, security entities cannot take actions in a timely manner. Therefore, actions can not be carried out adequately. For this reason, the design and implementation of a prototype for a panic alert system is developed in this work, which contributes with timely information of thefts events to the authorities. In the proposed system, a server is designed and a smartphone application is implemented in order to report the theft events. In particular, the notification is sent to two different servers. The person in distress sends the notification when press the lock/power button on the smartphone a certain number of times. The button activates a cluster of video cameras that share live feed to emergency officials in both entities.

Keywords: Panic system · Video surveillance · Mobile application · Android studio · NetBeans

1 Introduction

In the city of Quito, Ecuador, some statistics published in 2013 showed that during 2011 and 2013, reported assaults had an increase between 49% and 53% only in the north of the city [1]. In addition, according to the statistics, the rate of robberies to persons increased 1.18% in the first semester of 2018 (2648 cases) with respect to 2017 (2617 cases) [2]. As a worrying case, there are also several complaints that have been reported by the students being victims of robbery, violence or other events in the surroundings of universities campuses.

In the literature, some works have been studied panic events. As example, in [3], the authors analyze the causes, characteristics and response of public panic, then introduces the public panic measurement system based on the monitoring of opinions on network to provide an effective decision support for emergency management. Results show that it is necessary to have applications that

© Springer Nature Switzerland AG 2020
M. Botto-Tobar et al. (Eds.): ICAETT 2019, AISC 1067, pp. 303–312, 2020.
https://doi.org/10.1007/978-3-030-32033-1_28

allow those affected people to notify events immediately so that trained personnel can help them. Considering that almost all people have a mobile device, these applications can be implemented on these platforms.

In [4], the authors proposed a system that use the global positioning system (GPS) in order to take the location of the user in danger from Google maps. The main concept would be sending a voice message and a message template to the selected contacts and emergency numbers when in danger. Unfortunately, in robbery events it is very likely that the user will not be able to record a voice message. Another similar approach is presented in [5], where emergency medical services are equipped with a panic button aiming at reduction of the time of first aid provision. Nevertheless, this application is specific for medical scenarios.

Given the facts explained above, the idea of using a smartphone to generate a notification message to the authorities and activate a video surveillance system to help reducing the crime arose. However, it is also important to provide not only the victim locations but also images of the event if possible in order to have additional information to protect the victim and even be able to recover the stolen objects.

In this work we describe the designing and implementation of a system that operates as follows: An alert message is sent from an Android application to a server that hosts a Java application. Immediately, a video cameras system is activated in order to provide images or videos of the event. The Java application allows the access to the video surveillance system. As a test scenario, we consider that the information is stored into a server from a state security institution, such as the Integrated Security Service ECU911[1] and in a server located inside the educational institution outside of which the event is occurring.

The remainder of this paper is organized as follows. In Sect. 2, the logical operation, protocols and methods to perform the communication between an Android application and the mentioned servers is explained. The interaction between the server that host the Java Application and the video surveillance system is also described. Subsequently, the technical information and requirements for the panic system installation are addressed in Sect. 3. Some test results of the proposed system are shown in Sect. 4. Finally, some concluding remarks are presented in Sect. 5.

2   System Description

In this section, the proposed panic system is presented. Moreover, the Android Application and the Java Application server are described. For better understanding of the following, a block diagram of the system is shown in Fig. 1.

To give the user a huge advantage, it is configured a mobile Android application which is able to run in background mode. In addition, a quick method to trigger the panic system is implemented [6]. This method consists in sending

---

[1] The ECU911 is an Ecuadorian immediate response service to emergencies.

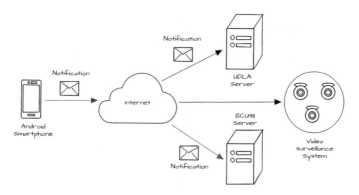

Fig. 1. Block diagram of the panic system.

the notification to the servers by pressing the smartphone power button 4 times consecutively[2].

When the application is installed and launching the first time on the mobile phone, it requests the user the next information: user name, user last name, ID/Passport, phone number, disabilities, blood type, allergies, emergency contact name, emergency contact last name, emergency contact phone number and emergency contact relationship. All this information is sent to the servers and in case of emergency, it is shown to people monitoring the panic system.

The main Android Application interface consists of a toggle button that allows to keep the application running in background mode. Hence, the application is waiting for the consecutively press of the power button event to trigger the panic system. In addition, the application has a button which allows to trigger the panic system directly, while remaining in the main interface of the application. A screenshot of the main Android mobile interface[3] is shown in Fig. 2.

Once the toggle button in the interface changes from OFF to ON state, a persistent notification will be created within the Android notifications, it allows the application to remain running in background mode. In this case, the application is waiting a power button event. If it occurs, an alert message is sent.

For the application, a power button listener method was created [7], which is responsible for interpreting the Android intent actions of SCREEN_ON and SCREEN_OFF. Each time the power button is pressed, an intent action of SCREEN_ON (if the screen is off) or SCREEN_OFF (if the screen is on) is detected by the power button listener. A counter is set in the power button listener method. It validates if an intent action (SCREEN_ON or SCREEN_OFF) is received within a time interval, which is set to 2 s (this time can be modified

---

[2] Obviously, a greater number of times will send the notification as well.
[3] As the application was used by Spanish speakers during the tests, it was configured in Spanish language. However, it is easy to change it to other languages.

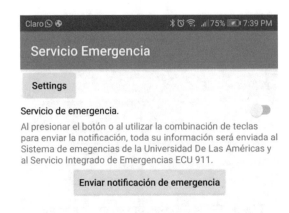

Fig. 2. Android application main interface.

at convenience). Once this time expires, the counter will return to zero and wait for the action to be carried out to start the account over again.

As soon as the panic system is triggered by the user, the application will send notifications to both ECU911 and university servers. Later, the screen in Fig. 3 is displayed in the mobile phone. This screen confirms the emergency event was correctly sent to the ECU911. It is important to mention that the sent notification includes the location of the victim in terms of geographical coordinates.

In order to give a brief summary of the Android Application paradigm, a workflow is shown in Fig. 4. This diagram explains the application behavior since the user opens it from the first time until the alert is sent to the servers.

After the notification is sent from the Android Application to the servers, the Java Application sends an email including the user information with the geographical coordinates of the event. Thus, the security staff receives an email like one in Fig. 5.

The video surveillance system is activated if the mobile phone is within a certain radius of coverage from each camera. The best way to know if a user from the mobile application is inside the coverage radius of the camera is by applying the Haversine formula [8] as an algorithm in the Java application to calculate the distance between 2 geographical points.

The information needed to calculate the distance is the latitude and longitude from two different locations, then replace those information in:

$$a = \sin^2\left(\frac{\Delta\phi}{2}\right) + \cos\phi_1 \times \cos\phi_2 \times \sin^2\left(\frac{\Delta\lambda}{2}\right), \tag{1}$$

where, ϕ_1 is the latitude of the fist geographical coordinate. ϕ_2 represents the latitude of the second geographical coordinate, $\Delta\phi$ is the variation (difference) between the latitude of the geographical coordinates and $\Delta\lambda$ is the variation

Fig. 3. Sent emergency screen.

between the longitude of the geographical coordinates. With the result of (1), then, it is calculated

$$c = 2 \times \arctan 2\left(\sqrt{a}, \sqrt{1-a}\right),$$ (2)

where $\arctan 2(x_1, x_2)$ is the element-wise arc tangent of x_1/x_2 choosing the quadrant correctly. The quadrant is chosen so that $\arctan 2(x_1, x_2)$ is the signed angle in radians between the ray ending at the origin and passing through the point $(1,0)$, and the ray ending at the origin and passing through the point (x_2, x_1). Finally, the distance between two geographical points is obtained as

$$d = c \times R$$ (3)

where R is the Earth radius, which is $6371\,\mathrm{km}$.

Consequently, if the application is inside the coverage area of the camera, the Java Application sends another email to the security staff with an uniform resource locator (URL) that gives access to the nearest camera from the incident.

3   System Design and Implementation

The system design and implementation are described in this section.

In order to guarantee an adequate functioning of the panic system, it is necessary that the server on which the Java application is implemented has the two main characteristics: public IP address and Java runtime environment. This

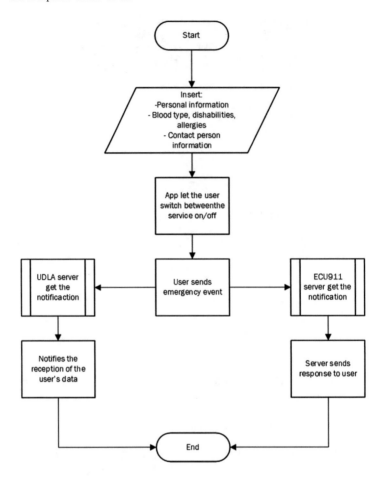

Fig. 4. Panic system workflow.

Fig. 5. User information details send by email.

guarantees the application can be used with a mobile network, without employing a Wi-Fi connection. The Java application is used to receive the notification sent from the user equipment.

The Android application communicates with the ECU911 server through a simple object access protocol (SOAP). Hence, the communication is secure sense and the connection is established beginning a session via a user and password. The user information, detailed in Sect. 2, is sent in a JavaScript object notation (JSON) string to the ECU911 server. Specifically, JSON is a lightweight data-interchange format [9]. In this project, it is used to send information from the Android application to the respective servers.

The Java application is also developed with Java Threads to perform the communication. This means that the transmission control protocol (TCP) is used in the connection. Java Threads supports multi-processes communications in order to create simultaneous channels (petitions) with different users [10]. Moreover, in order to establish a socket and to start the communication between the Android application and the university server, it is necessary to define a port, in this case, the port 24666 was used. In the Java Application, the user data is also received in a JSON chain.

When the user triggers the panic system in the Android application, the communication with Java Threads is started and the university server shows a text line indicating that the communication has begun, as show in Fig. 6.

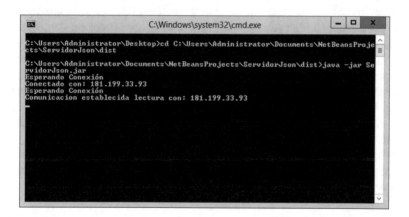

Fig. 6. Successful connection to the server.

After the Android application sends all the user data to the university server, the Java application displays all information, as shown in Fig. 7.

The Java Application implements a library called Javax Mail that allows to send emails through Gmail services. Hence, sender email address, sender password and the email recipient must be written in the code.

The video surveillance system is based on IP cameras, each one with a public IP addresses, so the university security staff can watch the video from every camera through any browser. Each camera and its respective geographical location must be defined in the Java Application, in this way, the expressions detailed in the Sect. 2 work adequately if a user is inside the camera coverage area. It

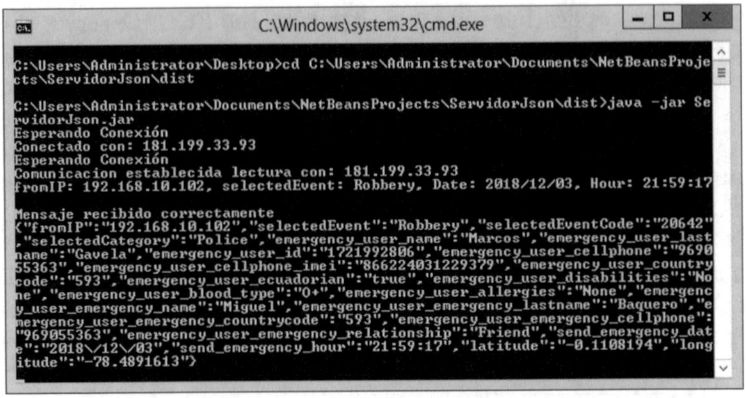

Fig. 7. Information received from the user application.

is recommended to use Pan, Tilt, Zoom (PTZ) cameras. This type of cameras provides a 360° coverage.

The video surveillance system is meant to be an outdoor system, so it is important to consider IP66 cameras, that is protected from total dust ingress and limited protection to high pressure water jets from any direction [13]. In order to get a better coverage area with a single camera, installing the video surveillance system in street intersections is an optimal way to reduce implementation costs and improve the effectiveness of each camera.

At least, the Android application needs to be installed on a smartphone running Android KitKat 4.4 (API 19) [11] or higher, in case that the device has a lower API, once the Android Package Kit (APK) [12] is installed, most of the Android application features can not work.

It is recommended for the smartphone to be a medium/high range device. This guarantees a more precise geographical location at the moment that the alert is generated.

4   Results

In order to test the system, multiple events are created from one device. This was performed to assure the response from the server depending on the configuration that was given as was explained in Sect. 2. The gadget used in this test is a smartphone Samsung Galaxy J5.

In total 19 tests were performed. The results obtained from all the tryouts are shown in Table 1. These results show that the server is capable of receiving a large amount of emergency notification. This aspect is decisive to ensure the viability for the application. All the results get the location with an error margin of 5 m in accordance to the data retrieved.

Table 1. User information displayed on the server each time the panic system is triggered from the smartphone.

Test number	Date (mm/dd/yyyy)	Time	Latitude	Longitude
1	12/19/2018	15:08	−0.1694631	−78.4710829
2	12/19/2018	15:12	−0.1680193	−78.4702626
3	12/19/2018	15:13	−0.1681101	−78.4702396
4	12/19/2018	15:14	−0.1680277	−78.470348
5	12/19/2018	15:16	−0.1684207	−78.4696808
6	12/19/2018	15:19	−0.1674689	−78.4708798
7	12/19/2018	15:20	−0.1675257	−78.4710067
8	12/19/2018	15:44	−0.1681628	−78.4728565
9	12/19/2018	15:48	−0.1667074	−78.4729565
10	12/19/2018	15:51	−0.1650741	−78.4726948
11	12/19/2018	15:52	−0.1651692	−78.4718446
12	12/19/2018	15:55	−0.1663824	−78.4719202
13	12/19/2018	15:57	−0.1679511	−78.4720165
14	12/19/2018	17:13	−0.1623699	−78.4616415
15	12/19/2018	17:15	−0.1623184	−78.4617591
16	12/19/2018	17:17	−0.1630898	−78.4626836
17	12/19/2018	17:18	−0.1632715	−78.4629283
18	12/19/2018	17:20	−0.1622993	−78.4616977
19	12/19/2018	17:35	−0.1628299	−78.4594461

5   Conclusions

In this work, a panic alert system to notify theft events has been proposed. The system sent information to different entities that can act to protect the victim from theft. The proposed system is able to send specific information about the victim and its location through email notifications. Additionally, the system incorporates a set of cameras that, depending on the location of the victim, can show the theft event in real time, which allows the authorities to know what actions to take.

In order to improve the impact of the panic system against robbery, audio-visual alarms can be added in order to thwart any attempt to steal and alert nearby people about the incident.

Finally, the proposed system is not meant to generate monetary profit, it generates social benefits preventing the lose of valuable artifacts, at the same time, this tool could be a lifesaver.

References

1. Secretaría de Seguridad y Gobernabilidad, Municipio de Quito: Delitos y Violencia del DMQ. http://omsc.quito.gob.ec/index.php/biblioteca-virtual/informes-mensuales.html. Accessed 28 Mar 2019
2. Ministerio del Interior: Indicadores de Seguridad Ciudadana. http://cifras.ministeriodelinterior.gob.ec/comisioncifras/inicio.php. Accessed 25 Mar 2019
3. Jimei, L., et al.: Urban mass panic monitoring and measurement system for incident based on the BBS. In: IEEE International Conference on Emergency Management and Management Sciences (2010)
4. KishorBabu, S., et al.: Exigency alert and tracking system. In: International Conference on Advances in Computer Engineering and Applications (2015)
5. Paramonov, I., et al.: Communication between emergency medical system equipped with panic buttons and hospital information systems: use case and interfaces. In: Artificial Intelligence and Natural Language and Information Extraction, Social Media and Web Search FRUCT Conference (AINL-ISMW FRUCT) (2015)
6. Android developers: Backgroud Executions Limit. https://developer.android.com/about/versions/oreo/background?hl=en-419. Accessed 25 Feb 2019
7. Android developers: Intent. https://developer.android.com/reference/android/content/Intent. Accessed 21 Feb 2019
8. Calculate distance and bearing between two Latitude/Longitude points using the haversine formula in JavaScript. https://www.movable-type.co.uk/scripts/latlong.html. Accessed 19 Feb 2019
9. Introducing JSON. https://www.json.org/. Accessed 29 Mar 2019
10. Thread (Java Platform SE 7). https://docs.oracle.com/javase/7/docs/api/java/lang/Thread.html. Accessed 29 Mar 2019
11. Android 4.4 API. https://developer.android.com/about/versions/android-4.4?hl=en-419. Accessed 22 Feb 2019
12. What is APK file (Android Package Kit file format)? https://whatis.techtarget.com/definition/APK-file-Android-Package-Kit-file-format. Accessed 22 Feb 2019
13. What is Ingress Protection? https://www.rainfordsolutions.com/what-is-ingress-protection. Accessed 31 Dec 2018

e-Learning

Towards an Improvement of Interpersonal Relationships in Children with Autism Using a Serious Game

Giancarlo Alvarez Reyes[1] , Valeria Espinoza Tixi[1],
Diego Avila-Pesantez[1(✉)] , Leticia Vaca-Cardenas[2] ,
and L. Miriam Avila[1]

[1] Escuela Superior Politécnica de Chimborazo, Riobamba, Ecuador
davila@espoch.edu.ec
[2] Universidad Técnica de Manabí, Portoviejo, Ecuador

Abstract. Autism Spectrum Disorder (ASD) or Autism is considered as a psychological neurodevelopment disorder which affects people. It is characterized by a wide range of conditions that difficult their social skills, as well as their verbal and non-verbal communication. According to the Centers for Disease Control, Autism affects an estimated 1 in 59 children in the USA. Technological advances seek to benefit the most vulnerable groups of society, creating tools that enhance the development of their abilities and skills. In this sense, Serious Games are structure-oriented videogames designed to learning or therapies in a fun environment. With this principle, "JOINME" game was developed using a natural user interface to benefit the progress of interpersonal relationships in children aged between 5 and 10, diagnosed with high-functioning ASD. For experimentation, a case study was applied. It was built from the adaptation of the KidsLife scale through a pre-test and post-test. As a result, a significant improvement of 8.17% in the development of interpersonal relationships was obtained in the kids. It reinforced their mathematical reasoning, recognition of emotions, and comprehensive reading.

Keywords: Autism Spectrum Disorder (ASD) · Autism · JOINME · Serious Games · SUM methodology

1 Introduction

Autism Spectrum Disorder (ASD) or Autism is considered as a psychological neurodevelopment disorder, which is characterized by a wide range of conditions that includes impairments in social interaction, repetitive behaviors, trouble understanding what other people think and feel [1–5]. People with Autism are cognitively different and show uneven skills development. Generally, ASD indicators appear at 2 or 3 years old. Some factors that indicate the presence of Autism are sensory sensitivities, seizures, sleep disorders, as well as mental health problems such as anxiety, attention problems, and depression. Today, according to the Centers for Disease Control, Autism affects an estimated 1 in 59 children in the USA [6]. Its etymological origin comes from

© Springer Nature Switzerland AG 2020
M. Botto-Tobar et al. (Eds.): ICAETT 2019, AISC 1067, pp. 315–325, 2020.
https://doi.org/10.1007/978-3-030-32033-1_29

the Greek < autos > meaning "oneself"; therefore, it is understood that Autism is self-enclosed, that is, self-absorbed [7, 8]. The first researcher of this disorder, Leo Kanner, determined four common features: preference for isolation, insistence on activities already known, resistance to change in their routines and specific skills that were paradoxical in people with such limitations [9, 10]. An individual with Autism has a different set of challenges and strengths. They learn, think, and solve problems in a variety of ways, which can be classified from highly qualified to people with serious difficulties. Some may require significant support in their daily lives, while others may need less help [6].

With the advance of science, currently, Autism can present seven behavioral traits, which are detailed in Fig. 1. One of the features that most affect their social development is their problem in identifying emotions in the people around them. It hinders the sense of empathy with others, complicating the recognition of facial features that show some emotion [11, 12].

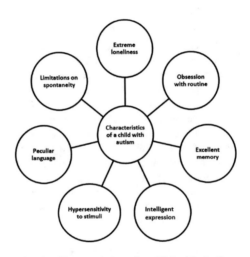

Fig. 1. Characteristics of a child with Autism

On the other hand, the expansion of communication technologies (ICT) has allowed the development of new tools that facilitate the lives of people with disabilities and learning problems [13–16]. These tools are a motivating, attractive, and highly versatile means of application. In this sense, Serious Games (SG) have been developed, which facilitate the teaching of various subjects through digital games that allow conveying knowledge to the participants, while having fun [16–19].

This paper is structured as follows: it begins detailing information related to Autism and the definition of SG. Section 2 describes the related works. In Sect. 3, it specifies stages of development and operation of JOINME. In Sect. 4, the statistical analysis of the results obtained is carried out. Finally, it presents the conclusions of the article.

2   Related Works

During the last decade, SGs development has employed a range of technologies (including smartphones, laptops, desktops, and console games) for ASD children. It includes different platforms, modes of interaction, and learning goals. For instance, ECHOES is an SG that seeks to develop communication skills in ASD children [14]. It works through a virtual character that becomes a companion for the patient. This character performs several activities in a two-dimensional sensory garden. Nevertheless, this game promotes only communication skills and does not reinforce other vital aspects of Autism. On the other hand, LIFEisGAME is a mobile application that focuses on developing the ability to recognize emotions, through this game children perform activities that induce them to quickly identify feelings such as surprise, fear, anger, among others [20], but it does not have any feedback. Another game analyzed is MOSOCO, a mobile application that uses Augmented Reality to improve social skills through real-life exercises [15]. Unfortunately, this application does not have feedback stimuli.

As can be seen in the examples, the most common interaction in these games is a keyboard/mouse and touchscreen. It motivated the development of an SG that helps improve the interpersonal relationships of the participants using a natural user interface (NUI) through the Kinect sensor. It could create a more significant impact on social interaction.

3   Serious Game "JOINME"

The developed SG is called "JOINME"; it was built using the agile methodology for the development of videogames known as SUM since it provides flexibility to define the software life cycle and proposes essential phases that are described below [21].

Phase 1 Concept: In this stage, the central concept of the game is defined, which must be validated by the involved parties (developers, graphic designers, pedagogues, psychologists, and therapists). This activity is carried out considering three components: the game aspects (narrative and setting), technical aspects (software & hardware necessary to run the game) and business aspects (the target audience to whom the play is directed).

Phase 2 Planning: Here, the primary goal is to plan the remaining phases that describe the methodology. Two activities are carried out in parallel: First, the central planning, which defines the objectives, creation of the development team, schedule, and the budget; and second, the game specifications, based on estimating and prioritizing the requirements.

Phase 3 Elaboration: It allows programming the game, it is done iteratively and incrementally until getting an executable version of the game at the end of each iteration. In this stage, the development tools and the game engine are selected.

Phase 4 Beta: It evaluates and fixes the different issues of the SG, eliminating the highest number of errors detected. In the same way, it works iteratively releasing different versions of the game to verify them with the help of professionals specialized in this field.

Phase 5 Closing: Offers the client a final version of the SG for evaluation. Two activities are carried out sequentially: a release of the SG (the final version of the game is built) and assessment of the project (relevant aspects that occurred during the development).

The risk management of the project is carried out during all the phases, mitigating the conflicts that may occur during the SG development. Two tasks are carried out simultaneously, the first identifies the risks at each moment of the project, and the second is responsible for the monitoring and application of the contingency plans.

JOINME includes activities that allow working with mathematical reasoning, the identification of emotions/good manners, as well as the development of comprehensive reading. Each of these activities was proposed by a group of psychologists/therapists who are experts in the area of Autism. This SG uses a natural user interface controlled by the movement of the player's hands, through the Kinect 2.0 device, which allows autistic children to interact with the activities of the game to improve their social abilities. For the creation of the application, the Unity 3D game engine was used, which has extensive documentation, it facilitated the development of several features within the game. As an integrated development environment (IDE), the Visual Studio 2015 tool was selected for its easy integration with C # scripts. Finally, to access the functionalities of the device, we worked with the software development kit (SDK) v2.0, which facilitated the integration of the SG with the Kinect 2.0 sensor. JOINME was implemented on a desktop PC under the Windows 7 operating system or higher. The game starts with one animation, and then the main menu is shown with the three modules that child can play within the SG.

Mathematical Module: this game has 2 options. "Mathematical Lake," whose goal is to complete 10 mathematical operations; with 5 addition and 5 subtraction calculations (see Fig. 2). Each activity will be shown on the poster, and children must choose the correct answer. If the participant succeeds, then the next action will be displayed. The other option is the "Jungle of numbers," whose objective is to complete 5 multiplication and 5 division operations (see Fig. 3). In this case, the action will be displayed on a poster, and 3 possible results will appear, the child must choose the correct result; when he does it, the next exercise will be presented. This game has 3 difficulty levels: basic, intermediate, and advanced.

Pictograms Module: this option contains 2 minigames. "Mountain of emotions" is an activity where children must identify the emotions represented in images (see Fig. 4). A poster with the name of the feeling will be shown, and children should be able to match it with a corresponding image. If the action is correct, they receive motivating feedback and continue playing until they reach 10 hits. The game has three difficulty levels. In the other activity called "Manners Beach," children must match the manners shown in letters (see Fig. 5). A grid of flipped cards is displayed, and each one will be visible whenever the participant selects it. If children cannot match, the cards will be flipped again. With each successful pair, feedback audio will be played. The game will have 3 difficulty levels (basic, medium, and advanced) depending on the number of cards to play. The game ends when all the pictures are paired.

Fig. 2. Screenshot of "Mathematical Lake" minigame

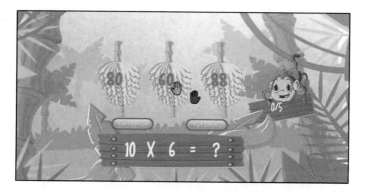

Fig. 3. The main screen of "Jungle of numbers" minigame

Fig. 4. "Mountain of emotions" minigame

Fig. 5. "Manners Beach" matching minigame

Story Module: this module tells the story of "The Adventures of Bender." In each scene, children listen and observe the description of an event and proceed to answer the questions about the established content (reading comprehension). It allows them to move between scenes until the end of the story (see Fig. 6).

Fig. 6. "The Adventures of Bender" game

4   Research Methodology

4.1   KidsLife Scale

The KidsLife scale is used to evaluate various dimensions related to the quality of life of children/adolescents with intellectual disabilities under 22 years of age [22, 23]. This scale is composed of 96 questions, which evaluate 8 dimensions: social inclusion, self-determination, emotional well-being, physical well-being, material well-being, rights, personal development, and interpersonal relationships. For the present case study, it was used only the dimension of interpersonal relationships that define relationships with different individuals, have friends and get along with people (neighbors, peers, and others) through indicators such as *Communication, Relationships relatives, Friends, Companions, and Society.* This instrument can be applied to individuals with high-functioning ASD, due to its general characteristics. For the evaluation of JOINME game, a survey of 12 questions was elaborated that corresponds to the dimension of interpersonal relationships (see Fig. 7), using a Likert scale of five values (5 = always, 4 = often, 3 = sometimes, 2 = seldom, 1 = never). The evaluation was performed through observation and completed by a person (therapist, family member, and teacher) who has known the patient for at least six months.

N=Nunca, R=Raramente, A=A veces, F=Frecuentemente, S=Siempre

Nº	Pregunta	N	R	A	F	S
1	¿Está integrado/a con sus compañeros, familiares, terapeutas?					
2	¿Realiza actividades de ocio con personas de su edad?					
3	¿Tiene una adecuada higiene e imagen personal?					
4	¿Aprende cosas que le hacen ser más independiente?					
5	Las actividades que realiza le permiten el aprendizaje de nuevas habilidades.					
6	Tiene oportunidades para demostrar sus habilidades					
7	Se estimula su desarrollo en distintas áreas (cognitiva, sensorial, social y emocional).					
8	Las tareas que se le proponen se ajustan a sus capacidades, ritmos y preferencias.					
9	Tiene oportunidades para iniciar una relación de amistad si lo desea.					
10	Busca oportunidades para estar a solas con sus amistades y personas conocidas.					
11	Se ha identificado la mejor forma de comunicarle información (visual, táctil, auditiva).					
12	Tiene facilidad para lograr identificar emociones a través de rasgos faciales.					

Fig. 7. Example of a survey filled out by the therapist to evaluate the interpersonal relationships

4.2 Participants

Children were selected by the therapists, considering their aptitudes and skills to use videogames. The population for this case study was 10 children diagnosed with high-functioning Autism, within 5 and 8 years old. They receive therapies at the "Autismo Ecuador" specialized center, located in the Guayaquil city - Ecuador. 60% of the participants are boys, and the rest are girls. The average age was 7.3 years (SD = 2.00). They have been attending the therapy center for more than 6 months. Their parents authorized their participation in the experimental intervention.

4.3 Procedure

For the evaluation of the results obtained, a quasi-experimental Pre-test/Post-test design was used with the group of participants. This type of design allowed the comparison of a variable in two evaluation times, before and after the participant's exposure to an experimental intervention with the use of Serious Game JOINME. In the pre-test evaluation, the participants received the traditional therapies for 30 days, then the therapists/psychologists proceeded to evaluate them using the adapted survey of the

KidsLife scale. For the second phase, the children used JOINME daily in their therapies for a month. In the same way, the evaluation instrument (Post-test) was applied through the therapists who worked with children (see Fig. 8).

Fig. 8. Therapies using JOINME in post-test

4.4   Results

The results obtained in pre-test and post-test were processed in the statistical software SPSS. In Table 1, the values obtained from the sum of the scores are displayed. Consequently, children evaluated in the pre-test got an average of 48.4/60 points, corresponding to 80,66%. In the post-test, the average was 53.3/60 points; it means 88,83%. So, the difference between both scores was 4.90 points equivalent to 8.17%. Furthermore, children showed great motivation, interest, and progress in the social interaction between children after using JOINME. Figure 9 reflects the difference in the scores of each participant in this experience.

Once the descriptive analysis was concluded, the inferential analysis was carried out. The t-Student was selected for paired samples to examine the differences to the same group of participants on two different occasions. This process raised the research question:

RQ: Is there any difference between traditional vs. JOINME therapies in the development of interpersonal relationships in children with Autism?

Two hypotheses derive from this question:

H_0: There are no differences in the development of interpersonal relationships perceived by children with Autism receiving traditional therapies versus the application of JOINME as a tool to support therapies.
H_1: There are differences in the development of interpersonal relationships perceived by children with Autism receiving traditional therapies versus the application of JOINME as a tool to support therapies.

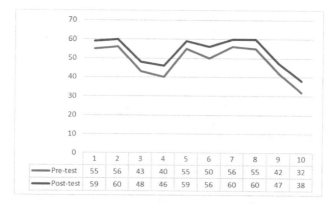

	1	2	3	4	5	6	7	8	9	10
━━Pre-test	55	56	43	40	55	50	56	55	42	32
━━Post-test	59	60	48	46	59	56	60	60	47	38

Fig. 9. Comparison between the scores obtained by participants in Pre and Post-test

Similar calculations were carried out; the results of the t-student method are presented (see Table 1).

Table 1. Students - t distribution

Experimental group	N°	Mean	Sample variance	Observations	Pearson correlation coefficient	Degrees of freedom	t Stat
Pre–test	10	48.40	73,155	10	0,997456279	9	−17.696
Post–test	10	53.30	62,455	10			
P(T <= t) two tail 0.00000002663							
t critical value (two tail) 2.26215716279821							

Several results were obtained, highlighting the value of p = 0.000000043 (where p < 0.05); therefore, the hypothesis (H₀) was discarded, and the H₁ was accepted. Hence, it is concluded that the therapies using JOINME SG as a support tool, achieve a significant change in the development of interpersonal relationships in children with Autism.

5 Conclusions

The development and use of Serious Games are an innovative strategy to support therapies in children with ASD. JOINME, as an assistive tool has a natural user interface, controlled by a motion sensor using the Kinect 2.0 device, which controls the game, through the movement of children's hands. This characteristic enables a better collaborative work with children in the multiple activities within the game. The research results showed that children improved recognition of emotions, good behavior, and a better level of commitment to others. In addition, the therapies that are promoted in the minigames motivate the participants to complete different activities, such as comprehensive reading and the execution of mathematical operations.

References

1. Alessandri, M., Mundy, P., Tuchman, R.: Déficit social en el autismo: un enfoque en la atención conjunta. Rev. Neurol. **40**(Supl 1), S137–S141 (2005)
2. Artigas, J.: El lenguaje en los trastornos autistas. Revista de neurología **28**(2), 118–123 (1999)
3. Boucher, J.: Research review: structural language in autistic spectrum disorder–characteristics and causes. J. Child Psychol. Psychiatry **53**(3), 219–233 (2012)
4. Grzadzinski, R., Huerta, M., Lord, C.: DSM-5 and autism spectrum disorders (ASDs): an opportunity for identifying ASD subtypes. Mol. Autism **4**(1), 12 (2013)
5. Ousley, O., Cermak, T.: Autism spectrum disorder: defining dimensions and subgroups. Curr. Dev. Disord. Rep. **1**(1), 20–28 (2014)
6. https://www.autismspeaks.org/what-autism. Accessed Mar 2019
7. Aguilar Jara, K.L.: Estrategias cognitivas conductuales para trabajar en el aula con un niño con trastorno del espectro autista (2016)
8. Myers, E.F., Follmer, R.K., Zinner, S.H.: Autism spectrum disorder and tourette syndrome: commonalities and connections. Curr. Dev. Disord. Rep. **3**(4), 210–212 (2016)
9. Güemes Careaga, I., Martin-Arribas, C., Posada de la Paz, M.: Evaluación de la eficacia de las intervenciones psicoeducativas en los trastornos del espectro autista (2009)
10. Kanner, L., Asperger, H.: Introduction to Autism (2016)
11. Berggren, S., Engström, A.-C., Bölte, S.: Facial affect recognition in autism, ADHD and typical development. Cogn. Neuropsychiatry **21**(3), 213–227 (2016)
12. Black, M.H., Chen, N.T., Iyer, K.K., Lipp, O.V., Bölte, S., Falkmer, M., Tan, T., Girdler, S.: Mechanisms of facial emotion recognition in autism spectrum disorders: insights from eye tracking and electroencephalography. Neurosci. Biobehav. Rev. **80**, 488–515 (2017)
13. Avila-Pesantez, D., Rivera, L.A., Vaca-Cardenas, L., Aguayo, S., Zuñiga, L.: Towards the improvement of ADHD children through augmented reality serious games: preliminary results. In: 2018 IEEE Global Engineering Education Conference (EDUCON), pp. 843–848. IEEE (2018)
14. Bernardini, S., Porayska-Pomsta, K., Smith, T.J.: ECHOES: an intelligent serious game for fostering social communication in children with autism. Inf. Sci. **264**, 41–60 (2014)
15. Escobedo, L., Nguyen, D.H., Boyd, L., Hirano, S., Rangel, A., Garcia-Rosas, D., Tentori, M., Hayes, G.: MOSOCO: a mobile assistive tool to support children with autism practicing social skills in real-life situations. In: Proceedings of the SIGCHI Conference on Human Factors in Computing Systems, pp. 2589–2598. ACM (2012)
16. Rego, P., Moreira, P.M., Reis, L.P.: Serious games for rehabilitation: a survey and a classification towards a taxonomy. In: 5th Iberian Conference on Information Systems and Technologies, pp. 1–6. IEEE (2010)
17. Avila-Pesantez, D., Vaca-Cardenas, L., Rivera, L.A., Zuniga, L., Avila, L.M.: ATHYNOS: helping children with dyspraxia through an augmented reality serious game. In: 2018 International Conference on eDemocracy & eGovernment (ICEDEG), pp. 286–290. IEEE (2018)
18. Roglić, M., Bobić, V., Djurić-Jovičić, M., Djordjević, M., Dragašević, N., Nikolić, B.: Serious gaming based on kinect technology for autistic children in Serbia. In: 2016 13th Symposium on Neural Networks and Applications (NEUREL), pp. 1–4. IEEE (2016)

19. Whyte, E.M., Smyth, J.M., Scherf, K.S.: Designing serious game interventions for individuals with autism. J. Autism Dev. Disord. **45**(12), 3820–3831 (2015)
20. Alves, S., Marques, A., Queirós, C., Orvalho, V.: LIFEisGAME prototype: a serious game about emotions for children with autism spectrum disorders. PsychNology J. **11**(3), 191–211 (2013)
21. de Lope, R.P., Medina-Medina, N., Paderewski, P., Vela, F.L.G.: Design Methodology for Educational Games based on Interactive Screenplays. In: CoSECivi, pp. 90–101 (2015)
22. Gómez, L.: Escala KidsLife-TEA: evaluación de la calidad de vida de niños y adolescentes con con trastorno del espectro del autismo y discapacidad intelectual (2018)
23. Morán Suárez, L., Gómez Sánchez, L.E., Alcedo Rodríguez, M.Á.: Relaciones interpersonales en niños y jóvenes con trastornos del espectro del autismo y discapacidad intelectual (2017)

Learning Style Identification by CHAEA Junior Questionnaire and Artificial Neural Network Method: A Case Study

Richard Torres-Molina[1] , Lorena Guachi-Guachi[1,2(✉)] ,
Robinson Guachi[3,6] , Perri Stefania[4] ,
and Francisco Ortega-Zamorano[5]

[1] Yachay Tech University, Hacienda San José, 100119 Urcuquí, Ecuador
lguachi@yachaytech.edu.ec
[2] SDAS Research Group, Yachay Tech, Urcuquí, Ecuador
[3] Department of Mechanical and Aerospace Engineering,
Sapienza University of Rome, Via Eudossiana 18, 00184 Rome, RM, Italy
[4] University of Calabria, via P.Bucci, 87036 Arcavacata di Rende, Italy
[5] University of Malaga, Avda. Cervantes, 2, 29071 Málaga, Spain
[6] Department of Mechatronics, Universidad Internacional del Ecuador,
Av. Simon Bolivar, 170411 Quito, Ecuador

Abstract. By the lack of personalization in education, students obtain low performance in different subjects in school, particularly in mathematics. Therefore, learning style identification is a crucial tool to improve academic performance. Although traditional methods such questionnaires have been extensively used to the learning styles detection in youths and adults by its high precision, it produces boredom in children and does not allow to adjust learning automatically to student characteristics and preferences over time. In this paper, two methods for learning style recognition: CHAEA-Junior questionnaire (static method) and Artificial Neural Networks (automatic method) are explored. The data for the second technique used answers from the survey and the percentage scores from mathematical mini-games (Competitor, Dreamer, Logician, Strategist) based on Kolb's learning theory. To the validity between both methods, it was conducted a pilot study with primary level students in Ecuador. The experimental tests show that Artificial Neural Networks are a suitable alternative to accurate models for automatic learning recognition to provide personalized learning to Ecuadorian students, which achieved close detection results concerning CHAEA-Junior questionnaire results.

Keywords: Learning Style · Automatic recognition · Artificial Neural Network

1 Introduction

Several people recognize that every individual acquires and processes information based on their learning styles and abilities. The term "Learning styles" refers to the comprehension that each person learns differently based on its cognitive, affective, and psychological factors, which determines how a person perceives, interacts, and responds to the learning environment. Learning styles are described in models, which

© Springer Nature Switzerland AG 2020
M. Botto-Tobar et al. (Eds.): ICAETT 2019, AISC 1067, pp. 326–336, 2020.
https://doi.org/10.1007/978-3-030-32033-1_30

based on specific scales (of perception and information processing) characterized by theorists in the fields of psychology and cognitive science, classify individuals according to conventional ways that they learn [1]. In this sense, some people have found that they have a blend of learning styles, where a learning style is dominant, with far less use of the other ones. Besides, other people found that they use different styles in different circumstances.

An established problem in the primary education is the lack of customized learning to students in different areas of knowledge, where the identification and use of a mix of learning styles is a challenge yet, due to only a few educators have started to identify them to improve their learning and teaching techniques. Notably, one of the difficulties is on the study of mathematics in students worldwide, where different studies suggest that learning math is complicated, and the outcome is being a "math hater" in the subject, being Latin America (Ecuador) with one of the lowest performance worldwide. To overcome this limitation, static and automatic approaches for identifying learning styles have emerged over time.

In the past years, questionnaires have been the most common approach to identify learning styles, which are characterized by its excellent reliability and validity [2]. In any case, they have also been subjected to some criticism considering that a questionnaire is a static approach, where their results are no longer valid over time, while learning styles change continuously. As well as, in the majority of cases the filling out a questionnaire produces boredom in children. Besides, students are not aware of the importance of the survey for the future uses, which may tend to pick answers self-assertively. Even in some cases, students can be influenced by the questionnaire formulation to give answers perceived as more appropriate. To overcome its difficulties such as boredom, recent proves had established a correlation between playing styles that match with learning styles applied on entertainment games in education [3].

The learning style identification has also been investigated in technical fields like mechanical engineering, for example in [4, 5] the impact of negative knowledge is discussed and implemented as a way to prevent and improve competency in computer-aided-design modeling. In [6] practical experience is linked to theory to help the novice to improve their theoretical results, and in [7] categories of skills and knowledge are defined for defining questions and related significant scores.

Besides, in the most recent years, some automatic approaches based on Artificial Intelligence have been introduced in the learning style identification, such as Artificial Neural Networks-based (ANN) [8], and Bayesian Networks-based (BN) [9]. Since automatic approaches tend to be more accurate and less error-prone, they focus on educational systems that adjust learning to student characteristics and preferences over time.

Although numerous static and dynamic approaches for learning style identification have been introduced with high accuracy, several primary educational systems in Ecuador, mainly for learning math is still a challenge. Therefore, in this work, we explore and compare CHAEA-Junior questionnaire and an automatic method based on ANN to determine their percentage achieved in each learning style identification.

The CHAEA-Junior questionnaire has been selected due to the test reliability in identifying learning styles in children [10], which has been translated into the Spanish language by the researchers and applied in different schools in Spain. Meanwhile, the ANN method was adopted by its demonstrated speed of execution, fast learning, and its efficient predictive capabilities to categorize and learn from specific examples data [2]. ANN as a data-driven method used data gathered from the CHAEA-Junior questionnaire, and the scores in the mathematical minigames based on ADOPTA playing styles to learning styles recognition. As a case study, we evaluated both approaches in students of 11 and 12 years old of seventh-grade primary education from the school "Teodoro Gómez de la Torre" (Ibarra-Ecuador).

The rest of the paper is organized as follows. Section 2 describes the most relevant related works. The explored methods are described in Sect. 3. Experimental setup and results are presented in Sects. 4 and 5, respectively. Finally, Sect. 6 deals with the concluding remarks.

2 Related Works

To learning styles identification, various theories typically focused on testing in teens and adults have been introduced since decades with its appropriate questionnaire, which assess and recognize with strong reliability the most aligned learning style of each learner. Kolb introduced the first theory of the Learning Style Inventory (LSI) based on experiential learning theory (ELT), which recognizes learners as Concrete experience (CE) or "feeling", Reflective observation (RO) or "observing", Abstract Conceptualization (AC) or "thinking", and Active Experimentation (AE) or "doing" [11]. Its LSI questionnaire has been used in accounting, psychology, nursing and business students [12].

Based on Kolb approach, Honey and Mumford introduced the Learning Styles Questionnaire (LSQ) [13], to remark that each individual has the sum up of all learning styles but with a main one. It describes four learning styles as follows: Activist (open-minded individual to new experiences eager to new challenges), Reflector (consider the problem in different perspectives by analysis to obtain conclusions), Theorist (logic is used to build relations and incorporate all details into an issue), and Pragmatist (applied in a practical way theories and techniques). The LSQ was modified and translated to Spanish by Alonso, Gallego, and Honey to use it in university students, which was named "Cuestionario Honey-Alonso de Estilos de Aprendizaje" (CHAEA). Additionally, considering reception and information process as two successive phases, Felder and Silverman developed a learning model with four dimensions (sensing/intuitive, visual/verbal, active/reflective, sequential/global), with its respective Index of Learning Style (ILS) questionnaire [14].

With the main objective of recognizing the learning styles in children, CHAEA-Junior [10], a reduced version of CHAEA was performed to students in elementary education level and first years of high school. This test allows verifying the learning style preference taken into consideration the psychological children characteristics between nine and fourteen years old from Spain. The results prove the reliability of the CHAEA-Junior and in a reciprocal way to the CHAEA itself.

Although static approaches as questionnaires have been the most common way to recognize learning styles with demonstrated precision, in the majority of cases produce boredom in children. For this reason, recent works have studied how the playing styles interfere in their learning styles due to nowadays children spend most of their time playing games. This correlation between playing and learning styles is known as ADOPTA (ADaptive technOlogy-enhanced Platform for eduTAinment) technique [15], which computes the results found in key performance metrics in the game (score, difficulty, and efficiency), and the answers recollected from the Honey and Mumford questionnaire. ADOPTA is based in games to education and describes the styles as: Competitors/Activists (players that like to take risk in the environment and they are fast in problem-solving), Dreamers/Reflectors (players that prefer to observe and listen to the arguments of others), Logicians/Theorists (players focused in logic analysis for task completion), and Strategist/Pragmatist (players interesting in resolving complex problems within a game in the most effective way).

Apart from the entertainment, certain affective, cognitive and psycho-social behaviors, automatic and dynamic detection of learning styles over time by gathering data about students' behavior is demanded. Therefore, in recent years, some methods to automatically predict learning styles have emerged in the Artificial Neural Networks area. Mainly, the ANN-based prediction often has been used with data from a conversational intelligent tutoring system (CITS) [8]. In this case, CITS gives the score of the learning style based in a set of rules based on Processing and Understanding dimensions from Felder-Silverman model taking into account the students' answers and behaviors in the platform. To improve the achieved accuracy, in [16] the four dimensions of the Felder-Silverman model were used, where the inputs were behavior data of a university course and the target was the learning style identified with the ILS questionnaire.

In order to find the most dominant learning style of a child, two alternative methods CHAEA-Junior questionnaire and ANN-based with ADOPTA playing styles are proposed and described in the following. It is notable that, while a child has one dominant style, the rest of the styles have itself contribution percentage to learn.

3   Methods

To identify the percentage of learning style in each individual, we apply and compare a static method: CHAEA-Junior questionnaire (CHAEA-JQ), and an automatic method: ANN. CHAEA-JQ has been selected by its demonstrated reliability by educators and academic counselors despite of being considered a boredom technique in the students [10]. For this reason, to provide entertainment in the assessment of learning style, in this work, video games to recognize the learning style based on ADOPTA playing styles have been used in conjunction with ANN. On the other hand, ANN was adopted by its attractive characteristic in the speed of execution and the updating of parameters [2, 17].

3.1   CHAEA-Junior Questionnaire (CHAEA-JQ)

CHAEA-Junior is a questionnaire focused on students of elementary education level and first years of high school. It is based on the theoretical foundations of Honey and Mumford, which is focused on experience aimed at academic improvement. The questionnaire identifies the learning styles preference of the students by a set of questions written in a way to be comprehended by the psychological children characteristics. It is characterized by its usability, speed, and ease, both in its application and in its correction by counselors and teachers.

The standard questionnaire is presented in a single folio sheet composed by 44 questions, distributed randomly, with the four groups of 11 items corresponding to the four learning styles: Active – Reflective – Theoretical – Pragmatic. The total score obtained in each Style is a maximum of 11, showing the level reached in each of the four Learning Styles (which will be between 0 and 11). The student needs to answer by drawing a circle in the item that he/she agreed; otherwise, it can leave the item without surrounding. On the back of the folio, four columns of numbers belonging to each of the four Learning Styles are presented to define the student's preferred learning profile.

For this work, it was selected randomly 24 questions from the standard questionnaire due to time constraints in the experiment and to avoid boredom in children, each group of 6 questions represents one learning style. It was not used a folio sheet as the standard version; instead, the questions were written in an interactive web system where the student needs to check in a checkbox if he/she is agreed with that question. The score in that similar learning style will be increased, otherwise, if he/she disagrees, the checkbox will be empty, and the score in that learning style will not be modified.

3.2   Artificial Neural Network

It is compounded by three kinds of layers: the input layer, hidden layers, and output layer as it is depicted in Fig. 1. This method comprehends two processes: training and validation. During the training process, the input layer contains the neurons that receive as entry data the CHAEA-JQ answers of the 24 questions. The output layer often also called target layer, holds the percentages of each learning style obtained from the mathematical mini-games based on ADOPTA styles (PLS_{ANN}). The PLS_{ANN} was computed after the student finished playing each mini-game. It measures the percentage

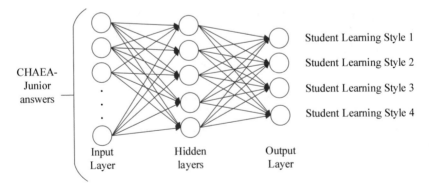

Fig. 1. Top-level architecture of the artificial neural network approach.

of each one by the division between the score obtained in each mini-game and the maximum score between them (which have been applied to the training students' group).

The network learning process takes the input neurons and the expected output neurons, to update the weights on the internal neurons of the one hidden layer until getting the most likely computed output neurons with respect to the target. This process uses the Back-Propagation (BP) algorithm to propagate back the derivative of the error function from the end to the start of the network. The difference between the result from the target and the output of the BP is used as a back-propagation error.

The BP main equation is given by Eq. (1), where o_i is the output of the neuron belonging from the hidden n_i, p represents the synaptic potential, w_{ij} are the synaptic weights between neuron i in the current layer and the neurons of the previous layer with activation $\hat{o}_j$. Therefore, the sigmoid activation function is computed as shown in Eq. (2).

$$o_i = s\left(\sum_{j=1}^{n} w_{ij} \cdot \hat{o}_j\right) = s(p) \tag{1}$$

$$s(x) = 1/(1 + e^{-\gamma x}) \tag{2}$$

The BP algorithm objective is to reduce the error obtained by modifying the synaptic weights, to get a minimum difference between targets and network outputs. The error is given by Eq. (3), where the first sum is computed on the p patterns of the data set and the second sum is calculated on the N output neurons. $t_i(r)$ is the target value for output neuron i for pattern r, and $o_i(r)$ is the response network output.

$$E = \frac{1}{2}\sum_{r=1}^{p}\sum_{i=1}^{N}(t_i(r) - o_i(r))^2 \tag{3}$$

The synaptic weights between two last layers of neurons are computed as shown in Eq. (4), where η is the learning rate and s' is the derivative of the sigmoid function o_i, and the other weights are modified according to deltas (δ) that propagate the error.

$$\Delta w_{ij}(r) = -\eta\left(\partial E/\partial w_{ij}(r)\right) = \eta[t_i(r) - o(r)]s'_i(p_i)\hat{o}_i(r) \tag{4}$$

In the validation process, the input layer receives the CHAEA-JQ answers; meanwhile, the output layer contains the neurons that provide the percentage of the student learning styles which are correlated with ADOPTA playing styles.

4   Experimental Setup

4.1   Data Preparation

Experiments were carried out over collected data of 100 students between 11 and 12 years old from seven-grade primary education of the school "Teodoro Gómez de la

Torre" (Imbabura-Ecuador), which were divided randomly into two data sets, 80% for training and 20% for testing phase. Both methods CHAEA-JQ and ANN were evaluated on the testing set.

4.2   Quality Metrics

The ability to identify the learning styles that have a student was measured through metrics such percentage for learning style recognition for CHAEA-JQ (Pls_{JQ}) and ANN (Pls_{ANN}). Pls_{ANN} corresponds to the output of the network in Eq. (1), while Pls_{JQ} for each learning style is given by Eq. (7), which takes into account the Sumatory of Each Learning Style ($SELS_L$), and the Sumatory of all questions related to the Learning Styles (SLS):

$$SELS_L = \sum\nolimits_{j=1}^{6} q_j^L \quad \begin{array}{l} q_j^L = 1 \ if \ it \ is \ check \ the \ checkbox \\ q_j^L = 0 \ if \ it \ is \ not \ check \ the \ checkbox \end{array} \tag{5}$$

$$SLS = \sum\nolimits_{j=1}^{24} q_j \tag{6}$$

$$PLS_{JQ} = SELS_L / SLS \tag{7}$$

4.3   Experiment Description

To assess the quality achieved by CHAEA-JQ and Artificial Neural Network approaches to identify students learning styles, several experimental tests have been performed in a testing group of children by answering the questions in the CHAEA-JQ and by answering mathematical questions in math mini-games based on ADOPTA playing styles.

To test CHAEA-JQ, each student answered 24 questions in an interactive web-system. The answers were captured in the system, with 1 corresponding to agree (checked checkbox) and 0 corresponding to disagree (empty/unchecked checkbox). An example of data captured in one student will be 1 0 1 ...... 11111, where each number represents the answer of the 24 questions.

By using (PLS_L), it can be found the percentage of learning style with the CHAEA-JQ. An example of one student is 0.00 | 0.50 | 0.17 | 0.33, where each value represents learning styles percentage in the questionnaire, like 0.00 it means the percentage of the Activist learning style, meanwhile 0.50 represent the Reflector learning style.

To determine the percentage of all the styles to learn by using ANN, each student played four mathematical mini-games based on the four ADOPTA playing styles with 32 mathematical questions each one. The mini-games are Competitor (shooting rockets to the correct answer), Dreamer (pressing a puzzle piece to the correct answer), Logician (pressing a card to the correct answer) and Strategist (jumping an avatar and shooting bubbles to the correct answer). The goal of the mini-games is to solve basic mathematical operations (sum, subtraction, multiplication, and division) according to the rules in each mini-game. If the student answers the question correctly, the score is increasing at one point in that mini-game; otherwise, there is not a score increment. After the student finished playing the four mini-games, the data was captured in the system. An example of the scoring results of one student could be 120 | 100 | 130 | 200, where each value represents the score in one of the mini-games. For instance, 120 represent the score of the Competitive style correlated to the Activist learning style; meanwhile, 200 represent the score of the Strategist style correlated to the Pragmatist learning style.

Then, it is calculated the percentage of the learning style based on the scores from the mini-games, dividing the score of each game by the sum of the scores of all the mini-games, obtaining results as 0.22 | 0.18 | 0.24 | 0.36. The binary answers from the CHAEA-JQ were used as input in the ANN, and the target was the percentage calculated from the mathematical mini-games based on ADOPTA playing styles as explained before. In that way, both techniques with a testing set of 20 students were evaluated finding the error between the percentages in the learning styles calculations. Therefore, the error (ε) between both methods for the four Learning Styles (N = 4) was calculated by the Eq. (8).

$$\varepsilon = \frac{\sum_1^N \left[abs \frac{(Pls_{ANN} - Pls_{JQ})}{Pls_{JQ}} \right]}{N} \tag{8}$$

For the training and testing processes, ANN uses $\eta = 0.05$ and $\gamma = 1/2$, using 1000 for the maximum number of iterations as the parameters that best fit the model in different experimental tests. Besides, different architectures were used with the calculation of the Mean Square Error (MSE), the best performance was achieved by the architecture with one hidden layer of 10 neurons.

5 Results and Discussion

This section depicts and discusses the achieved precision to recognize the learning styles for each analyzed method. The average of the results of the percentage reached by all styles that own every student of the testing group has been computed for the Pls_{JQ} and Pls_{ANN} metrics. The overall results are summarized in Table 1, where the existing relation between Learning Styles and ADOPTA Playing Styles is shown in the first column.

Table 1. Average of percentages learning styles based on CHAE-JQ and ANN

Learning Styles/ADOPTA playing styles	Pls_{JQ}	Pls_{ANN}
Activist/Competitive	18.82%	23.44%
Reflector/Dreamer	**32.45%**	**29.91%**
Theorist/Logician	22.33%	26.70%
Pragmatist/Strategist	26.39%	20.07%

The results depicted in Table 1 show that Reflector/Dreamer is the most dominant learning style in the students that have achieved the highest percentage value for both CHAEA-JQ and ANN approaches. It describes that the majority of the students follow the traditional educational system, and have a higher tendency to prefer to listen to the people opinion and consider the problem in different ways to obtain conclusions.

It is essential to mention that the results found in the CHAEA-JQ have a variance of $\sigma^2 = 34.25\%$. The variation in this technique states a considerable difference in the dispersion between the data. The reason is the highest percentage found in the Reflector/Dreamer learning style with a value of 32.45% in comparison with the other learning styles. This predominance in this learning style denotes the psychological evolution in children by following a traditional educational system, where the students need to adapt to that learning style since their early stages of life. Being this tendency to this learning style to preferred to listen first, and then act to conclude, to create solutions; with a tendency to be thoughtful and cautious. The students with a higher-level Reflector/Dreamer learning percentage do not learn when they are forced to take a leadership position in a group and doing tasks without prior preparation. The CHAEA-JQ could slight differ if the 24 questions were selected different from the 44 questions and if the questionnaire could be presented with graphical representation in each question. It could produce an increase or decrease in the learning style percentage in the student, but the main improvements could be made by boosting interactivity in the questionnaire to amuse the students to answer the question self-aware because even that the survey is the most common approach, the student could lie in it producing an unreliable classification. Also, the ANN approach has a variance $\sigma^2 = 17.91\%$, meaning that the spread of the data is more uniform as is seen in Table 1 starting from the lowest percentage style which is 20.07% in Pragmatism/Strategist learning style until the highest value with 29.91% which was stated before. It is related in the amount of data that was collected because according to the input in the network and the scoring in the mini-games, it could drastically improve the learning style percentage prediction that will be discussed in the next paragraph.

In Table 1 is shown that the results were not fit in the rest of the learning styles, by comparing the CHAEA-JQ with ANN approach. It is because of the data gathered from the mechanics in the game, such that in the Activist style related to the Competitive playing style, the students, attracted by the game, shoot to an answer randomly without trying to solve the question correctly. On the Pragmatist style related to the Strategist playing style, the students were confused in jumping the avatar precisely to throw away bubbles to the correct answer, and in the Theorist style, related to the Logician Style, the students were confused with pressing the card center in the right answer selection. Those behaviors generated outcome where the students could not answer some questions. In fixing it some components in the game, the data gathering could be more effective in the prediction of learning styles percentages. Also, the reliability in learning style detection can be enhanced by amplified the sample size.

However, it is interesting to note that the automatic method based on ANN is not too far from the CHAEA-JQ results, which in the past has been characterized by its excellent reliability and validity, with just an error $\varepsilon = 18.9\%$.

6   Conclusion

In Ecuador there is an inefficient personalized education in teaching math to students in the primary education level, being an indicator one of the lowest scorings in the international mathematical examination called PISA in comparison with developing countries. By lack of accessibility to excellent and personalized training in teaching math to children and the motivation that takes part in it. The use of ICT has been an advantage, but the customized content to each student according to its learning style is still a challenge. However, finding the most optimal method to learning style detection to the student it can be a useful tool to diminish this problem.

As an overall conclusion, by the comparison between the two methods: CHAEA-JQ questionnaire and ANN, it was found that the ANN provides percentages close to the most effective traditional one (CHAEA-JQ) to recognize the most dominant learning style, and obtain an average error $\varepsilon = 18.9\%$ for identifying all of them. To have better recognition and in the future could be used to provide content related to the learning style of the student by using mathematical mini-games based on ADOPTA playing styles to increase the mathematical skills in Ecuadorean students.

To obtain better results in the learning style recognition it must be considered the environment in which the experiment is taken place, i.e., the computational resources equipment because the CHAEA-JQ and the mini-games were implemented in a Web system. The sample size should be acknowledged to enhance the results in the experiment. The gathered data to enhancement in this work can be done by improving the game mechanics, adapting dynamic tutorials in the game and animated avatars in each question in the CHAEA-JQ. Some data that can be collected in future works is the emotional state for the student after it finishes to play each game to be used in the ANN.

Acknowledgements. We want to thank the director Juan Vázquez, the teacher Silvia Diaz, and to the administrative staff of the school "Teodoro Gómez de la Torre" (Ibarra-Ecuador) who have contributed in the data gathered in this work.

References

1. Cassidy, S.: Learning Styles: an overview of theories, models, and measures. Educ. Psychol. **24**(2), 419–444 (2004)
2. Feldman, J., Monteserin, A., Amandi, A.: Automatic detection of learning styles: state of the art. Artif. Intell. Rev. **44**(2), 157–186 (2014)
3. Bontchev, B., Vassileva, D., Aleksieva-Petrova, A., Petrov, M.: Playing styles based on experiential learning theory. Comput. Hum. Behav. **85**(0), 319–328 (2018)
4. Otto, H., Mandorli, F.: Integration of negative knowledge into MCAD education to support competency development for product design. Comput. Aided Des. Appl. **14**(3), 269–283 (2016)
5. Otto, H., Mandorli, F.: A framework for negative knowledge to support hybrid geometric modeling education for product engineering. J. Comput. Des. Eng. **5**(1), 80–93 (2018)

6. Baronio, G., Motyl, B., Paderno, D.: Technical Drawing Learning Tool-Level 2: an interactive self-learning tool for teaching manufacturing dimensioning. Comput. Appl. Eng. Educ. **24**(4), 519–528 (2016)
7. Villa, V., Motyl, B., Paderno, D., Baronio, G.: TDEG based framework and tools for innovation in teaching technical drawing: the example of LaMoo project. Comput. Appl. Eng. Educ. **26**(5), 1293–1305 (2018)
8. Latham, A., Crockett, K., Mclean, D.: Profiling student learning styles with multilayer perceptron neural networks. In: 2013 IEEE International Conference on Systems, Man, and Cybernetics, pp. 2510–2515, Manchester, United Kingdom. IEEE (2013)
9. Hasibuan, M.S., Nugroho, L.: Detecting learning style using hybrid model. In: 2016 IEEE Conference on e-Learning, e-Management and e-Services (IC3e), pp. 107–111, Langkawi, Malaysia. IEEE (2016)
10. Sotillo Delgado, J.: CHAEA-Junior Survey or how to diagnose elementary and secondary students' learning styles. J. Learn. Styles **7**, 182–201 (2014)
11. Kolb, D.: Experimental Learning. Prentice-Hall, Englewood Cliffs (1984)
12. Manolis, C., Burns, D., Assudani, R., Chinta, R.: Assessing experiential learning styles: a methodological reconstruction and validation of the Kolb Learning Style Inventory. Learn. Individ. Differ. **23**(73), 44–52 (2013)
13. Honey, P., Mumford, A.: The Learning Styles Questionnaire. Peter Honey Publications, Maidenhead, Berks (2001)
14. Dziedzic, M., de Oliveira, F., Janissek, P., Dziedzic, R.: Comparing learning styles questionnaires. In: 2013 IEEE Frontiers in Education Conference (FIE), pp. 973–978, Oklahoma City, OK. IEEE (2013)
15. Bontchev, B., Georgieva, O.: Playing style recognition through an adaptive video game. Comput. Hum. Behav. **82**(3), 136–147 (2018)
16. Bernard, J., Chang, T.W., Popescu, E., Graf, S.: Using artificial neural networks to identify learning styles. In: Conati, C., Heffernan, N., Mitrovic, A., Verdejo, M. (eds.) Artificial Intelligence in Education. LNCS, vol. 9112, pp. 541–544. Springer, Switzerland (2015)
17. Hmedna, B., Mezouary, A., Baz, O., Mammass, D.: A machine learning approach to identify and track learning styles in MOOCs. In: 5th International Conference on Multimedia Computing and Systems (ICMCS), pp. 212–216, Marrakech, Morocco. IEEE (2016)

Digital Learning Objects for Teaching Telemedicine at University UNIANDES-Ecuador

Eduardo Fernández[1(✉)], Paola Pérez[2], Víctor Pérez[2],
and Gustavo Salinas[2]

[1] Universidad Regional Autónoma de los Andes, Ambato, Ecuador
ua.eduardofernandez@uniandes.edu.ec
[2] Universidad Técnica de Ambato, Ambato, Ecuador

Abstract. The present investigative work arises from the existent problematic in the University Regional Autónoma de Los Andes, and that is related to the limited use of technological elements in the educative process that carry out the teachers of the Career of medical Sciences of this Institution. The problem also has to do with the limited knowledge that there is about Telemedicine. As a solution to this problem, the project proposed as general objective is: "To elaborate learning objects about Telemedicine so that they serve as a technological element in support of the educational process." To clarify the problem, a survey was conducted among teachers and students of the same career, is concluded that there is ignorance about objects of learning and telemedicine, also the researched point out that teachers do not use technology as an educational support element. The theoretical foundation of the work has to do with education, with objects of learning and with their computer tools for the development of them. Finally, several learning objects useful for Telemedicine teaching have been developed, in the eLearning tool, and since they are web applications they are published in a digital repository created for it.

Keywords: Learning objects · Telemedicine · Education

1 Introduction

The fundamental characteristic of the final decade of the XX century and early XXI century was the significant technological development, especially in three areas such as information systems, communications, and general electronics. The fusion of these devices emerges dispositive such as computers that have reached a degree of popularization that nowadays in every household is very reasonable to have one of them. The same technological advance and this fusion of technologies, the Internet emerges as a means of communication between computers and obviously between people [1].

The popularization of computer equipment begins in 1984 with the appearance of the first PC from IBM, from there, to date, have been raising the processing speed, storage capacity and the number of programs to be used in them, on the other hand, also they have been shrinking the size of the equipment and the price of the same. All these factors have combined to make each day computers more accessible and mostly used in the world for all people [2].

© Springer Nature Switzerland AG 2020
M. Botto-Tobar et al. (Eds.): ICAETT 2019, AISC 1067, pp. 337–346, 2020.
https://doi.org/10.1007/978-3-030-32033-1_31

The international computer network for civilian use taking off in the mid-90s, the connecting nodes were universities in the United States, he worked deficient, and its purpose was eminently educational. Over the years allowed the increase in rate as well as the emergence of new services within this network, the same that have been facilitating communication and use of computers in the international system [3].

As it is logical to assume, this new technological advance has been used by other areas of human knowledge, and that is why the technical aspect has been fully taken up in the business field, this means that there are not companies that do not have technological devices for all activities operational and commercial that carry out. Another area greatly benefited from this technological breakthrough is education; this means that technologies are influencing directly in the modern educational process; there are several tools that develop activities of teaching and learning process. Even now they have virtual platforms that simulate this educational process, thus solving problems of distance and time for students, allowing knowledge to reach more people and to remote sites [1].

The impact of technology in education has grown exponentially in recent years, and this has required a significant effort of teachers to manage and adapt to their classes and all teaching activities. On the other hand, it may also be noted that teachers try to go slowly, digitalizing all educational materials to share with their students and other teachers soon. Consider that the statement above had been met with a low percentage of teachers, there are a several factors that make that many teachers do not use neither technology nor academic content, digitize their materials or their classroom work [4].

2   Developing

For **UNESCO,** one of the basic requirements of the XXI century education is to prepare the population to participate in a knowledge-based economy, which includes social and cultural perspectives.

In the central region is the University: Uniandes, disposer of Medical schools. Initial research conducted in these centers of higher education, and they have to do with learning processes in these faculties, have encountered some difficulties, which may be mentioned, for example:

- Teachers define the contents of the materials; this creates an infinite number of variations in the modules of the same subject, either in the same or a different cycle.
- The modules are for a single learning style, and as it is known, learning depends on each student; therefore, there are different learning styles, and in the module we are only concentrated to one.
- The appearance of the syllabi is not very motivating for students; these require the use of more graphics and visual resources.
- The contents are not permanently available to students; this means that the student must necessarily be at home or take out the syllabus physical where they go.
- The use of Tics is not encouraged permanent, and teachers do not handle them properly.

- You cannot reuse syllabi or other educational activities that are not known because they are not digitized.
- The strategies collaborative between teachers does not apply, and neither are encouraged the same among students.

As described above, it can be concluded that University has trouble in learning processes due to the limited use of technology by teachers. As a solution to this problem the project "Telemedicine Learning Objects" arises, the following **general objective,** Develop several learning objects that are available in the digital repository of CEDIA and through its use can achieve improvement in the teaching and learning process for subjects as treatment of trauma and telemedicine in Medical Schools of the University Uniandes.

To achieve this overall objective, it has been considered necessary to meet the following specific goals:

- Basing theoretically learning objects, development tools, telemedicine, and education in general.
- Diagnose the level of knowledge and use of learning objects in the educational process of the Faculty of Medical Sciences at this university.
- Develop several learning objects on the subjects mentioned.

The project was planned, and developed during 2017, its budget came to $ 12,600 of dollars and in the benefits that generate may include:

- Digitalization lesson plans to create so-called learning objects in areas such as telemedicine and trauma care.
- Overall improvement of the educational process due to the strengthening of skills of both students and teachers.
- Availability of learning objects because being on the platform CEDIA can be reused and improved.

For these benefits, the project is fully justified. As follows, it is exposed the criteria issued by various authors and universities with learning objects.

The Polytechnic University of Valencia (2013) believes that the establishment of the European Higher Education Area (EEES) an activity of great importance is the implementation of a digital repository of learning objects with multimedia features and as a basis for creating interactive learning modules.

In this context, the University defines a learning object as "The minimum unit of learning, in digital format, which can be reused and sequenced." [5].

Therefore, conceived these small components (O.A.) as integrated or integrators elements of the teaching, offering the students the opportunity to improve their performance and satisfaction. The presence of learning objects has created a new way of thinking about learning content. The content is no longer a means of achieving a goal and becomes an object in its own right, capable of being reused [6].

A learning object is "a collection of content, exercises, and evaluations are combined based on a simple learning objective" is to reach a goal. The term is awarded to Wayne Hodgins, who used to create a working group in 1994 that includes its name, but the concept was first described in 1967 by Gerard [7].

On the other hand, the authors consider that a learning object is: An independent digital teaching unit, the structure contains a learning objective, content, a set of activities, and evaluation. It can be reused in different contexts technological and educational; also, it has metadata that allows their location within repositories [8].

When referring to "**Teaching Unit**" refers to integrated set, organized and sequential elements with its own sense that allows students to appreciate the result of their work, after study [4].

"**Reuse**" of a Learning Object is essential features that make it better and different versus other approaches to learning on the Web. From an educational point of view, the more context it has the immersed contained therein, the less chance of reuse to compose lessons, units or new modules and reused in other contexts will possess, while a more general contextual approach of content, allow that it can be reused in other educational settings [6].

Several authors have stated that **Learning Objects** should have specific characteristics, which are in agreement within the scientific community and are:

Interoperable: property that allows an OA to have the conditions and be able to be deployed on various recognized technological environments that enable full functionality and using uniform standards.

Education: quality that has OA through which meets or acquires an intention and/or educational purpose to facilitate the understanding, the representation of a concept, theory, phenomenon, knowledge or event, in addition to promoting individual development of skills, abilities and skills of a different order: cognitive, social, cultural, technological, scientific, among others [9].

Generative: the ability to build new lessons, units, modules, etc., from its assembly with other OA. Possibility of renovation or modification through community collaboration OA development, increasing their potential.

Publishable: facility to be identified, located through labeling with various descriptors (metadata) that enable cataloging and storage in the corresponding repository. It also refers to the form of "license" OA so that access to these its public or private domain [10].

Reusable: the quality that has an OA so it can be used for the purpose of create new levels or components of instructional design (course, lesson, etc.) and be used as a teaching resource in different technological and educational contexts [11].

Learning Objects need to be stored, located, and retrieved by some mechanism or system to perform this process can use the Learning Object Repositories (ROA). The ROA is application of information system that facilitate storing, searching, location and retrieval of OA. It makes possible to publish an OA in a repository of learning objects. The most commonly known ROA operate independently with a web interface and a search mechanism. By the way, resources are concentrated, it is possible to identify two types of ROA:

- Those containing learning objects and their metadata, being within the same system and even within a single server, as shown in the figure below.
- Those containing only the metadata and access the object through a reference to its physical location are in another system or object repository [7].

The general structure of a learning object depicted in the following scheme (Fig. 1):

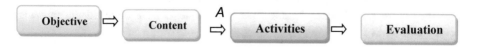

Fig. 1. Structure of a learning object

The starting point of design tasks is not only the content to be transmitted, but fundamentally the forms of organization of the activity are intended to develop the students interact with "educational material."

There are several methodologies for the development of so-called learning objects; one of the most popular is called DICREVOA, which has the following stages: **Analysis, Design, Implementation, Evaluation, and Publication**.

These phases are intended to guide the teacher in designing an OA, for which, given a set of guidelines that allow it to be this one to make crucial decisions regarding planning learning through OA. This methodological proposal is designed for those teachers who do not have in their institutions of a multidisciplinary team that is responsible for providing the necessary support for the creation of OAs, thus having that be the teacher, who is responsible for the entire process design, development and evaluation of O.A.

3   Methods and Results

After the respective theoretical foundation, we proceeded to the methodological part; for this, first, the population and sample defined as follows (Table 1):

Table 1. Population

Function	Number
Students of Medical Sciences Uniandes	800
Teachers of Medical Sciences Uniandes	80
Total	880

Based on this population and assuming an error of 5%, the respective sample was obtained.

$$Sample = 275$$

Investigative techniques adopted were: the survey by both teachers and students of universities and observation of teaching activities directly into classes.

The instruments associated with these research techniques were questionnaires and notebook. These instruments were designed based on the symptoms of the problem, leaving the main structured with the following questions:

After surveys conducted, both teachers and students have obtained the following results (Table 2):

Table 2. Results of the investigation

Questions	Answers				
1. Mark your level of knowledge about digital learning objects?	None	Very Low	Under	Medium	High
	167	75	16	11	6
2. Do you think, as a teacher of Medical Science must use digital learning objects as elements of educational support?	Never	Rarely	Frequently	Always	
	25	26	93	131	
3. Would you agree that there is a digital repository with learning objects related to medical science to help you serve your educational training process?	Yes	No	Does not matter to me		
	207	21	47		
4. Do you think that the formation of health care today should be strongly supported by technology-based pedagogical elements?	Yes	No	Does not matter to me		
	241	26	8		
5. Mark your level of knowledge about Telemedicine?	None	Very Low	Under	Medium	High
	19	69	93	81	13
6. Would you like to learn telemedicine supported by learning objects?	Yes	No	I have no idea		
	204	49	22		

Investigation to 275 people (teachers and students)

From the results, it follows that:

61% of those surveyed do not know about learning objects.

Approximately 50% of respondents believe that teachers of Medicine should use digital learning objects as support elements in the educational process.

The vast majority (75%) states that technology should support the teaching of medicine and that it would be good if there is a digital repository for learning objects.

There is a low level of knowledge related to Telemedicine, but it is considered that it will be essential to learn it using learning objects.

From the results, the teaching of Telemedicine can be recommended, but using digital learning objects. Based on this recommendation, we proceed to develop a digital learning object for the learning of Telemedicine.

For the development of learning objects assume the following criteria:

Development methodology.	DICREVOA
Software used:	exeLearning
Material:	Telemedicine

How is assumed the method DICREVOA requires the following initial conditions (Tables 3, 4 and 5):

Table 3. Matrix of needs

Matrix of needs	
Topic	Introduction to Telemedicine
Description	Initial concepts of telemedicine
Level	Higher Education
Student Profile	A student with primary knowledge of medicine and computer
Estimated time	8 h
Educative Context	Educational context Telemedicine is a new paradigm of health care is based on the use of the internet and computer, solves problems away
Type of license	Free

Table 4. Instructional design

Template for design of learning objects	
Instructional Design	
1	**Textual Description of the content**
	The object will contain the initial concepts of telemedicine, for it summarizes the evolution of the internet and communications in general, is required. It also incorporates some history related to remote of medical care. Then definitions of telemedicine are specified according to OMS. Based on the definitions the application areas of telemedicine are deducted. The concept of Telehealth is extended. Finally, ideas are evaluated
2	**Objective and Learning:** Understand the concepts of telehealth, regions and applications and development in the last decades
3	**Contents** • Explication about the term "tele" • History of telemedicine • Telecommunication, Evolution • Information Evolution • Internet development • Conceptualization about Telemedicine by world entities of health. Medical areas that can be used for telemedicine

(continued)

Table 4. (*continued*)

Template for design of learning objects

Instructional Design

4	**Activities** • Summary of evolution historical of telemedicine • Dispositive about the development history of information systems and internet • Map about telemedicine
5	**Auto-evaluation:** Evaluative Crossword of fundamental concepts developed

Designed template

1	**Designed template of Interfax** OA uses a simple and friendly Interface configured to 800 × 600 Pixel for the browser also uses a style of INTEF cascade (CSS) of E-Learning, it also looks beautiful in tone colors
2	**Structure of screen:** La structure of screen of OA, is based on the designer of left browser block
3	**Browser:** The organization of OA contends follows a sequence of hierarchical browse, going from the known to the unknown, the immediate to the mediate, concrete to abstractor and the easy to the difficult

Table 5. Internal structure

No	Structure of O.A.	Subject	I-devices de eLearning
Internal structure of O.A. Y LOS i-devices used			
1	Start	Cover page	Free text
2	Objectives	Description of previous knowledge and general objectives of O. A.	Information textual
3	Contents	Texts, videos, enlaces, résumés sobre la evolution historical del internet, de las telecommunications y de la telemedicine Conceptualization de la telemedicine por parte de entities mundiales de salud	Free text
4	Activities	Elaboration of templates, summaries and conceptualist maps and dispositive	Ardora, related to text, images, external sites as goconqr and emaze
5	Evaluation	Crossword	Ardora
6	Bibliographic	Bibliographical References	Free text

After the design proceeded to the development of the learning object in the exeLearning program, as follows some screenshots of this development (Fig. 2).

Fig. 2. Learning object of Telemedicine. Source: http://uniandesinvestigacion.edu.ec/repositorio/oa/telemedicina1/

4   Conclusions

Learning objects are essential element in the modern educational support top-level educational process; its application allows the achievement of meaningful learning.

The digitization of content of a syllabus, embodied in a learning object, implies a modernization in the educational process of any university, to which the improvement of digital skills by teachers is required.

Learning objects are not disseminated in the participating institutions of the project, which is supported the development of this research, needs to be reflected, from the classroom among the main actors of educational process teacher-student.

The incorporation of learning objects as elements of educational support generates a culture of teaching by technology and a positive impact among students, because most of them are digital natives and their typical development environment.

Learning objects are complemented by the use of other technological tools applicable to educational; they, can be used in different learning modalities.

Designed learning objects following stages of implementation and use are required, with the next step of assessment of their level of complexity, applicability, and importance, among others.

References

1. Burgos, J., Lozano, A.: Educational technology and learning networks. Trillas, México (2010)
2. H. I. University, Technologies, and Learning. Prieto, Miami (2014)

3. Alversond, R., Collins, A.: Rethinking Education in the Age of Technology: The Digital Revolution and Schooling in America. Teachers College Press, New York (2010)
4. Pérea, A.: Educate Yourself in the Digital Era. Morata, Madrid (2012)
5. Oscar, S.: Approach to the origin of the notion of learning object: historical and bibliographic review. INGE CUC **12**(2), 26–40 (2016)
6. Cameron, T., Bennett, S.: Learning objects in practice: the integration of reusable learning objects in primary education. Br. J. Edu. Technol. **41**(6), 897–908 (2010)
7. Klemke, R., Ternier, A., Kalz, M., Specht, M.: Implementing infrastructures for managing learning objects. Br. J. Edu. Technol. **41**(6), 873–882 (2010)
8. Tylor, P.: Learning Objects for Instruction. Design and Evaluation. IGI-Global, Miami (2007)
9. The University of Sydney School of Education and Social Work: Learning objects, 19 Marzo 2013. http://sydney.edu.au/education_social_work/learning_teaching/ict/theory/learning_objects.shtml. Accessed 29 Mar 2019
10. Benítez, J., José, G., Eulalia, H., Jaime, M.: Some learning objects to explain Kepler's laws. Comput. Appl. Eng. Educ. **21**(1), 1–7 (2013)
11. Frantiska Jr., J.: Creating Reusable Learning Objects. Springer, Cham (2016)

Virtual Education Based on Digital Storytelling as an Alternative for the Inclusion of the Microenterprise Sector and ICT

Francisco Jurado$^{(\boxtimes)}$ ⓘ, Tannia Mayorga ⓘ,
and Mario Oswaldo Basurto Guerrero

Universidad Tecnológica Israel,
Francisco Pizarro E4-142 y Marieta de Veintimilla, Quito, Ecuador
{fjurado,tmayorga,obasurto}@uisrael.edu.ec

Abstract. The constant advance of society is directly related to the evolution of Science and Technology, where the level of education contributes to the development of people and companies. This document presents the theoretical basis of the work related to the inclusion of virtual spaces based on digital stories to be able to include Information and Communication Technologies (ICT) in the microenterprise sector, because the virtual environments that have the particularity of adapting to the characteristics of the participants, the same that should be used to train even more in this era where technology is part of all the activities of the human being, the microenterprises that, due to their characteristics, do not take into account the training of personnel and the implementation of technology, as in other types of companies that have been developed with ICT, in this way the growth of this sector has been limited due to the ignorance of those involved and do not have adequate methodologies to these, they have developed the presumption of not requiring technology, evidencing in the limited development and short time of life in the market that these companies maintain.

Keywords: Storytelling · ICT · Education

1 The Evolution of Learning Models

The globalization of information, a consequence of the advancement of science and technology, currently allows to have all kinds of information at any time, what allows anyone to access any type of information, either out of curiosity or because you want to learn something new, it is through the use of technology that one could suppose it would be simple for people who were born surrounded by technology and more complex for those who must adapt to it. For this, in the first instance, access to technology is required, which, due to figures presented worldwide, about 40% of the world population has access to Information and Communication Technologies [1].

Access to ICTs generally indicates the number of people who have a technological device with an Internet connection, as indicated by the World Bank [2] in its 2016 World Development Report, called Digital Dividends, booklet of the "Overview", where it is highlighted that the growth in the number of mobile phones, Internet and

© Springer Nature Switzerland AG 2020
M. Botto-Tobar et al. (Eds.): ICAETT 2019, AISC 1067, pp. 347–356, 2020.
https://doi.org/10.1007/978-3-030-32033-1_32

other digital technologies can transform economies, societies and institutions, obvious results if we observe how technology is part of almost all human activities.

The growth in acquisition of technology for obvious reasons does not guarantee that its use is adequate, especially in the field of education, where every day the profiles of professionals required demand that people seek to specialize in order to opt for better jobs, a task that in today's society where capital is favored more than the welfare of the people [3]. It presents some limitations when a person is professional or does not seek to learn or specialize, one of these is to have time to attend a school, this limitation to develop academic activities related to completing studies, opting for a third or fourth level degree or to learn about a specific topic, becomes more complex if to complete this activity you must attend a classroom regularly within an educational center.

Due to this, it is nothing new to indicate that the advance in modern science goes hand in hand with the actions that are given on people and nature is here where progress in the fields of technology and science begins, main generators of changes in society and productivity, a country is considered advanced if its level of generation of new knowledge is high, this new form of categorization allowed the emergence and development of the knowledge society.

These advances with all the advantages that they present have also generated the illusion that technology allows an excellent quality of life and well-being in people, without taking into account the effects and the new problems that this entails, Beck explains this, a society based on knowledge is a society closely related to some type of risk that goes hand in hand with situations in which men, due to their historical inexperience, present them in most cases due to the globalization [4].

This globalization is also evident in the way in which people acquire knowledge, whose models have been evolving together with the development of technology, both for face-to-face and distance or virtual modalities [5], starting with the face-to-face modality in a general way, as indicated by Barrantes (1992), is characterized by the fact that the teacher is sharing a physical space with his students, using as a means of communication to transmit the information in order to impart class, in this space the teaching that occurs in this period of time is also consumed at that time [6].

This modality developed certain characteristics in the industrial era in which people, due to the appearance of industrialization, were forced to drastically change their work habits that before this time were developed on a farm, to then move on to work in the city, in this new stage of the evolution of society, people had to adapt to the requirements of labor that did not necessarily have to be qualified for mass production, this characteristic was evidenced in the educational model which had the function of assigning roles to both men and women [7], who, in order to develop in the professional field, had to choose between working in a trade, in a factory or in an office position, in this stage, getting to occupy managerial positions, administrative was only for people who for different reasons had access to professional training.

The education of this era looked for the standardization, the uniformity of knowledge in order to reach a mass production, where the few favored people reached a directive position, the objectives that were reached with the characteristics of education focused on being able to serve to society through a profession, the application of engineering and science was applied in order to contribute to industrial progress, becoming each individual compared to a piece within the production and distribution chain [8].

The aim was to develop specific skills in the largest number of people through literacy and to provide them with basic arithmetic skills, so that they could develop specific skills for manufacturing, commerce and industry, the role of the teacher was simplified to being a transmitter of knowledge, focused on dictating a master class, students took notes and had to read specific bibliography to complete the study of any subject [9].

The way in which society evolved gave way to people seeking new ways of learning to seek better opportunities, this is how the distance education model appears that, due to its flexible schedule characteristics, allowed students to attend tutorials in accordance to its availability either for study at night time, specific weekends or days, when reviewing several works around the definition of this model of education, the definitions provided by Miguel Casas (1982) stand out for those who are models based on strategies where the relationship between student and teacher is the same. leads through the exchange of printed information and mechanical or electronic means [10], in this line is also Victor Guedez (1984) who indicates that distance education is one that allows the exchange of cognitive information and messages for student training by means that do not require a face-to-face relationship in a specific place [11].

As a basis for this learning model, together with the advantages that technology has incorporated, virtual education is developed. In this new form of learning, technology acquires a very important role as a tool to facilitate professional growth through online work on campus virtual, in this way, face-to-face is not a requirement to learn. This new approach to education, given the influence of ICT, allowed the term "virtuality" to emerge [12], which refers to the use of electronic devices to manipulate the senses of: hearing, sight and touch, in order to create an artificial space where learning processes are developed.

2 Characterization of Virtual Education

Education has found in virtuality a new learning scenario using the Internet where network communication is developed that can be synchronous or asynchronous, reaching the flexibilization of temporality and the sequentiality of curricular activities.

The profound changes that have occurred in all areas of society in recent years require a new training for young people regardless of the level of ICT integration that schools have, teachers also require digital literacy and a didactic update that allows them to know, master and integrate technological instruments and new cultural elements in their teaching practice.

The advance observed in the last decade in terms of technology applied to education has generated changes in the paradigm of the way in which it is taught and learned. The educational institutions are the first ones invoked to participate, design and implement methodological strategies for the application of technology in all its educational processes, which tend permanently, for the improvement of the quality of education, this is also indicated by UNESCO in its document on ICT Competencies Standards for Teachers (2008) [13], where it mentions that:

"Thanks to the continuous and effective use of ICT in educational processes, students have the opportunity to acquire important skills in the use of these. The teacher is the person who plays the most important role in helping students acquire these skills. In addition, he is responsible for designing both learning opportunities and the enabling environment in the classroom that facilitate the use of ICT by students to learn and communicate".

On the other hand, in this same sense is the study called "Humans & Machines", conducted by The Economist Intelligence Unit and sponsored by Ricoh (2013), where the impact of technology on creativity and human intuition in organizations is addressed, here it is stated that 71% of managers have helped them to make sound decisions, 72% affirm that human/technology interaction brings great advantages to the economy.

The main challenge is to be able to use ICT taking into consideration that technology evolves faster than the internal processes that sustain it and that those technologies that can most favor imagination and intuition are data analysis, email, cloud computing (data repositories) and telepresence [14].

The e-learning platforms are virtual environments that intend to follow an educational model based on the use of ICT, these become a support within the entire learning process, allowing to give more importance to pedagogy in education, in these environments should be maintained as main elements the motivation and the follow-up, giving importance to the wealth of the differences of the individuals, to the quality and warmth and not to the amount of information and coldness when sharing it. It is here, where the social processes that motivate the criticality and analysis of the presented information are added to the teaching, getting to build knowledge. In this new process, the tutor does not limit himself to inform, expose and teach, but also helps to create, educate, guide and share the results, applying the instructional design.

The importance acquired by virtual education raises the obligation that the tutor in charge of teaching not only manages the knowledge of the subject to dictate, but also must have solid knowledge on the handling of electronic devices to be used as smartphones, Tablets, or computers, in addition to the software required for your virtual platform forcing the teacher tutor before starting to develop a specific topic in a virtual environment to be clear about the following points:

– Define the level of knowledge and skills of technology management, both hardware and software of students, to define the language and characteristics of the content to be presented in the virtual environment, which would even give way to technological literacy so as not to limit the learning process due to not being able to use a printer, a mouse or not knowing that it is a browser, a website or how to access them
– Adapt the content of the virtual environment to the characteristics of technological management of students and develop the content of the virtual classroom, under the characteristics of presence, that is, create the need, so that students enter the virtual classroom motivated by its use; of Scope, that is, define and clearly explain the objectives of the course; Finally, Interaction, which is basically the use of all technological resources available and appropriate to the level of student knowledge

If the tutor is trained in the management of Technology, he will be able to transmit this knowledge to his students, prior to starting the virtual course, so that during the teaching process there are no students left behind or who drop out of the course, taking the data reference on students in this new modality highlights the one presented in Ecuador by the "Secretaria Nacional de Educación Superior Ciencia y Tecnología" (SENESCYT) in an article published in El Universo where it is indicated that 9.60% of students access the distance modality and 3.8% in the blended modality, lower data to countries such as Colombia, Mexico or Brazil. In addition to there are more than 2280 careers at a general level, where 13 public universities offer 71 non-face-to-face courses and 17 private universities offer 205 careers in this modality. Which indicates that the demand for courses and academic programs in this modality [15].

This new teaching and learning process, like any new process, generates advantages that Cacheiro mentions [16], these are generally:

- Motivation for carrying out tasks
- There is no time-space barrier, each person can learn at any time, it only requires an electronic device and Internet access
- Constant interaction between people, achieving a constant exchange of data, contributing to the feedback of information, a necessary process to create knowledge.
- Incentive to the use of the creativity of the student, since it develops activities of autonomous form, looks for sources of information and decides what information is relevant and how to use it.
- Teamwork (online work), which is simpler than gathering several people in a certain place
- Students optimize their ability to produce documents more efficiently with the use of software
- Access to all kinds of information at any place and time; whether text, video or audio (Internet).
- Constant support to students by the tutor, to resolve and clarify doubts
- Greater agility in the administrative management, such as elaboration of tasks, taking lessons, and other follow-up activities to the student

Undoubtedly, what is mentioned is a great contribution to the learning process, but there are also disadvantages that become obstacles for the student, as they are:

- Lack of time for teachers to integrate technologies into the learning process.
- Student's distractions due to not being clear about the teaching process.
- Decide what information is appropriate, since the information available on the Internet may contain errors

Minimizing these disadvantages, must always be present during all learning process, at this point, it was wrongly sought that technology solve all the academic and organizational problems of educational institutions, since in the virtual spaces the operative and administrative processes were reproduced with the same deficiencies that the face-to-face educational modality maintains, the technology was abused, the pedagogical elements were neglected, methodologies were used that did not correspond to the reality of each institution, making that the teaching staff enter the computer area without a solid knowledge in relation to a Virtual Learning Environment (EVA), forcing the development of technological skills without pedagogy.

The visible result of a bad incorporation of virtual education because they were introduced in this face-to-face education characteristics resulted in static virtual classrooms, with a lot of content and boring, with educational processes without interaction, where only text with information is presented, also in many cases the educational entities simply dedicated themselves to investing in technology without being clear about the way in which it would be used or if it would be useful, this is how digital whiteboards were acquired to replace the traditional blackboard where the tracing with ink or chalk was simply changed to make them in this device that does exactly the same but with greater difficulty, also purchase online portals to improve communication with parents or students that only substitute a phone call or physical visits for a virtual location.

3   Application of "Storytelling" as a Tool of Virtual Education

Looking for innovation in the way of teaching appears a new form that is known as the "storytelling" that could be defined as telling a short story looking to influence people either to highlight a specific issue, as mentioned by Salmon & Roing, the "Storytelling is a machine to make stories and format minds", is a technique used in several organizations dedicated to trade since the 80 [17], this communication technique that searches the public to which the story is directed to identify with the content in a subjective manner, this development generated a radical change to the traditional business management converting promotional marketing into a more effective way of involving and convincing the commercial communication [18].

Given the great effectiveness shown by this technique, the interest of applying this technique in education is born due to the importance that the narrative has in the digital environment [19], this technique allows through storytelling to present ideas with the aim of transmitting knowledge by presenting information in a novel way and capturing the attention of students by presenting attractive scenarios where the teacher becomes a producer of audiovisual content where puts to the test the narrative and creative capacity of this [20].

This technique within the educational field is known as digital storytelling (DST) allows to enhance the learning of reading and writing through the use of digital tools to thereby encourage the ability of students to express and communicate [21], as evidence of the effectiveness of applying this technique in education, it is possible to find several works such as the one presented by Del Moral & López who used digital stories in classrooms made up of children from basic early childhood education in the Principality of Asturias, to analyze the development of basic skills when using the teacher this technique, as a result the students despite their young age were able to carry out various activities requested to solve different problems associated with everyday situations [22].

In this sense also highlights the work presented by Maddalena & Pavón who in their work highlight the experience of using digital stories in an online training course for teachers of Spanish in Brazil where the results determined that there is a great possibility in the pedagogical field use ICT to create digital stories as a tool to learn Spanish [23].

In the same way in the work presented by Herreros that details the experience of working with digital stories in the classroom supported in the classic narrative and film in cognitive and psychological processes presenting digital stories as a tool for students to reflect on their identity personal, where the reception of a story supposes a cognitive perception situating it in an experiential context that allows the receiver to interpret the story by associating their own experiences [24].

Within this line of evaluating the effectiveness of this technique, the work of Ruiz & Alvarado also highlights the importance of this tool during the teaching - learning process in pre - school students in the region of the coast of Oaxaca, this research presents as the main results the importance that the DST at preschool level acquires, in the same way the teachers indicate that applying this technique at the time of teaching produces good results in terms of the understanding of their students, strengthening bonds of trust between both parts [25].

4 The Use of Virtual Environments Based on DST to Incentivize the Microenterprise Sector in the Adoption of Certain ICT

Implementing the latest advances in technology, whether in the business or social sphere, has generated knowledge to become a very important factor in the production stages, in this way, ICTs are an important part of the development of the economy since companies that do not include them within the strategies of buying and selling tend to give competitive advantages to the companies that adopt them, causing a short period of time disappear for not being able to compete [26].

In this way, ICTs are an important part of the development of the economy since companies that do not include them within the strategies of buying and selling tend to give competitive advantages to the companies that adopt them, causing a short period of time disappear for not being able to compete, since in the current knowledge is the main element and input for production and through technology are applied in different fields where the human being intervenes, as indicated by Kaushink and Sigh [27], for whom the access and use of ICT generate higher economic income not only for those who make up the company because it gains efficiency by economizing resources in their production, but also for the economy in general because they generate higher consumption that results in higher income to the country creating new employment opportunities all the benefits that with the use of technology in many companies have been evident in its expansion and growth, but in almost all microenterprises these benefits are not achievable due to the low level of knowledge and management of the technology of the members of these, this ignorance is known as digital illiteracy and finally there is the lack of economic resources.

The main factors for this business sector that acquires great importance within the economy of the countries as evidenced by the data presented by the "Comisión Económica para América Latina y el Caribe" CEPAL, this sector represents 88.4% of formal companies [28], the importance of these companies lies basically in the ability to generate employment sources that allow the subsistence of the people associated with each microenterprise, especially in times of crisis in a country, now starting from different studies related to education, technology and microenterprises, it is proposed to develop a research to determine the effectiveness of using virtual environments based on digital stories focused on the microenterprise sector so that they choose to implement technology appropriate to the special needs of this sector related to information security, for this purpose, the use of a descriptive exploratory type study is proposed as the first phase in order to be able to define the effective content for this class of companies to be interested in implementing technologies for information security, in a second stage a case study to evaluate in greater detail the effectiveness of the message based on a DST and how to help define strategies to adopt technology related to information security.

5   Conclusions

This work presents the theoretical basis that allows to develop activities based on the technique of digital stories together with virtual environments so that the microenterprise sector analyzing the characteristics that define it can incorporate technologies and achieve sustainable development.

This relatively new technique in the educational field is generating good results so it would not be unreasonable to assume that applying it in virtual environments in an appropriate way would contribute to the reduction of the so-called digital divide by allowing more people to access comprehensible information with the that can identify both people and micro-enterprises, opting to implement technology.

ICT are a very important element that, if implemented properly, significantly improve the productivity and competitiveness of companies, thus obtaining greater income that favors this productive sector, so that the people in charge of carrying it out must handle the appropriate knowledge so that the technology is implemented in the productive processes.

The capacity of a company to produce and generate resources can not only be measured by the indicator of investment in research and development, it is also a very important part of the human resource capacity that is available which allows companies to improve efficiency levels.

The limitations related to the use and access to technology by the people who make up a microenterprise become an important and determining factor that prevents these companies from developing adequately along with the current demands of the business, so that to look for an alternative to diminish this problem part of the base presented in this document.

References

1. ITC Facts & Figures: ITC Facts & Figures 2017 (2017). https://www.itu.int/en/ITU-D/Statistics/Documents/facts/ICTFactsFigures2017.pdf
2. Banco Mundial: World Bank Group - International Development, Poverty, & Sustainability. Banco Mundial (2016). http://documents.worldbank.org/curated/en/658821468186546535/pdf/102724-WDR-WDR2016Overview-SPANISH-WebResBox-394840B-OUO-9.pdf. Accessed 8 June 2018
3. Costo, I.J.: El conocimiento: un reto para gerentes y especialistas del futuro (2014). http://hdl.handle.net/10654/12407
4. Giddens, A., Desbocado, U.M.: Los efectos de la globalización en nuestras vidas. In: Runaway World, pp. 1–39 (2007)
5. López, S.G., Gómez, L.L.H.: La universidad entre lo presencial y lo virtual. Universidad Autónoma del Estado de México (2006). Colección Pensamiento universitario. https://books.google.com.ec/books?id=vCuR6687bX0C
6. Barrantes, E.: Educación a distancia. aspectos teóricos. Universidad Estatal a Distancia, San José, Costa Rica (1992)
7. Ferrer, A.T., Sauter, G.O., Fernández, F.S., Benito, A.E., De Castro, F.G.R., Mmdelp, A., et al.: Historia de la educación (edad contemporánea). UNED. Libro Electrónico (2005). https://books.google.com.ec/books?id=FCceh3lAXkAC
8. Tiana, A., Ossenbach, G., Sanz, F., Escolano, B., Gómez, F., Del Pozo, M., et al.: Historia de la Educación Social. UNED, 350 p., Grado (2014). https://books.google.com.ec/books?id=ffE3AwAAQBAJ
9. Nishio, M.: Tecnologías de Información y Comunicación para el Desarrollo
10. Casas Armengol, M.: Ilusión y realidad de los programas de educación superior a distancia en América Latina. Proyecto especial, 37 (1982)
11. Guedez, V.: Las perspectivas de la educación a distancia en el contexto de la Educación Abierta y Permanente. en Boletín Informativo de la Asociación Iberoamericana de Educación Superior a Distancia, Madrid, vol. 3 (1984)
12. Gebera, O.W.T.: La docencia en la educacion virtual: concepciones, metodos y perspectivas
13. UNESCO: Estándares de competencias en Tic para docentes. UNESCO (2008). http://www.eduteka.org/pdfdir/UNESCOEstandaresDocentes.pdf
14. RICOH: The Economist Intelligent Unit (2013). http://thoughtleadership.ricoh-europe.com/es/humans-and-machines/education-insight
15. El Universo: Gobierno promueve programa de Educación Superior Virtual, ante la brecha de acceso a la Universidad (2018). https://www.eluniverso.com/noticias/2018/04/04/nota/6697693/senescyt-anuncia-programa-educacion-superior-virtual-bachilleres
16. Cacheiro, M.: Educación y Tecnología: Estratégias Didácticas para las Integración de las TIC. Editorial UNED, Madrid (2018)
17. Salmon, C., Roig, M.: Storytelling: la máquina de fabricar historias y formatear las mentes. Península Barcelona, Barcelona (2008)
18. Martín González, J.A.: La eficacia del Storytelling. MK Marketing+ Ventas **251**, 8–17 (2009)
19. Illera, J.L.R., Monroy, G.L.: Los relatos digitales y su interés educativo. Educação, Formação & Tecnologias, 2(1), 5–18 (2009). ISSN 1646-933X
20. Hull, G., Katz, M.L.: Crafting an agentive self. Case studies of digital storytelling. Res. Teach. Engl. **1**(41), 43–81 (2006)

21. Pérez, M.E.D.M., Martínez, L.V., Piñeiro, M.R.N.: Competencias comunicativas y digitales impulsadas en escuelas rurales elaborando digital storytelling. Aula Abierta **45**(1), 15–24 (2017)
22. del Moral Pérez, M.E., López, B.R.: Experiencia innovadora: realización de relatos digitales en el aula de educación infantil. DIM: Didáctica, Innovación y Multimedia (32), 1–16 (2015)
23. Maddalena, T.L., Sevilla, A.: El relato digital como propuesta pedagógica en la formación continua de profesores. Revista Iberoamericana de educación **65**, 149–160 (2014)
24. Herreros, M.: El uso educativo de los relatos digitales personales como herramienta para pensar el Yo (Self). Digit. Educ. Rev. **22**, 68–79 (2012)
25. Ruiz, N.S., Alvarado, A.V.: El storytelling digital como herramienta pedagógica para el docente en el proceso de enseñanza-aprendizaje de los alumnos de educación preescolar en la región de la costa de Oaxaca. EDUCATECONCIENCIA **11**(12) (2016)
26. Gonzalez, R.M., Alfaro-Azofeifa, C., Alfaro-Chamberlain, J.: TICs en Las PYMES de Centroamérica: Impacto de la Adopción de Las Tecnologías de la Información Y la Comunicación en El Desempeño de Las Empresas. IDRC, 270 p. (2005)
27. Kaushik, P.D., Singh, N.: Information technology and broad-based development: preliminary lessons from North India. World Dev. **32**(4), 591–607 (2004)
28. Dini, M., Stumpo, G.: Mipymes en América Latina: un frágil desempeño y nuevos desafíos para las políticas de fomento. CEPAL (2018). https://repositorio.cepal.org/handle/11362/44148. Accessed 16 Oct 2018

Development of a Social Robot NAR
for Children's Education

Nathaly A. Espinoza E.$^{(\boxtimes)}$, Ricardo P. Almeida G, Luis Escobar,
and David Loza

Universidad de las Fuerzas Armadas ESPE,
170121 Sangolqui, Pichincha, Ecuador
{naespinoza, rpalmeidal, dcloza, lfescobar}@espe.edu.ec
http://www.espe.edu.ec

Abstract. There are robots aiming to develop cognitive abilities through building and programming, but there are other kind of social robots oriented to education, this paper shows an example of this branch of robots where aspects as anthropomorphism, bidirectional interaction, movement applied to the development of educational robots are explored. The social robot NAR is focused on education of children around 3 to 5 years. The robot is a didactic and interactive tool which purpose is to strengthen acquired knowledge. The main goal is to be able of getting a great interaction child-robot minimizing details through a minimalist but at the same time anthropomorphic design, being able of bidirectional interaction and capable to emulate and communicate basic emotions using movements, sounds, lights and sensors. Based on the results of a completed study applied to technical students and held at "Universidad de las Fuerzas Armadas ESPE", was demonstrated that NAR has a friendly physical appearance, its interaction goes together with its level of Anthropomorphism and gives appropriate answers while interacting. Due the results on this first stage of development is highly encouraging to develop a protocol to improve and expand capabilities of the robot and make a more comprehensive study with infant subjects and education specialist.

Keywords: Teaching-assistant platform · Social robots · Human robot interaction · Robot design

1 Introduction

The evolution of robotics and technology in education, has led to the study of educational robots [1], as an alternative to traditional education. There are different aspects to analyze within the educational robots such as: child-robot interaction, duration of the interaction, teaching methodologies, continuous learning, didactic tools, appearance of the robot, etc. [2]. The use of robots in education serves to: promote learning, creativity and generation of new skills in children [3]. Robots have great psychological and social

Supported by Laboratorio de Mecatrónica y Sistemas Dinámicos, Universidad de las Fuerzas Armadas - ESPE.

© Springer Nature Switzerland AG 2020
M. Botto-Tobar et al. (Eds.): ICAETT 2019, AISC 1067, pp. 357–368, 2020.
https://doi.org/10.1007/978-3-030-32033-1_33

effects [4], principally they can be implemented in two ways. First, the child learns to program and build its own robot [5]. An example of this approach is Mindstorm robot [6], which can adopt different forms, and has several sensors so children develop creativity and skills to program many functions. Second, the robot interacts directly with the child, with the aim of capturing their interest and study the user's behavior. Some examples of robots that interact directly with children, for education or other related purposes are: Robovie [7], Irobi [8], iCat [9], Probo [10, 11], etc. Where the studies are based mainly on how to improve the interaction with children, and the effects this interaction produces in the learning process. NAR is a social robot, which interacts directly with children. NAR's goal is to be an educational tool of interactive reinforcement for the teacher. Where the set of features aim to achieve effective interaction. The results were obtained during and after the educational process with children, and the aspects of interaction and the usability as a educational tool were evaluated.

2   Characteristics of Social Robots in Education

One of the main characteristics that social robots oriented for education should have is the capacity to achieve a fluid interaction with the student to get good results in the stimulated area [12]. According to [13] in a social robot some facts to be taken are: appearance, emulated emotions and established behaviors [13]. Therefore, in the case being analyzed some important characteristics that can be mentioned are: Anthropomorphism, bidirectional interaction and movement.

2.1   Anthropomorphism

Anthropomorphism consists of endowing a robot, computer or animal with human characteristics. In social robots this characteristic is an important one, because it makes bidirectional interaction easier [14]. This interaction depends on the observers, in order to get a socially attractive robot, looking for an anthropomorphic aspect is recommended. Robots' tendency nowadays try to achieve anthropomorphic appearance as high as possible, although to obtain a high degree of anthropomorphism is complicated, while the robot resembles more a human being, there is more likely to fall in the uncanny valley proposed by Masahiro Mori [15]. In the uncanny valley graph proposed by Mori, there is a section where the appearance of the robot is bearable. It is accepted up to an approximate 75% of resembling human beings or animals, passing after this percentage the "Uncanny Valley" is reached, where the robot's appearance is not pleasant, causing discomfort or fear. However, if the similarity increases, it leaves the valley and increases the liking towards the robot again, improving the interaction. [15] An important point about increasing anthropomorphism is that when human resemblance is higher, expectations of complex behavior and fluid movements are higher as well. [14] Within the anthropomorphism we can list several Parameters such as: physical appearance, sound, behavior, etc. It is important to detail the impact that a robot can have in a child, to endow it with the needed anthropomorphism [9].

2.2   Bidirectional Interaction

According to research [18] carried out with Karotz robot, it shows that a social robot must respond to stimuli made by a human, being this reaction visual, auditory or physical through its movements. It is essential that the robot had an answer to such stimuli making the human being feels comfortable with it [18]. The Kasukiro Kosuge study [17] works with two types of bidirectional interaction, one in which the robot takes objects that the human wants to manipulate, and another where the robot has to dance. For the first case, it is necessary to take into account human variables such as strength, location, etc. In the second case, both the human and the robot must predict the future movement of their partner, in order to be able to dance satisfactorily [17].

2.3   Movement

The movement of this type of robot should be as natural and fluid as possible, that means that the robot must control their respective joints and links in such a coordinate manner and in a slow manner in order to resemble a living being, that is avoiding sudden movements that frighten the user.

3   Design of Nar

The social assistance robot called NAR is oriented to the education of children from 3–5 years old. It is a didactic and interactive tool to reinforce knowledge already acquired by children. In order to achieve an appropriate interaction, it has a friendly appearance. Below are the specifications and for the final design, the VDI 2206 mechatronic systems design methodology was used [23].

3.1   Technical Specifications

NAR has a height of 35 cm and has a modular architecture. It is connected to 220 V and has a total of 7 degrees of freedom (DOF). Its body, head and feet are made of plastic and the hands are made of flexible material. The model was made using actuators and material according to its mechanical design. Obtaining the model of Fig. 1. From the model obtained, the construction of the robot was carried out. In Fig. 1, the hands (4) of flexible material are observed, which have lights and piezoelectric sensors to feel when left or right hand are pressed. It also has a touch screen on front (6) to perform the interaction with children. It has a camera in its mouth (3), can move the head (1), eyebrows (2) and feet (5). Also a base made of synthetic grass is included to supports the entire structure.

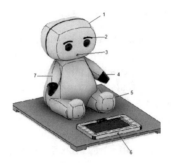

Fig. 1. Parts of NAR robot. 1. Head, 2. Eyebrow, 3. Camera, 4. Hands, 5. Feet, 6. Screen, 7. Arms

3.2   Social Specifications

According to the characteristics of social robots [15]. NAR has a friendly appearance, and is able to emulate basic emotions with the movement of eyebrows and neck with sounds and colors. It is white to transmit peace and tranquility to children, because a stronger or more striking color would be a limitation in the interaction [16]. It also has a haptic system (piezoelectric sensors) to emulate sensibility in its hands.

3.3   NAR Robot Functions

Movement. The NAR robot has a total of 7 DOF (Table 1), 3 DOF in its neck, 1 DOF in each foot and 1 DOF in each eyebrow. The neck uses a mechanism (Fig. 2a) which enables the up and down, left to right and turn clockwise and counterclockwise head movement. Movements needed to track faces. The eyebrows have a degree of freedom in each one Fig. 2c. The feet have a degree of freedom in each in Fig. 2b.

Table 1. NAR degrees of freedom

Part	DOF
Neck	Up-down Left-right clockwise Counter-clock wise
Eyebrows	Clockwise - counter-clock wise
Feet	Clockwise - counter-clock wise

Emotions. The emotions (Fig. 3) that NAR expresses are through the movement of the eyebrows, a distinctive sound and distinctive color lights. As shown in the Table 2.

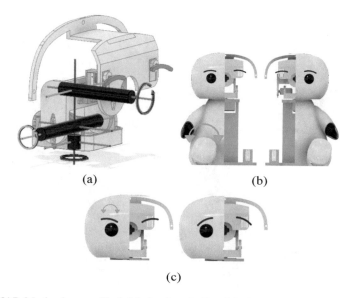

(a)   (b)

(c)

Fig. 2. NAR Mechanisms. a. Neck Mechanism, b. Feet Mechanism, c. Eyebrow Mechanism

Table 2. Level of freedom of NAR

Feeling	Hands color	Movement
Happy	Yellow	Feet
Sad	Blue	Head, eyebrow
Angry	Red	Eyebrow
Surprised	Orange	Head, eyebrow
Doubtful	Green	Head, eyebrow

Educational Module. Classes and teaching methods were classified through the advice of professional educators, according to the age and level of knowledge of an average child. Children from 3 to 4 years old learn a number and a color every day, while those from 4 to 5 years work with number and colors simultaneously in addition to identification of quantities.

Architecture. The robot program is done in ROS environment, with a total of 16 nodes and 13 channels (Fig. 4) to communicate these nodes. Here the behaviors and responses are established during the interaction with children. Some important nodes that can be mentioned are:

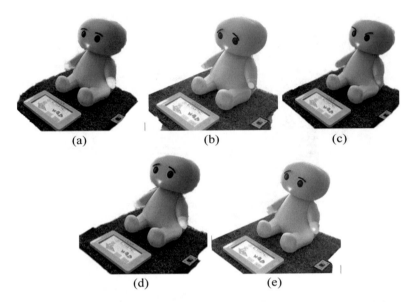

Fig. 3. NAR Emotions. a. Happy, b. Sad, c. Angry, d. Surprised, e. Doubtful

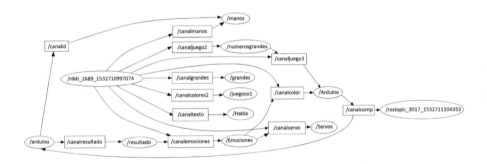

Fig. 4. Programming architecture

- Servo node
  Sends the angles of the servomotors to the SSC-32 card
- Emotion node
  Receives the behavior that the user selected, send the movements to the node servos and the colors that should be attached to node colors.
- Color node
  The color node receives the color that the hands of the robot should light.
- Node speaks
  Receives the text for the text to speech.
- Game node1
  Receives the number to be send to the touch screen and play on the screen.

- Hands node
  It receives the color of the theme that is selected, send which colors should hands light and finally gives the result if the child press it.
- Color game node
  The node is for the color detection game, it receives the color and compares it with the color of the card and sends the result.
- Comp node
  Compares if the user win or lose.
- Result node
  Receive if you won or lost in any of the games and sent the result to node emotions the behavior of winning or losing.

Other Functions. In addition to reinforcing the behavior of the NAR robot when interacting, more functions have been added such as: Text to Speech, artificial vision and a haptic system (Table 3).

Table 3. Other function

Text to speech	The robot uses TTS of Google [21], to speak what the operator writes
Detection and face tracking	This function works with the implementation from the DLIB bookstore [22], and the follow-up is done through of the inverse robot kinematics.
Color detection	This is done by QR codes located on each color card, which are interpreted and the robot shows the color through auditory perception
Haptic system	A piezoelectric sensor is used in each hand to feel when it is touched

4   Tests and Results

The main characteristics that NAR robot has to accomplish are: anthropomorphism, bidirectional interaction and movement which are evaluated through test and observation. One of the objectives is to show that a correct prototype was obtained from the specifications shown in Sects. 3.2 and 3.3. And a good interaction is achieved through emotions and games. As mentioned before, evaluation is made by doing a survey to a group of 25 people around the ages of 22 and 27 years old, this method had been used in other investigations [20, 21]. The quiz has 4 parts:

- Evaluation of resemblance to a human or a machine. (Anthropomorphism)
- Evaluation of appearance.
- Evaluation of correct interaction.
- Evaluation of correct emulated emotions.

Part of the evaluation is to gather information about the behavior and reactions from users during the surveys through observation records.

4.1   Process

First, the user is placed in front of the robot and an emotion is shown for the user to be identified, the same process is repeated for all cases of Fig. 5. Second, another survey is given to the user, explaining each question and their weighting. The first part of the survey evaluates the NARs robot anthropomorphism. The second part evaluates users acceptance. Finally, the user plays with the robot and the third part of the survey about interaction is completed. During all the process, information is acquired from the users behavior about the robot with observation records. For example, during the execution of emotions, it is observed if the user feels empathy according to their reaction.

4.2   Results

According to the design in its mechanic and physic structure mentioned in Sect. 3. The building is complied according to the requirements and the following prototype is obtained (Fig. 5).

Fig. 5.   NAR robot

In Fig. 5, a white robot made of plastic is presented. It makes different movements and its hands contain color lights and sensors. Another result obtained is a minimalist and simple design without great number of freedom degrees.

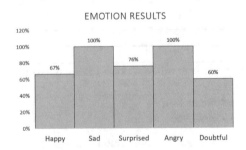

Fig. 6.   Emotions quiz results Happy, sad, surprised, angry, doubtful.

The results obtained in the survey for the emotions is shown in Fig. 6, there is more than 50% of success of emulated emotions. Being the most successful sad, angry and surprised. Data about appearance, movement and interaction are presented below in Table 4, where weighting is from 1 to 5, being 1 the least and 5 the higher, or according to the shown description, for example: 1 unpleasant and 5 pleasant.

Table 4. Results appearance, movement and interaction

Description	Weighting from 1 to 5
False - Real	3,40
Machine - Human	3,08
Artificial - Alive	3,28
Mov. Rigid Fluid mov.	**3,52**
Average	**3,32**–66,40%
Unpleasant - Pleasant	4,64
Not friendly - Friendly	4,68
Not kind - Kind	4,68
Average	**4,67**–93,33%
Robots response	4,52
Expressiveness	4,08
Objectives - Expectations	4,44
Average	**4,35**–86,93%

5 Results Analysis

According to the results, the main objective from the investigation was accomplished. NAR is an interactive and low-cost robot because of its minimalist design with friendly appearance, and good interaction levels were accomplished. The results were validated through data collected in surveys and observation records.

Comparing the virtual model Fig. 1 and the built prototype Fig. 5, the main objectives were complied about the robots construction. It has an appropriate size and friendly appearance. NAR is functional, meaning that from its technical specifications such as color, size and movement; social specifications are achieved, for example it emulates 5 emotions that will be analyzed later on, its piezoelectric sensors work appropriately during interactions in the games and color detection. Validated results through observation of the robot's operation.

The Nar robot emulates 5 emotions, Fig. 3: happiness, sadness, anger, surprise and doubt. According to the results of the first survey observed in Fig. 6, 67% of the users manage to detect that NAR robot shows happiness, 100% detect sadness and anger. 76% and 60% detect surprise and doubt, respectively. The highest results are obtained with sadness, surprise and anger, because NAR emulates them with the movement of the eyebrows and sounds, very similar as a human does. On the other hand, happiness and doubt are expressed through the mouth which NAR does not have. To compensate it sounds and movements are used. However, 100% was not achieved. The results obtained from the emotions tests are satisfactory because NAR emulates its emotions from its head movement, auditory stimuli and lights, in order to depend not only from facial expressions which would increase the robots DOF and consequently minimalist design would be lost making it more complex.

In Table 4 we have the results from the second survey. The first part shows the anthropomorphism and naturalness of movements, obtaining a score of 3,32 in the rank of 1 to 5, which is equivalent to 66,40%. In other words, it means that the robot does not completely resemble a human being, it only has some human attributes and does not have rigid movements but fluid and not in a human being level. In the second part of the table, results from the aesthetics of the robot are observed, reaching a 93.33% meaning that the robot is pleasant for the user. Finally, in the third part of the table results to establish the level of satisfaction with the interaction are found. It evaluates if the robot gives an adequate answer, its expressiveness and if it complies with the users expectation, getting an 88.93%. With that result, it can be said that a good interaction between the robot and the user was achieved. When talking about interaction, it can be mentioned that the robots automation level or its autonomy, placing it in level 5 of automation in the scale of [20], because the system, responds and acts when the human asks, procedure used during the games. The level is suitable for the purposes of the NAR robot, because from level 6 and above, the robot will take decisions for itself, often without the human being able to know what is happening, but for educational reasons of the user (children) they must know what is happening in order to react satisfactorily.

Finally, after analyzing acquired data from the users behavior and reactions through the observation records, it can be said that the user felt comfortable and excited after seeing the robot. Another fact that can be highlighted is the empathy from the user towards the robot when it emulates emotions during games.

6   Conclusions

NAR is a social robot oriented to education. Its characteristics act according to the area (education) where the NAR robot will be applied, being the most important anthropomorphism, movement and bidirectional interaction. The last one is the most relevant because it depends on the previous ones and will make the difference of a social robot from other technological tools oriented to education.

According to the results obtained in Sect. 4.2 and its respective analysis about interaction and robots appearance, it can be concluded that the robot has a good appearance according to the results in Table 4, and a level of intermediate anthropomorphism which are two important characteristics for interaction. If the anthropomorphism level were higher, the users expectations would also increase significantly and they would expect more complex behavior from the robot obtaining a bad interaction.

One of the objectives of the NAR robot was to achieve satisfactory results with a minimalist aspect. This objective was achieved because the user managed to read the emotions emulated by the robot through the movement of the eyebrows and the head because according to the survey more than a half of the users identified each emotion from NAR. It is important to spot that emotions are not only based in facial expressions but sounds and colors also help interpretation. Anthropomorphism, movement and emulated emotions are part of a good bidirectional interaction, because the user managed to feel empathy for the robot and it reacts to stimuli.

References

1. Han, J., Lee, S., Hyum, E., Kang, B., Shin, K.: The Birth Story of Robot IROBIQ for Childen's Tolerance (2009)
2. Meltzof, A.: Science. AAAS, New York (2012)
3. Ruiz, E., Snchez, V.: Educatrnica, Madrid (2007)
4. Saldien, J., Goris, K., Yilmazyildiz, S., Verhelst, W., Lefeber, D.: On the design of the Huggable Robot Probo (2008)
5. Barrera, N.: Uso de la robtica educativa como estretegia didctica en el aula. Boyac (2014)
6. LEGO, Midstorms EV3. www.lego.com/es-es/mindstorms/products/mindstorms-ev3-31313. Accessed 24 Sept 2018
7. Ishiguro, H., Ono, T., Imai, M., Maeda, T., Kanda, T., Nakatsu, R.: Robovie: an interactive humanoid robot. Ind. Rob. Int. J. **28**(6), 498–504 (2016)
8. Jeonghye, H., Miheon, J., Sungju, P., Sungho, K.: The educational use of home robots for children. In: IEEE International Workshop on Robot and Human Interactive Communication, pp. 378–383 (2005)
9. Saerbeck, M., Schut, T., Bartneck, C., Janse, M.: Expressive robots in education: varying the degree of social supportive behavior of a robotic tutor. In: Proceedings of the SIGCHI Conference on Human Factors in Computing Systems, pp. 1613–1622 (2010)
10. Goris, K., Saldien, J., Vanderborght, B., Lefeber, D.: Mechanical design of the huggable robot Probo. Int. J. Humanoid Rob. **08**(03), 481–511 (2011)
11. Saldien, J., Goris, K., Yilmazyildiz, S., Verhelst, W., Lefeber, D.: On the design of the huggable robot Probo. J. Phys. Agents **2**(2), 3–12 (2008)
12. Human Health, Carnegie Mellon Today (2018). https://www.cmu.edu/cmtoday/issues/dec-2004-issue/feature-stories/human-health/index.html. Accessed 24 Sept 2018
13. Fridin, M., Belokopytov, M.: Acceptance of socially assistive humanoid robot by preschool and elementary school teachers. Comput. Hum. Behav. **33**, 23–31 (2014)
14. Barteneck, C., Kanda, T., Ishiguro, H., Hagita, N.: Is the uncanny valley an un-canny cliff?. In: The 16th IEEE International Symposium on Robot and Human In-teractive Communication Jeju, pp. 368–373 (2007)
15. Duffy, B.R.: Anthropomorphism and the social robot. Rob. Auton. Syst. **42**(3–4), 177–190 (2003)
16. Heller E.: Psicologa del color (2004)
17. Fong, T., Thorpe, C., Baur, C.: Robot, asker of questions. Rob. Auton. Syst. **42**(3–4), 235–243 (2003)
18. de Graaf, M., Allouch, B., van Dijk, J.: What makes robots social? A users perspective on characteristics for social human-robot interaction. In: International Conference on Social Robotics, pp. 184–193 (2015)
19. Parasuraman, R., Sherida, T.: A model of tupes and levels of human interaction with automation. IEEE Trans. Syst. Man Cybern. Part A Syst. Hum. **30**(3), 286–297 (2000)
20. Bartneck, C., Kanda, T., Ishiguro, H., Hagita, N.: My robotic doppelganger - a critical look at the uncanny valley. In: The 18th IEEE International Symposium on Robot and Human Interactive Communication, RO-MAN 2009, pp. 269–276 (2009)

21. Cloud Text-to-Speech - Speech Synthesis, Cloud Text-to-Speech API, Google Cloud, Google Cloud (2018). https://cloud.google.com/text-to-speech/. Accessed 25 Sept 2018
22. Mallick, S.: dlib Learn OpenCV - Part 2, Learnopencv.com (2018). https://www.learnopencv.com/tag/dlib/page/2/. Accessed 25 Sept 2018
23. VDI RICHTLINIEN. VDI 2206. Design methodology for mechatronic systems. (Junio de 2004)

Security

Developing an Information Security Management System for Libraries Based on an Improved Risk Analysis Methodology Compatible with ISO/IEC 27001

María José Bravo Ramos and Sang Guun Yoo[(⊠)] [iD]

Departamento de Informática y Ciencias de la Computación,
Escuela Politécnica Nacional, Quito, Ecuador
{maria.bravor,sang.yoo}@epn.edu.ec

Abstract. This paper describes a new risk analysis methodology for libraries based on steps filtered from existing methodologies that are compatible with the ISO/IEC 27000: 2013 standard. After analyzing MAGERIT, OCTAVE and NIST 800-30 risk analysis methodologies, the most important steps were identified and those that do not fit in library type of organization were discarded. Once the methodology was created, it was tested through a real implementation in the Library system of a university to verify its benefits.

Keywords: Information security management system · Information security · Risk management methodology

1 Introduction

Due to the unending evolution of technology, information security has become a requirement for all organizations. It is because information is the most important resource for making sound decisions and implementing new business strategies. Therefore, the need for considering information security implementations based on protocols that guarantee confidentiality, integrity, and availability has surged.

In this circumstances, currently, the majority of libraries lack of security management protocols because the administrators do not understand their importance, even though libraries manage sensitive information, such as user accounts, management of fees, and the bibliographic data of their publications which constitute the raw materials for their business functions.

MAGERIT, OCTAVE and NIST 800-30 are risk analysis methodologies that are compliant to ISO/IEC 27001:2013 standard [2, 4, 12] which can be applied to any type of organizations [7, 9–11, 15]. Any one of the aforementioned methodologies can be adapted for security management of a library. However, due to the fact that these methodologies were not created exclusively for libraries, their steps may not cover all library functions, and additional steps may be necessary. Additionally, a systematic review of the literature revealed that there is no a risk analysis methodology that is exclusively designed for libraries [1, 5]. In this circumstances, this paper intends to

© Springer Nature Switzerland AG 2020
M. Botto-Tobar et al. (Eds.): ICAETT 2019, AISC 1067, pp. 371–379, 2020.
https://doi.org/10.1007/978-3-030-32033-1_34

design an optimized risk analysis methodology that can be applied to university library systems. The proposed methodology incorporates steps that adjust to the library business model, and discards unnecessary steps that are not adoptable to this king of organizations. Moreover, the proposed methodology obeys to ISO/IEC 27001:2013 requirements [6, 14].

In summary, the present paper analyzes the different risk analysis methodologies and creates a new methodology exclusive for libraries. Once created the new methodology, it is implemented in the library system of Escuela Politécnica Nacional (EPN), which is one of the most important universities of Ecuador, in order to test its effectiveness.

The rest of the paper is organized into 4 sections. Section 2 analyzes existing risk methodologies and explains the proposed risk analysis methodology for libraries. Then, Sects. 3 and 4 describe how the new methodology was implemented and analyzes the implementation results. Finally, Sect. 5 presents the conclusions.

2 Development of an Optimized Risk Analysis Methodology for Libraries

As mentioned before, this study has examined MAGERIT v.3, OCTAVE v.2, and NIST 800-30 risk analysis methodologies since they are the most popular worldwide. MAGERIT is very popular in Spanish speaking countries since its documentation is available in Spanish. On the other hand, OCTAVE is popular since it offers very simplified documentation in regard to how an organization's assets are identified and does not classify excessively, while NIST has become popular since it was one of the first methodologies and because it is focused on evaluating risks that information technology systems face.

Through the analysis of previous works, we could notice that the aforementioned risk analysis methodologies were created to be applied for all kind of business. In this sense, since the methodologies are general, organizations use one or more methodologies to adapt them to their needs. For example, R. Castro uses MAGERIT as a risk management methodology for his company (Xintiba) [8], while A. Susanto uses MAGERIT for assets' identification and OCTAVE for risk analysis in sight that these two methodologies are those that best suit your business model [9]. Finally, the National Institute of Information Technology [17] recommends that, in order to implement the ISMS in companies, some of these risk analysis methodologies should be used: Magerit, OCTAVE and NIST 800-30.

Following the same idea of previous works, which combine different risk analysis methodologies, the present study is intended to create an optimized methodology for libraries based on the popular methodologies (i.e. MAGERIT, OCTAVE and NIST). This section will describe the most important characteristics of each methodology that are applicable for risk analysis of library systems.

2.1 Steps of MAGERIT for Libraries' Risk Analysis

The ISO/IEC 27001:2013 standard requires analysts to create an asset inventory, since that is the best way to begin this analysis. However, it is difficult to complete this step when an organization is poorly understood. Therefore, the inventory of assets should be executed periodically.

Once a detailed asset inventory has been described, it is important to identify interlocking threats, to later determinate their risks. Threats to an organization depend a lot on its environment, positioning, and activity. But, there is a set of common threats that affects to most of assets such as fire, climatic factors, user errors, water damage, electrical overloads, and non-authorized accesses.

Once an organization's risks are understood, safeguards must be established which function is to reduce the level of the risk. In summary, in order to execute a complete risk analysis process, the expert needs to make an inventory of the assets of the organization. Then, the expert also has to identify the threats and measure the risks that the assets face. Finally, the expert must suggest safeguards that will counteract the identified threats and reduce the risks.

Among the different steps of MAGERIT, we have selected the following three specific steps that would be very useful in analyzing risks in library systems:

(1) *Assets:* MAGERIT proposes a wide list of predefined asset types which would be very useful in identifying assets of the library system. Once these assets are identified, they will be protected from the dimensions of integrity, confidentiality, availability, authenticity, and traceability.
(2) *Threats:* MAGERIT provides a catalogue of pre-identified threats. From this catalogue, we will select all those that are applicable to libraries. Once identified the list of threats, we will assess the influence that each threat might pose on any given asset and its probability of occurrence.
(3) *Safeguards*: MAGERIT also provides a list of possible safeguards which will be used for delivering security measures to the library systems [2].

2.2 Steps of OCTAVE for Libraries' Risk Analysis

The risk analysis methodology called OCTAVE is focused on identifying vulnerabilities and developing security plans. This is an important contribution, since MAGERIT methodology does not conceptualize the detection of organization weaknesses as an independent activity. Therefore, in order to complement the three steps adapted from MAGERIT (assets, threats, and safeguards), we have selected OCTAVE's steps for identifying vulnerabilities and developing security plans to create a more robust and optimized risk analysis process for university libraries. Below, we give a brief explanation of the steps taken from the OCTAVE methodology:

(a) *Vulnerabilities:* We will identify and evaluate the critical assets using a tool that help organizations evaluate their weak points in their Information Technology systems.

(b) *Developing security plans:* We will also prepare a library security plan to tackle identified risks and to develop a protection strategy for our organization and its critical assets. This document will contain the following items: introduction, development of security policies for various library domains, compliance verification, budget, and results.

2.3   Steps of NIST 800-30 for Libraries' Risk Analysis

Unlike MAGERIT and OCTAVE, NIST offers a way to communicate results. It is important to communicate the results obtained from the risk analysis process to the people to make them aware of the findings. Therefore, we have selected the "communicate results" step of the NIST methodology as the last step in our risk analysis methodology. Below, we give a brief explanation of the step adapted from the NIST methodology.

(a) *Communicate results:* The key activity of the step is to determine the appropriate method to communicate the risks uncovered to the pertinent parties.

By combining features from MAGERIT, OCTAVE, and NIST, we have developed a new methodology for risk analysis for the library entities. It combines the different steps of existing methodologies and eliminates steps that are not necessary for library systems.

2.4   Proposal of a New Risk Analysis Methodology for Libraries

Our proposal of risk analysis methodology focuses on qualitative analysis i.e. it uses a valuation grading scale. The qualitative scale allows users to make a ranking of assets based on their importance and involved risks. In order to implement an Information Security Management System (ISMS) in a university library, its personnel needs to identify the current state of security level of the system. The definition of the initial state helps to define the structure for the ISMS model. For this purpose, in the new methodology, we propose the execution of the following steps:

(1) *Identify Assets:* In this step, current assets of the library system will be detailed by specifying the type of the asset, its quantity, location, and responsible. For this step a specific form[1] will be used. The identified assets will be grouped into four categories taken from MAGERIT i.e. information, computing hardware, information applications, and digital memory devices. Later, a quantitative evaluation of all assets will be executed based on the 4 security levels using the Form B (See footnote 1) .

(2) *Identify Threats:* In this step, all threats that might materialize are identified. Threats will be grouped using the 4 categories proposed by MAGERIT i.e. natural disasters, industrial disasters, errors and non-intentional mistakes, and intentional attacks. Later, a quantitative evaluation of threats are executed taking into account

---

[1] Appendices and Forms associated with this study can be downloaded from: https://tinyurl.com/newriskmethodology.

its impact and its probability of occurrence. From this evaluation, each threat will be classified according to their potential risk. The valuation of threats is executed using the Form D (See footnote 1).

(3) *Identify vulnerabilities:* This step is executed using the Microsoft Security Assessment Tool (MSAT). This tool includes of a series of questions to detect vulnerabilities which is grouped in different categories such as infrastructure, applications, operations, and personnel. The usage of such tool will help to recognize the current weaknesses of the organization.

(4) *Select Safeguards:* In this step, the safeguards that are suitable to the assets and risks are identified. The selected safeguards are compliant to ISO/IEC 27002:2013 [13]. In this process, the Form E (See footnote 1) is used.

(5) *Develop the security plan:* After identifying the security problems and corresponding actions, a security plan is developed.

(6) *Communicate results:* Using a results matrix, the findings of the risk analysis process is shared with the people in charge of the library. Such document includes the list of assets with their risk ranges, threats, vulnerabilities, and safeguards associated with them.

3 Implementing the New Methodology

To verify the benefits of the proposed methodology, it was implemented in a real library i.e. library of Escuela Politécnica Nacional (EPN) which is one of the most important university of Ecuador. The library system of EPN is comprised of a central library and ten satellite libraries located in different faculties. The libraries at EPN are divided into the following departments: Acquisition, Technical Services, Circulation, Reference Services, and Binding and Conservation [16]. The application of the proposed methodology was executed under the implementation of the ISMS based on the Deming Cycle (Plan-Do-Check-Act phases) according to the ISO/IEC 27001:2013 standard.

Plan: Establish Your ISMS

Risk Management. This section describes the methodology used for risk management that supports valid decision-making based on information technology tools along with the organization's inventory of assets and their valuation. This process considers the confidentiality, integrity, availability, and authenticity of the information. It carries out a threat analysis and evaluates the identified threats in order to estimate all the risks that the organization can be exposed to. Below are the steps that were executed for the risk analysis process.

(1) *Compile assets:* In the first step, organization's assets were collected as detailed in Table 1 based on the different categories. Continuing with the methodology, the different levels of Asset Valuation used in this step were: 0 = irrelevant, 1 = low impact, 2 = moderate impact, 3 = severe impact, and 4 = extremely severe impact which were analyzed from different security dimensions i.e. Integrity (I), Confidentiality (C), Availability (D), and Authenticity (A).

Table 1. Assets inventory

Table of asset inventory	
Category	Asset
Information	User's personal data, Bibliographic data of materials currently in the library collection
IT Applications	Koha, Database Management System (MySQL), Digital repository (Dspace)
IT Hardware	Printers, Scanner, Router, IP Phones, PCs, Laptops, In-production virtual server
Digital Memory Devices	CD-ROM

(2) *Identify threats:* Threats can cause potential damage to either the information system or the organization. After analyzing the assets, the threats shown in Table 2 were detected using Form C (See footnote 1). Continuing with the methodology, we have used the Threat Valuation in conjunction with the Impact Scale (0 = Very Low, 1 = Low, 2 = Medium, 3 = High, and 4 = Very High) and Probability of Occurrence Scale (0 = rarely, 1 = non-frequently, 2 = frequently, and 3 = often). The threat valuation can be found in Appendix 4 (See footnote 1).

(3) *Identify vulnerabilities:* The next step is to identify technological vulnerabilities in the library using MSAT. Using the threat evaluation, analysts can obtain detailed results that describe weakly protected points through which threats can enter and exploit the system. The results reveal information that complements the risk analysis. Since the report generated by the tool contains extensive data, an example of the resulting report was included in Appendix 5 (See footnote 1).

(4) *Select safeguards:* After identifying threats and vulnerabilities, analysts must consider safeguards that help to reduce the identified risks. In the present case, we expect a considerable reduction of the impact. Safeguards was selected in conjunction with the controls described in ISO 27002:2013 using Form E (See footnote 1). An example of the resulting matrix can be found in Appendix 6 (See footnote 1).

(5) *Develop a security plan:* After determining the potential risks for each identified threat, the analysts have developed the security plan for the library.

(6) *Communicate the results:* In order to communicate the results, the different reports were delivered to the interested parties. The results matrix is detailed in Appendix 7 (See footnote 1).

Do: Implementing and Using the ISMS
Risk Treatment Plan. This step allowed implementers to create a detailed response plan. The Risk Treatment Plan can be found in Appendix 8 (See footnote 1).

Check: Monitoring and Revising the ISMS
The combination of a risk treatment plan with a monitoring plan allows analysts to measure how effective the implemented strategies are, and it allows to execute actions

Table 2. Identified threats

Area	Threat
Natural disasters	Water damage, Fire, Earthquake damage, Electrical storms
Industrial disasters	Explosions, Electrical overload, Dirt, Inappropriate conditions of temperature or humidity
Errors and non-intentional mistakes	User errors, Incomplete records, Virus propagation, Accidental alteration or loss of information, Information system overload
Intentional attacks	Identity theft, Abuse of access privileges, Intentional virus propagation, Alteration of data, Unauthorized access, Data traffic analytics, Information interception

timely to anticipate problems and guarantee project continuity. An example of a Monitoring Plan can be found in Appendix 9(See footnote 1).

In addition, internal audits were executed in order to analyze the level of maturity of the implemented ISMS according to the security controls stipulated in the ISO 27002:2013 standard. To accomplish this, CMM were used as the analytical methodology, since this model focuses on continuous improvement of processes and helps to identify current and desirable components. Our implementation used the evaluation found in Appendix 10 (See footnote 1). The Evaluation of ISMS' maturity, based on the interview with the ICT Library Assistant, can be found in Appendix 11 (See footnote 1).

Act: Maintaining and Improving the ISMS

ISMS implementation is a process that involves time, work, and economic resources. Nevertheless, its benefits growth and fulfills organizational objectives if it is managed in an appropriate way within a cycle of continuous improvement. Consequently, we suggest the implementation of training programs and monitoring activities and adoption of automatic controls and internal audits. We recommend the improvements found in Appendix 14 (See footnote 1).

4   Results

This section presents the results of the maturity evaluation of each control stipulated in ISO/IEC 27002:2013. Figure 1 shows a general view of the current state of EPN's library system in terms of ISMS implementation.

Outstanding factors from the ISMS maturity evaluation in the EPN library system include:

- Our ISMS implementation falls into maturity level 1, corresponding to "Initial."
- Even though the EPN libraries reflect a low level of maturity, it can be noticed that there is progress in several domains such as implementation of security policies, since before implementing the ISMS, the majority of information security controls had never been deployed.

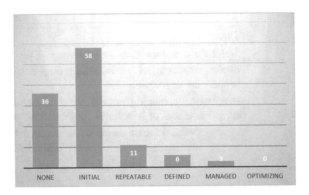

Fig. 1. Results of the maturity evaluation

- 51% of the controls specified in ISO/IEC 27002:2013 are in the "Initial" phase, 10% fall within the "Repeatable" phase, 5% are in "Defined" phase, and 3% of the controls are in "Managed." Phase.
- None of our controls achieve the highest level of maturity - "Optimized". This fact indicates that the library system lacks of automatic tools for ISMS control.
- ISMS implementation has strengthened EPN's information security via formal documentation and standardized procedures.

5   Conclusions

We have developed an optimized risk management methodology that can be effectively adapted to a business model of public university libraries. It incorporates 6 steps which were extracted from advantageous features of MAGERIT v.3, OCTAVE v.2, and NIST 800-30.

Implementation of ISMS in the EPN library system endowed it with resource optimization mechanisms and a continuous improvement process that facilitates achievement of the university's objectives and goals.

A risk analysis using our optimized methodology for public university libraries allows accurate identification of assets, threats, and potential risks to which the organization may be exposed. It also leads to the development of protection measures that mitigate or even avoid altogether detected risks.

The implemented improvements in EPN's library have allowed the entity to reach higher maturity levels, creating the possibility to apply for a certification based on the ISO/IEC 27001:2013 standard in the future.

The EPN libraries still have security policies that have not yet been completely formalized or implemented. As a result, it is important to maintain focus on this project. Finally, it is important to indicate that this project had a positive impact and fulfilled the expectations of the EPN library personnel.

References

1. Bayona, S., Chauca, W., Lopez, M., Maldonado, C.: Implementación de la NTP ISO/IEC 27001 en las Instituciones Publicas: Caso de Estudio. In: Web of Science, p. 6 (2015)
2. Esquema Nacional de Seguridad, MAGERIT – versión 3.0 Metodología de Análisis y Gestión de Riesgos de los Sistemas de Información Libro I - Método, Madrid: Ministerio de Hacienda y Administraciones Públicas (2012)
3. Alberts, C.J., Dorofee, A.J., Allen, J.H.: OCTAVE Catalog of Practices, Version 2.0, October 2001. https://resources.sei.cmu.edu/asset_files/TechnicalReport/2001_005_001_13883.pdf
4. National Institute of Standards and Technology, NIST Special Publication 800-30, Septiembre 2012. https://nvlpubs.nist.gov/nistpubs/legacy/sp/nistspecialpublication800-30r1.pdf
5. Vivancos Cerezo, M.E.: La seguridad en bibliotecas universitarias: normas y auditoría, 06 Agosto 2018. http://eprints.rclis.org/16220/1/Seguridad_informacion_en_BU_v1.pdf
6. Livshitz, I.I., Nikiforova, K.A.: The Evaluation of the Electronic Services with Accordance to IT-security Requirements Based on ISO/IEC 27001, pp. 128–131. IEEE (2016)
7. Narvaez Barreiros, I.R.: Aplicación de la norma ISO 27001 para la implementación de un SGSI en la Fiscalía General del Estado, 07 August 2018. http://repositorio.puce.edu.ec/bitstream/handle/22000/9780/TESIS_SGSI.pdf?sequence=1&isAllowed=y
8. Rodal Castro, P.: Implementation Plan for an ISMS according to ISO/IEC 27001:2013, 30 Diciembre 2016. http://openaccess.uoc.edu/webapps/o2/bitstream/10609/59325/8/prodalTF-M1216mem%C3%B2ria.pdf
9. Susanto, A.., Nurbojatmiko, Shobariah, E.: Assessment of ISMS based on standard ISO/IEC 27001:2013 at DISKOMINFO Depok City. In: Web of Science (2016)
10. Susanto, H., Nabil Almunawar, M., Chee Tuan, Y.: Information security management system standards: a comparative study of the big five. Int. J. Electr. Comput. Sci. **11**(5), 23–29 (2017)
11. Valencia-Duque, F.J., Orozco-Alzate, M.: Metodología para la implementación de un Sistema de Gestión de Seguridad de la Información basado en la familia de normas ISO/IEC 27000. Revista Ibérica de Sistemas y Tecnologías de Información (2017)
12. INTERNATIONAL STANDARD ISO/IEC 27001, 1st edn. (2013)
13. INTERNATIONAL STANDARD ISO/IEC 27002, 1st edn. (2013)
14. Aginsa, A., Matheus Edward, I.Y., Shalannanda, W.: Enhanced Information Security Management System Framework Design Using ISO 27001 and Zachman Framework a Study Case of XYZ Company. IEEE (2016)
15. Disterer, G.: ISO/IEC 27000, 27001 and 27002 for information security management. J. Inf. Secur. **4**(2), 92–100 (2013)
16. Bravo, M.J., Portilla, M.F.: Desarrollo de una interfaz biometrica para el modulo de prestamo del sistema de bibliotecas de la Escuela Politecnica Nacional, 22 May 2015. http://bibdigital.epn.edu.ec/handle/15000/10528
17. Instituto Nacional de Tecnologías de la Comunicación. Implantación de un SGSI en la empresa. https://www.incibe.es/extfrontinteco/img/File/intecocert/sgsi/img/Guia_apoyo_SGSI.pdf

SIDS-DDoS, a Smart Intrusion Detection System for Distributed Denial of Service Attacks

Luis Álvarez Almeida and Juan C. Martinez-Santos[✉]

Universidad Tecnologica de Bolivar, Cartagena, Colombia
jcmartinezs@utb.edu.co
http://www.utb.edu.co

Abstract. In the last few years, the Digital Services industry has grown tremendously, offering numerous services through the Internet and using a recent concept or business model called cloud computing. For this reason, new threats and cyber-attacks have appeared, such as Denial of Service attacks. Their main objective is to prevent legitimate users from accessing services (websites, online stores, blogs, social media, banking services, etc.) offered by different companies on the Internet. In addition, it produces collateral damage in host and web servers, for example, exhaustion of network bandwidth and computer resources of the victim. In this article, we will analyze the information contained in NSL-KDD data-set, which possesses important records about the several behaviors of network traffic. These will be selected to present two methods of selection of features that allow the selection of the most relevant attributes within the data set, to build an Intrusion Detection System. The attributes selected for this experiment will be of great help to train and test various kernels of the Support Vector Machine. Once the model has been tested, an evaluation of the classification model will be performed using the cross-validation technique and we finally can choose the best classifier.

Keywords: Machine learning · Data set · Classification model · DoS attacks · Support Vector Machine · Feature selection

1 Introduction

Nowadays, there are many services that depend on Internet access. These services are delivered to end users through internet connection. All this is possible due to the rising and developing network technology. The Internet cloud is a configurable computing resource environment that offers flexibility and helps reduce hardware and maintenance costs. This is what we call cloud computing [7]. In this environment exist different vulnerabilities that deliver opportunities for web attackers to break the security policies [1]. Distributed Denial Of Service (DDoS) is one of the most powerful type of attacks. DDoS has the power of disrupting

© Springer Nature Switzerland AG 2020
M. Botto-Tobar et al. (Eds.): ICAETT 2019, AISC 1067, pp. 380–389, 2020.
https://doi.org/10.1007/978-3-030-32033-1_35

and degrading the normal performance of web servers. Moreover, it can prevent the user from legitimately accessing a specific service [14]. The DDoS attack decreases available machine resources, so it cannot respond to any request made by a legitimate user [12]. This attack has been known by the network research community since 1980. In the summer of 1999 the Computer Incident Advisory Capability published the first Distributed DoS attack incident [5].

Figure 1 is a report delivered by the company Dyn DNS. It shows how DDoS have increased in the last few years. One of the launched attacks had an attack strength of 1,2 TB (1.200 Gbps) per second, and the attacker had included 100.000 malicious agents. Thus, the average of attack strength has been increasing [12].

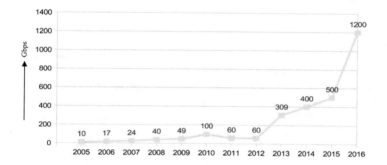

Fig. 1. Strength of DDoS attacks in Gbps.

Currently, the tools used to detect this type of attack are known as intrusion detection system (IDS). Systems that in many cases are based on signature. These are systems used to monitor and analyze the activities of a user and the system, examine the user's violations of network policies, and detect abnormal activities [6]. For this project, we will implement a classification model based on Machine Learning techniques that can help us to detect, identify or classify the anomalous network traffic inside a network data.

2 Related Work

During the review of the State of the Art performed throughout this research, several authors have proposed methods that allow to mitigate the effect of cyber-attacks. The Internet of Things (IoT) technology is one of the most recent researched fields because the number of devices, connecting to the Internet have increased. Thus, these become potential victims of Dos attacks. In this article, it is proposed to apply five machine learning algorithms to distinguish normal IoT packets from DoS attack packets, such as: K-nearest neighbors KDTree algorithm (KN), Support vector machine with linear kernel (LSVM), Decision Tree (DT) using Gini impurity scores, Random Forest using Gini impurity scores (RF) and

Neural Network (NN). The main aim of this implementation is to show evidence that packet-level machine learning DoS detection can accurately distinguish normal and DoS attack traffic from consumer IoT devices [8]. Others authors have proposed to use different classification models to create an Intrusion Detection System and prevent DDoS Attacks in Software Defined Networks (SDN) Using Machine Learning. SDN is an emerging networking model that is intended to change the limitations of current network infrastructures by separating the control logic of network from the underlying switches and routers, suggesting logical centralization of network control and allowing to program the network [2]. On the other hand, classification models are not used only for analyzing network traffic, but also for identifying infected mobile devices. The method presented here is a Machine Learning Classifiers for Anomaly-Based Botnet. The purposes of it, is to identify mobile devices that have been infected by malware. For the evaluation of the study, we used different machine Learning Models, such as: Nave Bayes, K-Nearest Neighbor, Decision Tree, Multi-layer Perceptron, and Support Vector Machine. The evaluation was validated using malware data samples from the Android Malware Genome Project [10].

In the State of the Art presented above, it is possible to evidence the boom in the implementation of classification models based on Automatic Learning Machines. In addition, the results obtained in each of these projects are highly relevant. We can observe how the application of classification models based on Automatic Learning techniques are being used today to counteract problems in the field of computer security. However, the works explained above do not describe the process or technique of features selection on which they were based to choose, train, and test their models. In this paper, we focus on the pre-processing steps previous to train the machine learning model. This process is an important step in the construction of the models, because it allows us to perform a suitable data transformation and standardization to select the appropriate attributes and improve the performance of system.

3   Methods and Materials

3.1   Methodology

The methodology implemented in this work will be based on Knowledge Discovery in Databases (KDD) technique. This approach will allows analyze and process information large amounts of information to identify relevant patterns in data set selected [9]. The steps applied in this methodology are shown in the Fig. 2 and will be developed below in this article .

Select Dataset: To construct a model there needs to be a set of data available and pre-processed. This will help us train and test our classification model. For this reason, NSL-KDD[1] is the data set suggested to train our classification

---

[1] https://www.unb.ca/cic/datasets/nsl.html.

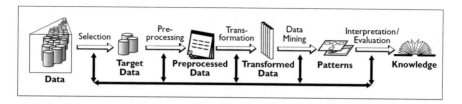

Fig. 2. Graphic schema of KDD process:

model. It has relevant and important information about many types of cyber-attacks, among them are: Denial of Service Attack (DoS), User to Root Attack (U2R), Remote to Local Attack (R2L) and Probing Attack. On the others hand, it contains a standard set of data collected of network traffic. NSL-KDD is the dataset chosen by many researchers to train classification models to classify network traffic based on anomaly detention methods. Moreover, it is an improved version of the KDD cup99 data set, which presents some weaknesses with regard to NSL-KDD, such as redundant records in the train set and duplicate records in the proposed test sets [13].

Preprocessing Data: Selecting the necessary information is important for our model to achieve high performance. Perform basic operations such as: removing redundant or duplicate records, analyzing missing data are of great importance.

Data Transformation: Basically NSL-KDD, is a data-set that contains 22.544 records and 43 attributes. Before starting to train and test our model is necessary to identify attributes with categorical variables which are showed in the Table 1. Therefore, these must be encoded to binary values [3]. The transformation of categorical variables will be performed using a get_dummies module of python. It will identify the categorical values in the attributes and create a binary column for each category. Once the encoding is done the number of columns in the dataset will increase from 43 to 115 columns.

Table 1. Categorical attributes

No.	Attribute name	Description	Sample data
0	protocol_type	Protocol used in the connection	TCP
1	service	Destination network service used	http
2	flag	Status of the connection Normal or Error	SF

The attribute 'type' contains several types of DoS such as: *neptune, back, land, smurf,* and *teardrop*. Also normal traffic is part of our target variables. These will be transformed to using binary value, so that the values for normal traffic will be identified with zero (0) and attacks traffic with one (1). This will allow the model to identify with integer values the different types of traffic. Thus we have only two classes in 'type' attribute.

Data Set Normalization: This is a common requirement for many machine learning estimators, because it is extremely important to improve the performance of the system. In this step, we try to get all attributes formatted to the same scale. To achieve this, we use the Min-Max method of standardization [3]. It scales the features so as to lie between a given minimum and maximum value. Generally, between zero and one. The motivation for using it, is due to its robustness, maintaining very small standard deviations of features and preserving zero entries in sparse data.

3.2   Model

For the development of the experiment, the algorithm known as Support Vector Machine (SVM) will be implemented. It is a popular algorithm used for creating Instruction Detection System. This method was proposed in 1998 by Vladimir Vapnik, and his principal objective is finding the better hyper-plane that will be able to separate the data-set into different classes, taking into account that the better hyper-plane found is the one that allows it to trace the distance or maximum margin between the hyper-plane and the data point. To accomplish this, the Support Vector Machine (SVM) takes the training data obtained from the basic input space, into a higher dimensional feature space using kernels and then gets the most favorable isolating hyper-plane or a decision boundary in the form of support vectors. [11]. The Support Vector Machine used in this paper will use the following kernels: Gaussian Kernel (Radial Basis Function), Polynomial and Linear.

3.3   Features Selection

To carry out this phase, we use *VarianceThreshold* and *SelectkBest* feature selection methods, which are functions provided by Scikit-learn library[2]. The method of feature selection presented in this article, to reduce the number of features, is called filtering method. It is based on the use of variable ranking techniques as first criteria. To remove attributes that have low correlation between them, it is necessary that we define a suitable range and threshold. This technique must be applied before we perform the test of classification model [4].

4   Results

4.1   *VarianceThreshold* Method

The first method exposed here is applied using the *VarianceThreshold* function. This function makes it easier to evaluate the $p - value$ parameter to select

---

[2] Scikit-learn Machine Learning in Python, https://scikit-learn.org.

the most relevant features in all data-sets. This method considers that Boolean features are Bernoulli random variables, and the variance of them is defined by:

$$VAR(X) = p * (1 - p) \tag{1}$$

where p is the probability of the variable to be on the samples.

Table 2. Results of changing p value parameter - $VarianceThreshold$ method.

p VALUE				
0.6	0.7	0.8	0.9	Metrics
0.820000	0.823922	0.921961	0.982745	Accuracy
0.687623	0.694466	0.954043	0.978087	Precision
0.934330	0.939873	0.837162	0.974960	Recall
0.825926	0.823995	0.926876	0.980777	Cross validation

Table 3. Models performance using classification metrics - $VarianceThreshold$ method.

Models				
No.	Linear	Polynomial	GaussianNB	Metrics
0	0.916667	0.921961	0.923725	Accuracy
1	0.949294	0.951297	0.962629	Precision
2	0.810411	0.824110	0.818630	Recall

Since the goal is to select the best features for the model, we iterated the parameter p between a value of 0.6 and 0.9 to get the Table 2. This table has four important metrics: accuracy, precision, recall, and finally cross-validation score that will help us validate the system. In this part of the process, we will be able to identify the appropriate $p-value$ for selecting the best attributes. To generate the p-value a polynomial kernel will be used. Moreover, it will be evaluated using the classification metrics mentioned above. The box graph represented in Fig. 3 we'll help us to display in a graphic way the performance of the model.

Once the $p-value$ is generated, a cross-validation shall be performed using several SVM kernels, including: Linear, Polynomial and Gaussian. These will be trained and evaluated using classification metrics. Table 3 shows the results. A box graph represented in Fig. 3 is also used to visualize the performance of each model.

The achieved result shows us that when the Variance method iterated with p equal to 0.9 was implemented, the model could suffer an over-fitting, because the score obtained was 0.980777, and it is very close from 1 which is a perfect score. Thus, the model could miss classify the new entries. Therefore to reduce the risk

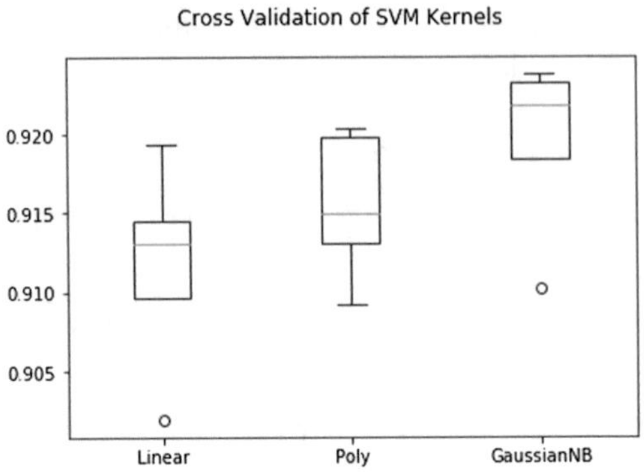

Cross Validation of SVM Kernels

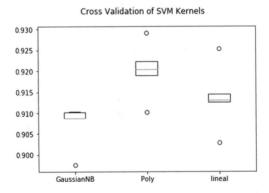

Fig. 3. Box plot of cross validation score of each model - $VarianceThreshold$ method

of over-fitting we selected p equal to 0.8, because the measurements obtained with each of the metrics, show a better behavior of the model. The attributes obtained iterating with $p - value$ equal to 0.8 are showed in the Table 4.

Figure 3 shows the performance of the three kernels using the features selected in Table 4 are very good. In addition, looking at the Fig. 3 which has a cross validation of three models, it is evident to note the differences in performance of the there model. Thus, it clear to say that the better kernel is GaussianNB.

4.2　$SelectkBest$ Method

Cross Validation of SVM Kernels

Fig. 4. Box plot of cross validation score of each model - $SelectkBest$ method

The second method implement is the $SelectkBest$ function, which uses chi-squared stats to compute the stats between each non-negative feature and class and retrieve only the two best features. Thus, this score will be used to choose the k features with the highest values for the test chi-squared statistic from the

Table 4. Table of features selected from *VarianceThreshold*

No.	Attribute name	Description
1	Logged_in	Login Status: 1 if successfully logged in; 0 otherwise
2	Serror_rate	The percentage of connections that have activated the flag (4) s0, s1, s2 or s3, among the connections aggregated in count (23)
3	Srv_serror_rate	The percentage of connections that have activated the flag (4) s0, s1, s2 or s3, among the connections aggregated in srv_count (24)
4	Same_srv_rate	The percentage of connections that were to the same service, among the connections aggregated in count (23)
5	Dst_host_srv_count	Number of connections having the same port number
6	Dst_host_same_srv_rate	The percentage of connections that were to the same service, among the connections aggregated in dst_host_count (32)
7	Dst_host_rerror_rate	The percentage of connections that have activated the flag (4) REJ, among the connections aggregated in dst_host_count (32)
8	Dst_host_srv_rerror_rate	The percentage of connections that have activated the flag (4) REJ, among the connections aggregated in dst_host_srv_c ount (33)
9	enco_host_name	Destination network service used
10	enco_printer	Destination network service used
11	enco_S3	Destination network service used

Table 5. Models performance using classification metrics - *SelectkBest* method.

Models				
No.	Linear	Polynomial	GaussianNB	Metrics
0	0.912157	0.916078	0.912157	Accuracy
1	0.952866	0.956329	0.952866	Precision
2	0.800000	0.808021	0.800000	Recall

data-set. The attributes obtained with this function are shown the Table 6. The only difference between the *VarianceThreshold* and *SelectkBest* methods is, the latter generates two attributes that are not included in *VarianceThreshold* which are: enco_RSTR and enco_whois. Once the features have been chosen we will test and train the model.

Table 5 shows the results of the metrics (accuracy, precision, recall, cross validation) score obtained after evaluating the model. The experiments indicate

Table 6. Table of features selected from *SelectkBest* method.

No	Attribute name	Description
1	Logged_in	Login Status : 1 if successfully logged in; 0 otherwise
2	Serror_rate	The percentage of connections that have activated the flag (4) s0, s1, s2 or s3, among the connections aggregated in count (23)
3	Srv_serror_rate	The percentage of connections that have activated the flag (4) s0, s1, s2 or s3, among the connections aggregated in srv_count (24)
4	Same_srv_rate	The percentage of connections that were to the same service, among the connections aggregated in count (23)
5	enco_whois	Destination network service used
6	enco_RSTR	Destination network service used
7	Dst_host_rerror_rate	The percentage of connections that have activated the flag (4) REJ, among the connections aggregated in dst_host_count (32)
8	Dst_host_srv_rerror_rate	The percentage of connections that have activated the flag (4) REJ, among the connections aggregated in dst_host_srv_c ount (33)
9	enco_host_name	Destination network service used
10	enco_printer	Destination network service used
11	enco_S3	Destination network service used

that the moment when the model works better is when it is evaluated with the polynomial kernel, unlike the *VarianceThreshold* method mentioned above. To visualized the results, a box plot is drawn as defined on Fig. 4.

5   Conclusions

To construct a robust model that permits classifying new entry samples or identifying the anomaly network traffic takes a lot of time. Thus, it is necessary to perform several tests and apply different steps, that permit the improvement of the performance of the models selected. For this reason, it is important to have an ideal data-set and then carry out a phase of pre-processing data. This last one, allowed eliminate redundant and duplicate records, encode attributes, normalize data, and select the most important features. In this paper we were able to reduce the number of features considerably from 115 to 11 attributes. Which will reduce computational expense. Once the most relevant attributes are selected and the model is tested, our Intrusion Detention System (IDS) could be capable of identifying anomalous traffic in the network. It is important to mention that Support Vector Machine work better when the values of target variables are transformed from labels to binary values. As we can see at Fig. 3

of the *VarianceThreshold* method, the performance of three kernels using the features selected in Table 4 show good results using polynomial and Gaussian Kernels. In addition, using the *SelectkBest* features selection method shows us that Polynomial Kernels has better performance. Finally, we can say that the Gaussian Kernel model is a good candidate to build a Intrusion Detection System. since, in both filtering methods it shows excellent results, as showed in the Fig. 4.

References

1. Ajagekar, S.K., Jadhav,V.: Study on web DDoS attacks detection using multinomial classifer. In 2016 IEEE International Conference on Computational Intelligence and Computing Research (ICCIC), pp. 1–5, Chennai, India, December 2016. IEEE (2016)
2. Ashraf, J., Latif, S.: Handling intrusion and DDoS attacks in software defined networks using machine learning techniques. In: 2014 National Software Engineering Conference, pp. 55–60. IEEE (2014)
3. Bhavsar, Y.B., Waghmare, K.C.: Intrusion detection system using data mining technique: support vector machine. Int. J. Emerg. Technol. Adv. Eng. **3**(3), 581–586 (2013)
4. Chandrashekar, G., Sahin, F.: A survey on feature selection methods. Comput. Electr. Eng. **40**(1), 16–28 (2014)
5. Criscuolo, P.J.: Distributed denial of service: Trin00, Tribe Flood Network, Tribe Flood Network 2000, and Stacheldraht CIAC-2319. In: Lawrence Livermore National Laboratory, p. 18, February 2000
6. Deokar, B., Ambarish, H.: Intrusion detection system using log files and reinforcement learning. Int. J. Comput. Appl. **45**(19), 28–35 (2012)
7. Deshmukh, R.V., Devadkar, K.K.: Understanding DDoS attack and its effect in cloud environment. Procedia Comput. Sci. **49**, 202–210 (2015)
8. Doshi, R., Apthorpe, N., Feamster, N.: Machine learning ddos detection for consumer internet of things devices. In: 2018 IEEE Security and Privacy Workshops (SPW), pp. 29–35. IEEE (2018)
9. Fayyad, U., Piatetsky-Shapiro, G., Smyth, P.: The kdd process for extracting useful knowledge from volumes of data. Commun. ACM **39**(11), 27–34 (1996)
10. Feizollah, A., Anuar, N., Salleh, R., Amalina, F., Maarof, R.R., Shamshirband, S.: A study of machine learning classifiers for anomaly-based mobile botnet detection. Malays. J. Comput. Sci. **26**, 251–265 (2013)
11. Gyanchandani, M., Rana, J.L., Yadav, R.N.: Taxonomy of anomaly based intrusion detection system: a review. Int. J. Sci. Res. Publ. **2**(12), 1–13 (2012)
12. Kaur, P., Kumar, M., Bhand, A.: A review of detection approaches for distributed denial of service attacks. Syst. Sci. Control Eng. **5**(1), 301–320 (2017)
13. Tavallaee, M., Bagheri, E., Lu, W., Ghorbani, A.A.: A detailed analysis of the KDD cup 99 data set. In: 2009 IEEE Symposium on Computational Intelligence for Security and Defense Applications, pp. 1–6. IEEE (2009)
14. Zargar, S.T., Joshi, J., Tipper, D.: A survey of defense mechanisms against distributed denial of service (DDoS) flooding attacks. IEEE Commun. Surv. Tutor. **15**(4), 2046–2069 (2013)

Trusted Phishing: A Model to Teach Computer Security Through the Theft of Cookies

Germán Rodríguez[1,2(✉)]⑩, Jenny Torres[1], Pamela Flores[1],
Eduardo Benavides[1,2]⑩, and Paola Proaño[2]⑩

[1] Systems Engineering Faculty, Escuela Politécnica Nacional, Quito, Ecuador
{german.rodriguez,jenny.torres,pamela.flores,diego.benavides}@epn.edu.ec
[2] Computer Science Department, Universidad de las Fuerzas Armadas ESPE,
Santo Domingo de los Tsáchilas, Ecuador
{gerodriguez10,debenavides,pmproano2}@espe.edu.ec
https://www.epn.edu.ec/
https://www.espe.edu.ec/

Abstract. Social engineering is a common practice to obtain information through the manipulation of users' trust; while phishing refers to a computer attack model that is executed through social engineering. Combined with Cross-Site scripting (XSS), users' curiosity to access their cookies and steal information from their sessions could be abused. The objective of this proposal was to teach concepts about cookie theft through vulnerable blogs. Our idea was to develop a blog vulnerable to XSS attacks to steal information from a test cookie that was created on the computer of users who accessed this site. Subsequently, the information corresponding to the attack was organized to present a new publication on the blog in order to explain to users how, with great care, we stole their cookies. Our goal was to challenge the trust and curiosity of our contacts in the social network Facebook and in the WhatsApp messaging application, so that they were tempted to visit this compromised blog whose content was false information. The results show that 182 contacts accessed the compromised blog and 100% of the users assumed that the blog was reliable. It was also shown that through this controlled attack, all 182 contacts learned about the theft of cookies that can be produced through fake blogs.

Keywords: XSS · Social Engineering · JavaScript · Vulnerable blog · Cookies

1 Introduction

The art of hacking people is called Social Engineering [1], its purpose is to cheat the user. Its popularity has increased due to the growth of social networks, emails and the online communication. In other words, is the psychological manipulation

© Springer Nature Switzerland AG 2020
M. Botto-Tobar et al. (Eds.): ICAETT 2019, AISC 1067, pp. 390–401, 2020.
https://doi.org/10.1007/978-3-030-32033-1_36

of users [2]. For example, an attacker could send a malicious script to an user's browser and will execute it. This attack is known as Cross-Site Scripting (XSS) and is one of the most present vulnerabilities [3] in web applications.

If we combine social engineering with XSS the attacker could steal victims' cookies. Cookies play a fundamental role in our Internet experience. With it, a server can remember the times we visit the website [4]. In consequence, the compromised information could be passwords, credit cards, bank information or among others. The hacker known as phisher [5] uses this techniques, pretending to be a *trustworthy person or company* using an email, some instant messaging system, social networks, SMS/MMS, or even phone calls.

In this scenario, there are proposals developed by the scientific community to mitigate this type of attacks [6–12], but their problem is that they are aimed at educating users with knowledge in computer security, our proposal is an incentive to awaken the curiosity of users without knowledge of computer security and, therefore, create awareness about the danger that exists when someone uses social networks to publish false information that directs them to false blogs.

With this background, we proposed a model to teach tips to avoid the theft of their cookies through fake blogs, through a controlled attack using Social Engineering and Cross Site Scripting (XSS). The most important phase is the development of a blog vulnerable to XSS attacks. Through the HTML and JavaScript codes we were able to demonstrate that the Blogger application does not offer any control for the safe development of blogs. This allowed that through our social networks on Facebook and WhatsApp exploit the curiosity and trust of our contacts to access this fake blog. Our contribution was to demonstrate that users' cookies could be stolen without their knowledge. With this model we were able to successfully execute an XSS attack and at the same time explain to the user how this attack was achieved. Later, this information was published as a new blog note to teach them about the theft of cookies through fake blogs.

The rest of the document has been organized in the following way: in Sect. 2 a brief review of the existing literature is made to guide the proposed solutions and their limitations, Sect. 3 discusses the proposal, the techniques and methods used, in Sect. 4 we analyze the results obtained with our proposal and finally, Sect. 5 talks about the conclusions and future work.

2   Related Work

This section offers a general description of the proposals that we have found in the scientific literature and that relates to teach of computer security through the theft of cookies.

In [6] the development of an XSS attack and defense laboratory has been proposed. Its objective is to allow university students to learn this type of attack, that is, to offer them an environment to practice with a behavior similar to that of hackers. For this, they use a dynamic website and a virtual environment, where they include a website for the theft of cookies and a file to store stolen cookies.

As the same way, in [7] a CSRF (Cross Site visit request) attack and defense practice laboratory environment for students to dominate these skills, has been proposed, that is, to attack as a hacker and defender with a web administrator. As tools, they use a computer and two virtual machines, in addition to a web application with Asp + Access. This application will extract the value of the user name of the cookie sent by the browser.

On the other hand, in [8] a laboratory environment for the spoofing attack of cookies has been developed. Its goal is to allow students to dominate the attack and defense using cookie counterfeiting. In the same way, they use an application where they deliberately include cookies forgery vulnerabilities. This application has an interface to start a session, stores this data within a cookie and then uses this cookie to authenticate the user when accessing a web page.

In [9] they show how to train browsers by developing a proven prototype to more than 50 vulnerable web applications in the real world. During the training, the website administrators must create all the fingerprints on the server side, however, this affects the effectiveness of the approach because they depend on the creation of all fingerprints. Its advantage is that it analyzes all the scripts that come from each website.

In [10] a learning platform has been proposed that is portable so that network applications are tested in a secure way. It is oriented to universities and faculties and its objective is to prepare students in the safe development of mobile applications through the application of practical laboratories. This works through an online repository, affordable and adoptable learning materials for further dissemination, however, there is no cookie analysis.

In [11] a model has been presented to design and implement social engineering awareness programs aimed at encouraging behavioral change in schools. This research began looking at the problem of social engineering when it comes to schools. Its advantage is that it is complemented with evaluation and continuous reinforcement approaches. Its disadvantage is that this program runs only once a year and mentions Social Engineering very briefly, so it cannot change the behavior of users in a short time.

In [12] they develop teaching ideas for two issues in computer security: by analyzing the web's cookies and public access to government databases. Both graduate and undergraduate students participate in this development. They propose a method of game through conferences for students to investigate the history of cookies on the web and thus expose by analogies with other types of transactions.

As shown in Table 1, the weakness of the proposals that they are oriented to teach students with knowledge in computer security [6–8, 10–12], only the proposal [9] is not oriented to the students. Three of these are laboratories that are controlled environments for attacks, our proposal makes a real attack without altering the susceptibility of people, so too, not many studies were found that teach computer security through the theft of cookies, but, we rely on those who did use this methodology to create an attack base through a real scenario using social networks.

Table 1. Classification of solutions

Tool	Objective	Oriented	¿Theft of Cookies?	Documents
Laboratory	XSS attack and defense	University students	Yes	[6]
Laboratory	CSRF attack and defense	University students	Yes	[7]
Laboratory	Spoofing attacks of cookies	Students	Yes	[8]
Prototype	Train browsers	Not specified	No	[9]
Portable platform	Test applications	University and faculty students	No	[10]
Model	Design and implement social engineering awareness	Students	No	[11]
Game	Web cookies and public databases analysis	Students	Historic of cookies	[12]

3   Trusted Phishing Structure

No matter how sophisticated our security is, the biggest risk will always be the same: users click on the wrong links and send their passwords to the wrong websites. So in Fig. 1 we have proposed a model to teach computer security tips through a controlled attack of XSS. Our model is composed of the following phases:

(A) **Recognition or Contact**: The objective of this section is to select the people that will be the target of our analysis

(B) **Collection of information**: The objective of this phase was to stimulate the curiosity of this people.

(C) **Compromised blog**: Using the Blogger application to create a blog vulnerable to XSS

(D) **Steal of cookies**: Create and steal a test cookie that is created on the computers of the victims

(E) **Teach about the executed attack**: Create a blog post to show what happened with the attack

Due to the sensitivity of the users, an intrusive attack was not carried out in order to not affect the privacy of the users.

3.1   Recognition or Contact

In this initial phase we will test the confidence and curiosity of our contacts. For this, we have applied a social engineering technique that arouses the curiosity to review links with relevant information. In this case, their intention is to

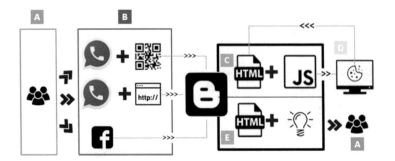

Fig. 1. Proposed model to teach computer security through XSS attacks

discover something they do not know (discover how to steal data using a What-sApp status), but as mentioned in [15], curiosity does not always coincide with intelligence, so we can take advantage of it to Encourage our contacts who offer information that is really fictitious and that acts as a distraction through the following blog [16].

3.2   Collection of Information

Three real scenarios were proposed to play with the trust of our contacts in the messaging application WhatsApp and in the social network Facebook. The objective of this phase was to stimulate the curiosity of the people. To fulfill this first objective, 3 attack vectors were used:

3.2.1   WhatsApp Status that Contains a QR Code

As seen in Fig. 2a this QR code was generated with the Social Engineering Tools (SET) software, which allowed us to camouflage a link to our compromised website. However, as the sensitivity of our contacts cannot be affected, our blog only contains information that specified the methodology to obtain their preferences by stealing their cookies. This website was implemented as a personal blog.

Fig. 2. (a) QR Code generated as a WhatsApp status, (b) Status of WhatsApp with link to the compromised blog

In the implementation of this blog, a complement was used as -Visitor Counter-, as shown in Fig. 3, that acted as a collector of the following characteristics of the visitors:

– Country of origin
– Type of browser used
– Type of operating system
– Date and time of visit

Fig. 3. Data collected with the hidden visitor counter

The objective of this vector was to count how many contacts were persuaded to use some medium or application that could decode the QR code. In Table 3 it is observed that only 3 contacts managed to decode the QR code and access to the compromised blog.

With this scenario it was found that the vector is not very intrusive since the QR Code acts as a static figure and users need to download a QR Reader application to be able to decode it. We must remember that our contacts have basic computer knowledge. However, 3 contacts managed to decode it and access to the implemented blog.

3.2.2   Status of WhatsApp with a Link to the Compromised Blog

The objective of this second scenario was to create a link to the compromised blog and put it as a WhatsApp status, as shown in Fig. 2b. However, the vector was not very intrusive either; there were only 5 contacts who accessed the blog after reviewing our WhatsApp status, as detailed in Table 3.

3.2.3   A Post on Facebook with a Fake Note

In this third scenario, showed in Fig. 4, a post was published on the Facebook social network that read: *¿Do you want to know how to steal information through a WhatsApp status? More information here ...* The same one was linked to our compromised blog.

3.3   The Compromised Blog

This section explains how the compromised blog was implemented using the Blogger application that allows to create and publish online blogs. According to the Blogger website [13], for to publish content, the user does not have to write any code or install server or scripting programs. However, here we show that it

Fig. 4. Post on Facebook with a false note to redirect to the compromised blog

is not true, and that a blog could be implemented with HTML and JavaScript code that steals a cookie was created by the same blog.

As mentioned before, we cannot affect the sensitivity of our contacts, however we abuse their confidence and curiosity to access this blog implemented with information that showed them how their information could be obtained through the theft of their cookies. This means that, we do not steal the cookies from the sites they have visited; we only obtained information from the cookie created with the visited blog itself.

3.4 Steal of Cookies

Figure 5 presents the scheme proposed, which objective was to steal a cookie using JavaScript codes. For this, we made a test to obtain the cookie created by the blog itself and we documented the action to then add an entry in the blog with these results.

Fig. 5. Proposed scheme to implement the compromised blog

In summary, this scheme proposes the implementation of our blog with XSS codes in 3 steps:

3.4.1 Create the Test HTML Blog

For this, an entry called Phishing Test was implemented using the Blogger application, and JavaScript codes to create a cookie and to obtain its information, as shown in Fig. 6.

3.4.2 Create a Test Cookie on User's Computer

The JavaScript code was developed within the blog to create a cookie on the user's computer with the information shown in Table 2.

Fig. 6. Compromised blog design using JavaScript code

Table 2. Content of the cookie that will be created on user's computer when visiting the compromised blog.

Parameters	Value
Name	Username
Content	ResearcherCookies
Domain	pentabyteblog.blogspot.com
Root	/
Send to	Any type of connection
Created	Monday, June 4, 2018, 16:34:39
Expires	Saturday, December 18, 9999, 07:00:00

3.4.3 Steal the Content of the Cookie Created

For this, the JavaScript code was embedded in the HTML code of the blog to retrieve the content of the cookie. When a user visits the website, the same page will show the content of the cookie that was created on his computer. This can be improved by designing a script that redirects the information of all cookies obtained from each user who visited the blog, however, this would be a more intrusive attack and violate the privacy of our contacts.

Fig. 7. Viewing the content of the stolen cookie

As shown in Fig. 7 the result is shown in the same entry of the blog with the label STOLEN COOKIE = ResearcherCookies, we show the content of the cookie that created the page to demonstrate how can get its content.

3.5   Teach About the Executed Attack

According to the analysis of the scientific literature, these proposals are not intended to educate the end user (without knowledge), on the other hand, there are proposals aimed at educating the user, but its limitation is that its object of study are advanced users [14].

Our goal was to educate users who do not have knowledge about computer security. For them, we use tips and suggestions, through the same blog we use to steal the test cookie from their computer, to teach them how to prevent all actions that users execute in this controlled attack.

With the first and second phases of our model, we verified that our contacts used their curiosity to access the compromised blog. These are some of the tips published in the blog post.

(A)   **¿Do you know how we managed to access our compromised blog?** We gently abuse your curiosity and confidence. Hackers exploit this because they know that the first movement of attack is always trust.

(B)   **¿Did you know that the information we publish on the blog was false?** The majority of users are interested in how to hack the Facebook or WhatsApp of their partners.

(C)   **¿Did you know that you cannot steal information through the status of our WhatsApp?** Users without knowledge of computer security access any type of information that catches their attention.

4   Analysis of Results

In this section we analyze the results obtained during the data collection in the execution of the 3 attack vectors.

Analyzing Table 3 we found that there were 3 contacts who accessed the compromised blog through an application to decode QR codes, this was verified by the visits counter programmed to register them during a day of publication of the status in the WhatsApp application. On the other hand, only 5 contacts accessed the compromised blog, checking for curiosity the URL link that was set as WhatsApp status. The highest number of visits were obtained through the Facebook status with the false note, in total 174 contacts clicked on the post and were redirected to the compromised blog.

As shown in Fig. 8a, our visitor counter recorded that the operating system most used to access the blog was Android (32% - 58 users) followed by Windows (31% - 56 users) and iPhone (30% - 54 users). On the other hand, the most used browsers were Google Chrome (57% - 103 users), followed by Mobile (30% - 5 users) and Firefox (9% - 1 user), as shown in Fig. 8b. It was also recorded that the largest number of visitors was in Ecuador (140), followed by the United States (29), Peru (6), Spain (3), Italy (2), the Philippines (1) and Chile (1). So, we verified that all the contacts that accessed the blog were interested in more information about these tips, through their comments on the blog.

Table 3. Summary of users who visited to the blog

–	Through QR Code in WhatsApp	Through Link from WhatsApp	Visiting Facebook Post	TOTAL
				–
Contacts that accessed the blog	3	5	174	182

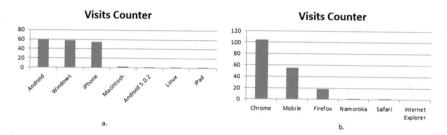

Fig. 8. Statistics by Operating System obtained (a) and Browsers (b) from the visitor counter

5   Conclusions and Future Work

This work presents a controlled attack executed through 3 vectors of information collection: (1) A status in WhatsApp with a QR Code, (2) a status in WhatsApp with a URL link, (3) a Facebook post with a false note.

For this purpose, a personal blog was implemented with false notes that talked about the theft of information through the WhatsApp status. These vectors redirected the victims to our blog that was implemented with the Blogger application. In this way and through JavaScript codes a test cookie was created and we stole the content of that same cookie, also installed a plugin to collect information of: Country of origin, Type of Operating System, Type of Browser, and Date/Time of visit.

It was shown that 100% of contacts did not ask about the objective of our status and the post in social networks and we took advantage of their confidence and curiosity. As a result of this attack it was obtained that 95.7% of contacts accessed through the Facebook post, 1.6% use a QR Reader application to access through our WhatsApp status and 2.7% reviewed our WhatsApp status with the URL link to the blog. In addition, it was found that the Blogger application does not have any type of control for the execution of JavaScript codes, giving the possibility of developing a blog vulnerable to XSS attacks that allow stealing user's cookies.

With this attack, our contacts have been socialized in the way that their curiosity and trust was abused through a social engineering attack combined with an XSS attack. Our goal was to teach users without knowledge of computer security through tips and suggestions. With this, the user learned how it is possible to steal cookies through fake blogs using social networks as a means to visit these types of sites.

As future work, we propose a similar attack, but including a credential collector that will supplant our identity to demonstrate that trust is the first factor required for a phishing attack to be executed with satisfaction and without affecting the user's sensitivity. A server will be implemented that will collect the information of all the cookies of all the users, using a mechanism that is not intrusive and that does not affect the privacy of the users.

References

1. Kaspersky Lab: Ingeniería social, hackeando a personas (2018). https://www.kaspersky.es/blog/ingenieria-social-hackeando-a-personas/2066/?slow=1/
2. BBVA: Ataques de ingeniería social: qué son y cómo evitarlos (2018). https://www.bbva.com/es/ataques-ingenieria-social-evitarlos/
3. Backtrackacademy: XSS-Capturando Cookies de sesión (2018). https://backtrackacademy.com/articulo/xss-capturando-cookies-de-sesion
4. LaCroix, K., Loo, Y.L., Choi, Y.B.: Cookies and sessions: a study of what they are, how they work and how they can be stolen. In: 2017 International Conference on Software Security and Assurance (ICSSA), pp. 20–24 (2017)
5. ESET: Cómo opera un phisher (2018). https://www.welivesecurity.com/la-es/2010/08/25/como-opera-phisher/
6. Zeng, H.: Research on developing an attack and defense lab environment for cross site scripting education in higher vocational colleges. In: 2013 International Conference on Computational and Information Sciences, pp. 1971–1974 (2013)
7. Zeng, H.: Research on developing a lab environment for cross site request forgery: attack and defense education in higher vocational colleges. In: Proceedings of 2013 3rd International Conference on Computer Science and Network Technology, pp. 56–60 (2013)
8. Zeng, H.: Research on developing a lab environment for cookie spoofing attack and defense education. In: 2013 International Conference on Computational and Information Sciences, pp. 1979–1982 (2013)
9. Dua, M., Singh, H.: Detection and prevention of website vulnerabilities: current scenario and future trends. In: 2017 2nd International Conference on Communication and Electronics Systems (ICCES), pp. 429–435 (2017)
10. Lo, D.C., Qian, K., Chen, W., Shahriar, H., Clincy, V.: Authentic learning in network and security with portable labs. In: 2014 IEEE Frontiers in Education Conference (FIE) Proceedings, pp. 1–5 (2014)
11. Mohammed, S., Apeh, E.: A model for social engineering awareness program for schools. In: 2016 10th International Conference on Software, Knowledge, Information Management Applications (SKIMA), pp. 392–397 (2016)
12. Miller, K.W.: Computer security and human values interact. In: 2016 10th Proceedings Frontiers in Education 1997 27th Annual Conference. Teaching and Learning in an Era of Change, vol. 2, pp. 1025–1029 (1997)

13. Blogger.com: Escribe sobre lo que te apasiona y a tu manera (2018). https://www. Blogger.com/
14. Mahdi, A.O., Mohammed, I., Abu, S.: An intelligent tutoring system for teaching advanced topics in information security (2016)
15. Villanueva, T.: No siempre la curiosidad coincide con la inteligencia (2014)
16. Anonymous: PENTABYTE - No instales, No configures..Desarrolla (2018). http:// pentabyteblog.blogspot.com/

Analysis of Cryptographic Technologies Based on ISO27001 Standards

Henry Rodrigo Vivanco Herrera[✉] [iD]
and Aurelio Gregorio Camacho Reina [iD]

Universidad Tecnológica Israel,
Francisco Pizarro E4-142 y Marieta de Veintimilla, Quito, Ecuador
{hvivanco, acamacho}@uisrael.edu.ec

Abstract. With advances in information technology, more and more processes are automated; consequently, the information generated increases. The centralization of information becomes a necessity, so that companies are forced to send all this information from their branches to a single repository.

The transport of information throughout the network must be done in a safe and fast way, and it is necessary to use mechanisms that support this first factor, one of which is cryptography.

The present work describes particularities of some encryption methods and compares their performance in an environment with reduced hardware characteristics.

Keywords: Encryption · Cryptography · ISO27001 · Raspberry

1 Introduction

Currently, technological advancement offers new and better solutions for all areas of different types of business and agriculture is no exception. The agricultural companies of Ecuador use Electronic Scale Systems (SBE) that; through a web interface, it allows registering data of fruit packing process from workstations. This information collected daily in the client application must be stored in repositories that could be physically located even in other cities, so it is necessary to ensure that such information arrives quickly and safely.

The above shows that companies require a mechanism to ensure that the information has not been modified during the transmission and reception process and at the same time without causing consumption of intranet resources, causing other users of the network to delay their work. Hence the importance of choosing an adequate mechanism that maintains the perfect balance between security and resource consumption for the transfer of information.

There are many methods of securing existing information among which the encryption of data stands out. On the Internet there is a lot of information about the characteristics, advantages and disadvantages of each of the different encryption mechanisms, implemented in computational equipment with robust characteristics and the results are very satisfactory.

© Springer Nature Switzerland AG 2020
M. Botto-Tobar et al. (Eds.): ICAETT 2019, AISC 1067, pp. 402–411, 2020.
https://doi.org/10.1007/978-3-030-32033-1_37

Currently there is a wide variety of cryptographic systems and knowing how to choose the right tool requires an exhaustive analysis of its characteristics, knowing its classification and fundamental concepts in order to be able to use it properly and obtain the greatest possible benefit.

Generally Raspberry devices have been used for educational development due to the ease of acquisition and maintenance, but this type of equipment is currently being implemented in projects in agricultural companies [1].

In itself each encryption mechanism has a different operation from the others, which means different workloads. The Raspberry devices still in their most current version have processing characteristics that limit their functions, so it is necessary to perform tests in their most extreme environments to consider them suitable before their release to production.

That is why, in the present investigation, encryption tests will be carried out with 2 algorithms, each with different structures, in Raspberry Pi2 equipment. The particularities of the different encryption systems and their performance on platforms will be identified, with reduced hardware characteristics, reaching harmony between the process load and response times.

2 Theory and Context

2.1 Encryption Algorithms

With the arrival and development of the Internet and the mass use of the computer, it is necessary to use tools that protect documents and information stored in them [2]. Some of these utilities are: firewalls, Intrusion Detection Systems and the use of cryptographic systems; These allow to protect both the information and the computer systems in charge of managing it.

3DES

This algorithm descended from the DES algorithm; designed by IBM and published in 1975, initially standardized for financial institutions. It is currently one of the most used algorithms for credit cards and other electronic payment methods; however, it is slowly disappearing due to its encryption process that tends to be relatively slow and being replaced by the AES algorithm, of which no vulnerabilities have been discovered to date.

This algorithm is based on doubling the effective length of the key to 112 bits, but due to the need to triple the number of necessary operations the total length of the key will be 168 bits, without modifying the DES algorithm. [3].

Figure 1 presents graphically how the 3DES algorithm works with respect to the management of its public key, and the process of encryption and decryption.

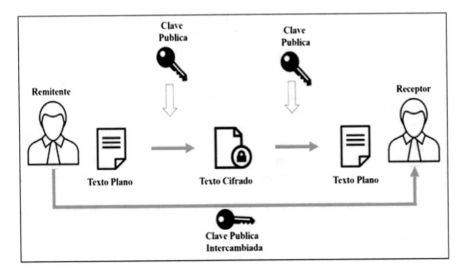

Fig. 1. Symmetric encryption operation [4]

On the other hand [5] describes that 3DES uses a method called 3DES-Encrypt-Decrypt-Encrypt (3DES-EDE) to encrypt plain text. First, the message is encrypted using the first 56-bit key, called K1, then, the data is decrypted using the second 56-bit key, called K2 and finally, the data is again encrypted with the third key of 56 bits called K3.

In the following graph, Fig. 2 is shown and this time the functioning of the 3DES algorithm with respect to the triple encryption process is presented, both to encrypt the document and to decipher it.

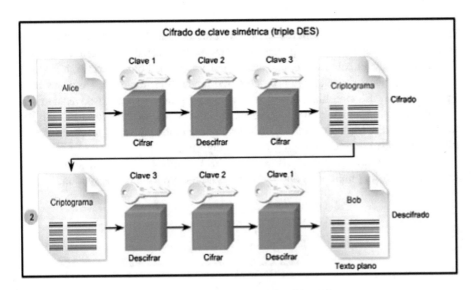

Fig. 2. 3DES encryption algorithm [6]

Rivest, Shamir y Andleman (RSA)

The RSA algorithm is known by the initials of its 3 discoverers (Rivest, Shamir and Adleman) and it is stipulated that this algorithm has resisted all attempts to break them for more than a quarter of a century, it is considered very robust [7]. Its biggest disadvantage is that it requires keys of at least 1024 bits to guarantee greater security, compared to 128 bits of the symmetric key algorithms, which makes it relatively slow.

The RSA algorithm consists of three basic steps: Key generation, encryption and decryption. Figure 3 presents graphically how the RSA algorithm works with respect to the management of its public key and its private key, and the process of encryption and decryption.

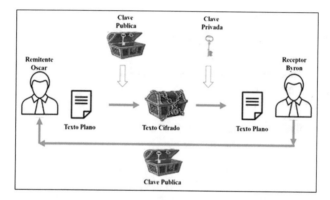

Fig. 3. Asymmetric algorithm operation [7]

3 Methodology

The present work is based on a methodology with the following characteristics:

Descriptive Research: The analysis that will be given, is focused on evaluating variables in independent environments to highlight the favorable and unfavorable characteristics of each, in each of its processes.

Experimental Design: It aims to culminate with real data, demonstrating in an explanatory way the cause-effect relationship.

Quantitative Approach: Because they are going to measure variables that will determine at the end of the investigation and important features of each of the algorithms in their respective scenarios, highlighting their weaknesses and strengths in regards to variables such as assignment of security, processing speed, cost of loading, among other factors that will be measured in teams with very limited resources.

3.1 Population and Sample

Population

The population is an agricultural company, with 83 processing plants and Raspberry PI equipment, which are in process of fruit packing process. In each process plant there are 3 workstations with their respective wireless network access points, from which data is sent to the central repository located in the city of Guayaquil.

Sample

The sample was selected through a stratified sampling with proportional affixation, using the zones as the stratification variable. The different strata nh, were obtained by the formula:

Where:

N: 249 Size of the target global population
Nh: Proportional population size
n: 25 Size of the global sample to be obtained
k: 0.1 Ratio coefficient
L: 6 Number of layers

From which the distribution shown in Table 1 is obtained:

Table 1. Distribución de la muestra

Estr.	Identification	Host	Proportion	Sample
1	North Zone 1	39	15,70%	4
2	North Zone 2	48	19,30%	5
3	Zone Center 1	66	26,50%	6
4	Zone Center 2	57	22,90%	6
5	South Zone 1	27	10,80%	3
6	South Zone 2	12	4,80%	1
Total		**249**	**99,80%**	**25**

3.2 Project Presentation

For the realization of the present investigation, a technological platform was used with limited features devices, single-board microcomputers that are part of wireless networks that connect to a central repository, where the information sent is verified.

With the protocol analyzer the data will be verified while they are being sent by the transport layer, checking the status in which it is found, both in encrypted format and in plain text format, and its content.

The data collection tests were analyzed in each one of the respective packers, of each hacienda, in each zone; making 3 shipments from the Raspberry PI3 device to the central repository, then after averaging, record that value.

The time taken to send the file from the device to a repository in the cloud was measured, both as plain text, and as encrypted with 3DES and RSA. In the transport, the file data, header and user and password values were verified with Wireshark.

In addition to the aforementioned, the quality of the encryption of the algorithms subjected to the test was analyzed and, finally, the percentage of growth of the file after having been encrypted in relation to the original file.

The results are presented individually, by area and globally, both in numerical values and in statistical graphs that facilitate their understanding and analysis.

In order to achieve a geolocation of the zones and how they are distributed nationwide, their locations are presented below along the 5 provinces of the country, in addition to the distribution according to the provider of the last mile link service.

Although the internet service provider is one, "Claro", there are 2 different providers that are providing the last mile links service (physical level, antennas, radios, etc.), in Fig. 4 you can see that the North 1 and North 2 areas are covered by the "Skyweb" service provider, while the other zones are with "Transdatel".

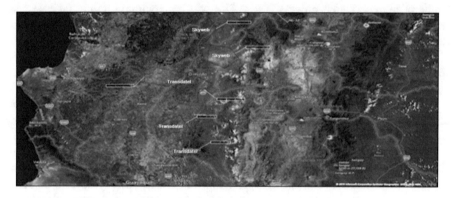

Fig. 4.

It is also important to indicate that the North 1 and North 2 zones maintain a band gap of 2 MW, while the other zones are coming out with a band gap of 6 MW, and even though they did not previously have link problems in the North 1 and 2 areas, since you are working with the SBE there have been some problems with several users regarding this issue.

The file sent from the Raspberry PI3 devices to the central repository are in two different formats:

(a) Plane text. In this format, the information contained will be readable, being possible to read the information in the file.
(b) Encrypted text In this format, the information contained will be encrypted with both RSA and 3DES, which will make it impossible to understand the data sent.

In the original file, unencrypted, there are 2,500 records of heavy boxes with all its component parts, which is an average of the boxes that are processed in all the packing houses on a daily basis. In addition, it has a file size of 572 Kb, and was sent from each of the hosts that are in the packs, from where they are sent to the central repository.

The file encrypted with 3DES, is the same file in plain text that after being processed with the 3DES algorithm becomes an encrypted file, which is illegible during its passage through the transport layer, and its size varied depending on the results of the encryption process.

Similarly, the file encrypted with RSA, is the original file that is encrypted with the algorithm RC4, which is a version of the RSA, and its size also varied depending on the results of the encryption process. These files will be monitored during their passage through the transport layer by the Wireshark protocol analyzer, with which their content was verified, in addition to other parameters.

According to the policy of using the cryptographic controls defined by ISO 27001: 2013, the protection of keys or encryption keys is implemented.

One option for key management is KeePass, because it is under free code and GNU GLPv2 license is available for Linux platforms, you can install the KeePass portable edition on a USB disk drive and keep it in your pocket. Do not write any data outside that unit, so you can use it on any computer.

KeePass uses an unusual composite master key system that can use any or all three different authentication methods: master password, key file and Windows user account.

Passwords must be long and complex. They do not necessarily have to be memorized, most password managers include a password generator, but many of them use defaulted defaults. Norton uses eight-character alphanumeric passwords by default; Dashlane offers 12-character passwords by default, and En-pass Password Manager 5 has a maximum of 18 characters. KeePass offers a default password of 20 characters, which makes it a good option.

4   Results and Discussion

Next, the comparative analysis between the encryption technologies under study is presented.

4.1   Size File

Generally after encrypting any document, it is normal that there is an increase in size in the original file and that increase in the size of the document is important to be considered because the larger the file the longer it will take to be transferred from the host to the repository. In the Fig. 5 the results are exposed in terms of the original file size and after being encrypted.

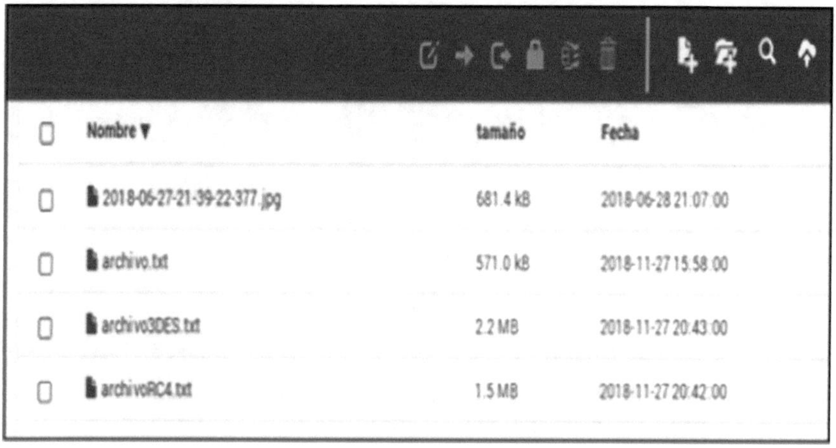

	Nombre ▼	tamaño	Fecha
☐	2018-06-27-21-39-22-377.jpg	681.4 kB	2018-06-28 21:07:00
☐	archivo.txt	571.0 kB	2018-11-27 15:58:00
☐	archivo3DES.txt	2.2 MB	2018-11-27 20:43:00
☐	archivoRC4.txt	1.5 MB	2018-11-27 20:42:00

Fig. 5. Size file

You can notice the sizes of the files in the repository, both unencrypted and after having gone through the encryption process. The file encrypted with 3DES has a larger size than the encryption with the RSA method.

In addition to a smaller size, RSA provides an asymmetric encryption which implies a better level of security when compared to the symmetric 3DES encryption.

The original file that had a size of 571 Kb, after the encryption process with the RSA algorithm, has increased to 1,523 Kb, that is to say a growth of 167%; while the same plain text file after the encryption process with the 3DES algorithm has reached 2,285 Kb, that is, a growth of 300%.

4.2   Upload File Time

The 25 hosts; object of study, are distributed among the 6 areas in which the present investigation was conducted. Figure 6 shows the average delay time in the transfer of the file to the central repository and; As you can see, when you encrypt the original file (text) with the 3DES method, it considerably increases its size, which directly affects the time of transfer of information, far exceeding what it takes to transmit the same file encryption with the RSA method. It should be mentioned that the hosts from which the files originated are Raspberry devices.

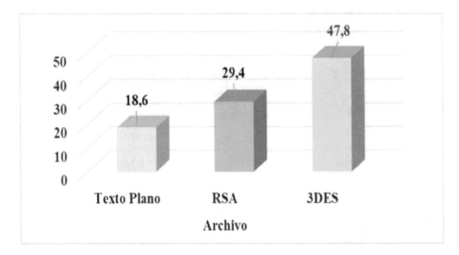

Fig. 6. Average file transfer time to the central repository

5   Conclusion

After the analysis of the information obtained during the investigation, the following can be concluded:

On the Internet there are many sites, either in online libraries, research forums, and in many other portals of scientific interest, where information about research is found, and a variety of encryption algorithms are compared, in which not only the process speed, but also encryption vulnerability, and even the ability to withstand the attacks of various programs to break the security of encryption, and it is always done in robust computing with memories of up to 32 Gb of RAM and with state-of-the-art processors, however, these same investigations have not been carried out on devices with limited resources such as the single-board Raspberry Pi or Arduino devices.

In the tests performed on the 25 hosts, evidence that the RC4 asymmetric encryption algorithm the last variant at the time of this research, exceeds by 100% with 29.4 s the encryption with the 3DES algorithm, which took 47,8 s, with respect to the process time of encryption of the flat text file that contains the information of the daily process of the haciendas where the tests were carried out, compared both with the sending of the file in plain text that took 18.6 s in be delivered to the central repository of the company.

It is important to clarify that, in the time of loading the file, it is necessary to consider the time it takes the algorithm to carry out the encryption process, which shows that the RSA algorithm is faster in this process than the 3DES algorithm.

In addition to having obtained a better time during the shipment, RC4 also achieved an encrypted file size smaller than 3DES, that is, RC4 increased by 167% with 1.5 Mb, while 3DES increased by 300% its size with 2.2 Mb, both in relation to the size of the plain text file that weighed 0.571 Mb.

Another important point that should be noted is that, the 3DES algorithm uses keys of 112 bits in length while RSA uses keys of at least 1024 bits which makes it clear that it is more robust in terms of security.

However, despite the information obtained in this investigation it must be clarified that in any case IT management will ultimately decide on this or that encryption system to apply for the work area.

6 Recommendations

Perform tests on other platforms such as HummingBoard devices which has hardware features similar to Raspberry Pi like BeagleBone Black and Odroid U3 that reaches 2 GB of RAM to confirm the results, these being also microcomputers unique, the results could be contrasted with those obtained in the present investigation.

In order to achieve a greater degree of security, incorporate other encryption algorithms such as AES and combine them with those used in the present investigation and compare the results.

Use other types of larger files during the testing process, such as images, pdf, audios and videos, which are larger than those used in the present investigation.

References

1. González, D.U.W.G.K.: Revista Electrónica de la Facultad de Ingeniería (2014). [En línea]. http://revistas_electronicas.unicundi.edu.co/index.php/Revistas_electronicas/article/view/164. Último acceso: 2018
2. Granados, G.: Introducción a la criptografía, Revista Digital Universitaria, pp. 1–17 (2006)
3. ISO: «www.iso27000.es, [En línea]. http://www.iso27000.es/iso27000.html. Último acceso: 20 septiembre 2018
4. Arenas, X.: Sistemas distribuidos y seguridad web, 16 Octubre 2016. [En línea]. http://sistemasdistribuidosyseguridadweb.blogspot.com/2016/10/cifrado-simetrico-firmas-sells-y.html
5. Pousa, A.: Plataforma Virtual Educativa ITCA-FEPADE (2011). [En línea]. https://virtual.itca.edu.sv/Mediadores/cms/u63_esquema_general_de_un_sistema_de_cifrado_simetrico.html
6. R. A., Rojas, J.: Implementación de Protocolo de Cifrado TLS para mejorar la Seguridad de la capa de Transporte, Universidad Señor de Sipan (2017)
7. Lucena, M.: Criptografía y Seguridad en Computadores, Openlibra (2014)

Author Index

© Springer Nature Switzerland AG 2020
M. Botto-Tobar et al. (Eds.): ICAETT 2019, AISC 1067, pp. 413–414, 2020.
https://doi.org/10.1007/978-3-030-32033-1

Printed in the United States
By Bookmasters